Birkhäuser

Nečas Center Series

The Nečas Center Series aims to publish high-quality monographs, textbooks, lecture notes, habilitation and Ph.D. theses in the field of mathematics and related areas in the natural and social sciences and engineering. There is no restriction regarding the topic, although we expect that the main fields will include continuum thermodynamics, solid and fluid mechanics, mixture theory, partial differential equations, numerical mathematics, matrix computations, scientific computing and applications. Emphasis will be placed on viewpoints that bridge disciplines and on connections between apparently different fields. Potential contributors to the series are encouraged to contact the editor-in-chief and the manager of the series.

All manuscripts are peer-reviewed to meet the highest standards of scientific literature. Interested authors may submit proposals by email to the series editors or to the relevant Birkhäuser editor listed under "Contacts."

Miroslav Bulíček • Eduard Feireisl • Josef Málek •
Milan Pokorný
Editors

New Perspectives in Mathematical Fluid Mechanics

Birkhäuser

Editors
Miroslav Bulíček
Mathematical Institute of Charles University, Faculty of Mathematics and Physics
Charles University
Praha 8, Czech Republic

Josef Málek
Mathematical Institute of Charles University, Faculty of Mathematics and Physics
Charles University
Praha 8, Czech Republic

Eduard Feireisl
Institute of Mathematics
Czech Academy of Sciences
Praha 1, Czech Republic

Milan Pokorný
Mathematical Institute of Charles University, Faculty of Mathematics and Physics
Charles University
Praha 8, Czech Republic

ISSN 2523-3343 ISSN 2523-3351 (electronic)
Nečas Center Series
ISBN 978-3-032-02318-6 ISBN 978-3-032-02319-3 (eBook)
https://doi.org/10.1007/978-3-032-02319-3

This book is published under the imprint Birkhäuser, www.birkhauser-science.com by the registered company Springer Nature Switzerland AG
The registered company address is: Gewerbestrasse 11, 6330 Cham, Switzerland

Preface

This volume contains five surveys dedicated to various aspects of mathematical fluid dynamics. The contributions are based on lecture series presented during two editions of the EMS School in Applied Mathematics, "Mathematical Aspects of Fluid Flows", held in Kácov, Czech Republic, in 2019 and 2024. The first part of the volume comprises three lecture notes from the most recent school, held from 12th May to 17th May 2024. The second part includes two shorter notes originating from the 2019 school, which took place from 26th May to 31st May.

The period following the 2019 school was significantly impacted by the global COVID-19 pandemic, which disrupted the tradition of these schools for several years. Consequently, only two lecture notes from the 2019 edition could be collected. After consulting with the authors, we decided to include these contributions in their original form, as submitted in early 2020, without updating references or content. Although reflecting the state of the field at that time, these surveys remain valuable resources, especially for doctoral students and early-career researchers interested in mathematical fluid mechanics and thermodynamics.

The lectures forming the basis of these surveys were delivered by the following experts: Paolo Antonelli (Gran Sasso Science Institute, Italy), Noemi David (Laboratoire de Mathématiques Raphaël Salem, CNRS & University of Rouen, France), and Vincent Giovangigli (CMAP, CNRS, École Polytechnique, France)—all from the 2024 school—as well as Vlad Vicol (Courant Institute, New York University, USA) and Emil Wiedemann (University of Erlangen, Germany) from the 2019 school.

The core mission of these schools remains unchanged: to present modern methods and recent advances in the mathematical theory of incompressible and compressible fluids as well as more complex systems of partial differential equations. Special emphasis is placed on physically relevant models, rigorous mathematical analysis, and, where possible, their connections to numerical methods. Each survey in this volume offers new insights into fundamental questions in mathematical fluid dynamics.

The 2024 lectures, which form the basis for the first part of this volume, address very general aspects of fluid flow models and their analysis, extending far beyond classical models such as the Navier–Stokes equations, and aiming to understand complex but physically relevant phenomena. Thus, this section represents a broad spectrum of recent trends and research directions in fluid flow analysis.

Paolo Antonelli presents a concise review of the mathematical theory of quantum hydrodynamics (QHD) equations, which describe the evolution of a compressible, inviscid, barotropic flow influenced by quantum effects on a macroscopic scale. These fluid systems are closely linked to nonlinear Schrödinger (NLS) type equations. To handle singularities near vacuum states, an alternative approach based on the polar decomposition of wave functions, rigorously connecting NLS evolutions to QHD systems, is introduced. The notes provide existence results for finite-energy weak solutions, including collisional models without NLS analogues, and study their stability and asymptotic behaviour via entropy-based functionals and dispersive estimates.

Noemi David explores singular limit problems motivated by mechanical models of tissue growth, where living tissues exhibit fluid-like behaviour. The analysis focuses on the porous medium equation, Hele–Shaw problems, and cross-diffusion systems. These models relate pressure laws to velocity fields through Brinkman's or Darcy's laws, distinguishing compressible from incompressible regimes through pressure stiffness. The survey rigorously describes how different mechanical tissue models relate through singular limits such as the inviscid limit towards the porous medium equation and the incompressible limit towards free-boundary Hele–Shaw type problems, including systems with convective effects.

The survey based on Vincent Giovangigli's lecture series focuses on fluid models derived from the kinetic theory of gases, addressing multicomponent flows with detailed transport and chemical reactions, internal energy relaxation phenomena, and diffuses interface models for non-ideal fluids exhibiting capillarity effects. Central themes include symmetrisation, hyperbolic-parabolic structures, and the asymptotic stability of equilibria. These approaches and techniques yield local existence results for models with stiff source terms and introduce an augmented formulation for capillary fluids that reveals a hyperbolic-parabolic-dispersive structure with antisymmetric coupling terms.

The second part of the volume features two surveys from the 2019 school, focusing primarily on regularity and singularity issues in fluid flow, especially concerning the Navier–Stokes equations and related topics. The key techniques discussed in both surveys include convex integration and renormalisation. Vlad Vicol investigates intermittent weak solutions of the three-dimensional Navier–Stokes equations using convex integration techniques, constructing novel solutions with significant implications for turbulence theory. The lecture notes discuss convex integration methods alongside background on weak solutions to the Euler and Navier–Stokes equations and their relation to turbulence models.

Emil Wiedemann's lectures address conservation laws, renormalisation, and statistical solutions in fluid dynamics, bridging classical results and modern challenges such as Onsager's conjecture and the construction of counterexamples via convex integration. The survey explores the subtle interplay between regularity and conservation of physical quantities such as energy or entropy, reviewing classical commutator methods and recent advances in the field.

We sincerely thank all lecturers, authors, and participants for their enthusiasm, collaboration, and the stimulating atmosphere they created during the Kácov school. We hope that this volume will serve as a valuable resource for researchers and students interested in the analysis of fluid flows. We also warmly invite all interested colleagues to future editions of the EMS School on Mathematical Aspects of Fluid Flows, which will continue to be held in Kácov in 2026. Further information, including lecture notes and an overview of the school's history, can be found at https://ems-maff.cuni.cz/.

Praha, Czech Republic
June 2025

Miroslav Bulíček
Eduard Feireisl
Josef Málek
Milan Pokorný

Contents

Contributors

Paolo Antonelli Gran Sasso Science Institute, L'Aquila, Italy

Tristan Buckmaster Courant Institute of Mathematical Sciences, New York University, New York, NY, USA

Maria Colombo École Polytechnique Fédérale de Lausanne, Faculté des Sciences de Base EPFL PH A2 364 (Bâtiment PH), Lausanne, Switzerland

Noemi David Laboratoire de Mathématiques Raphaël Salem Avenue de l'Université, BP.12, Saint-Étienne-du-Rouvray, France

Vincent Giovangigli CMAP, CNRS, École Polytechnique, Palaiseau, France

Pierangelo Marcati Gran Sasso Science Institute, L'Aquila, Italy

Vlad Vicol Courant Institute of Mathematical Sciences, New York University, New York, NY, USA

Emil Wiedemann Department of Mathematics, Chair of Analysis, Friedrich-Alexander-Universität, Erlangen-Nümberg, Germany

Part I
Lectures from Kácov School 2024

The first part contains three contributions based on three series of lectures presented at the EMS school Mathematical Aspects of Fluid Flows held in the Czech Republic, in Kácov, in May 2024.

Chapter 1
An Introduction to the Mathematical Theory of Quantum Fluids

Paolo Antonelli and Pierangelo Marcati

Abstract These lecture notes provide a concise review of the mathematical theory for Quantum Hydrodynamics (QHD) equations, describing the evolution of a compressible, inviscid, barotropic flow subject to a stress tensor depending on the mass density and its derivatives. Such fluid dynamical systems describe several phenomena where quantum effects remain relevant also on a macroscopic scale. Most of these models are deeply interconnected with nonlinear Schrödinger (NLS) type equations through the so-called Madelung transformation.

Our approach exploits this analogy to construct and analyze global-in-time, finite energy weak solutions to the QHD system. It is known that the Madelung transformation becomes singular close to vacuum, where the mass density vanishes. We introduce an alternative approach based on the polar decomposition of a wave function that rigorously establishes the relation between NLS evolutions and QHD systems. After briefly reviewing the literature concerning NLS equations, we introduce the main relevant objects for quantum fluid mechanics. We then discuss notions of weak solutions defined through physically relevant entropy functionals.

We first analyze the polar factorization method and its inverse given by the wave function lifting. By exploiting these tools, we provide some existence results for solutions to various QHD systems, including the collisional model that does not have an analog for NLS-type evolutions. We then study the stability properties of general weak solutions. We derive dispersive-type estimates that provide information on the asymptotic behavior for large times. Moreover, by introducing a novel entropy functional that controls a generalized chemical potential, we identify a class of weak solutions for which it is possible to prove a compactness result for the one-dimensional QHD system with large data.

P. Antonelli (✉) · P. Marcati
Gran Sasso Science Institute, L'Aquila, Italy
e-mail: paolo.antonelli@gssi.it; pierangelo.marcati@gssi.it

M. Bulíček et al. (eds.), *New Perspectives in Mathematical Fluid Mechanics*,
Nečas Center Series, https://doi.org/10.1007/978-3-032-02319-3_1

1.1 Introduction

These notes aim to present the rigorous analytical background to study a class of hydrodynamic systems arising in the description of quantum fluids.

Strictly speaking, a quantum fluid is a system of interacting particles characterized by the fact that quantum statistics effects are also appreciated on a macroscopic scale. Typical examples comprise superfluid Helium II [76, Chapt. XVI], or Bose-Einstein condensates (BEC) [89], where the extremely low temperature causes the thermal de Broglie wavelength to be comparable with the interatomic distance.

However, the term *quantum fluid* is now used more broadly. Similar phenomena are observed not only in the realm of low-temperature physics but also in different contexts, such as semiconductor physics [85], exciton-polariton condensates [33], or neutron star cores [55].

From the fluid dynamics standpoint, three prominent aspects distinguish a quantum fluid from classical ones. First, its *superfluid* behavior, namely the ability to freely flow through narrow capillaries without experiencing dissipation due to viscous forces [67]. Second, their local angular momentum is constrained to infinitesimally thin vortex lines (*quantized vortices*), whose characteristics such as circulation and core thickness are determined by quantum mechanical constraints, and thus only attain discrete values [43, 50, 88]. Finally, at non-zero temperature, quantum fluids exhibit a *two-fluid* behavior [71, 92].

Usually, such models are derived through different approaches, also depending on the specific phenomena under consideration. In the context of BEC, for instance, they typically arise as the fluid dynamical formulation of the Gross-Pitaevskii [54, 88] equation. This analogy is also exploited in superfluid Helium II models, even though the complete derivation in this context is given using thermodynamic considerations [71].

On the other hand, quantum hydrodynamic equations for semiconductor device modeling are also derived from kinetic equations, for instance, by performing a moment closure for the Wigner-Boltzmann equation [44]. A similar kinetic approach is also exploited for finite temperature BECs [53], where the hydrodynamic limit leads to the Landau-Khalatnikov two-fluid model of superfluidity [71]. For more details, we address the interested reader to [39, 63, 64].

In what follows, we briefly review the derivation of the quantum hydrodynamics equations as the fluid dynamical formulation of nonlinear Schrödinger type equations in the spirit of Madelung [81]. As a physically relevant example, we focus on the celebrated Gross-Pitaevskii (GP) equation modeling a BEC. This will be useful in introducing some basic notions and aspects related to the model and presenting some formal arguments that will be rigorously justified in the subsequent sections.

Finally, we remark that Madelung's original approach [81] was motivated to provide an alternative formulation of the quantum theory, reintepreting quantum mechanics through fluid dynamical concepts.

1.1.1 The Gross-Pitaevskii Model and Its Hydrodynamic Formulation

The most common model for a dilute Bose gas at zero temperature is the so-called Gross-Pitaevskii equation [54, 88], namely

$$i\hbar\partial_t\psi = -\frac{\hbar^2}{2m}\Delta\psi + V\psi + g|\psi|^2\psi, \tag{1.1}$$

defined in the whole three-dimensional space $\mathbf{R}^3$. This is a nonlinear Schrödinger (NLS) type equation describing the gas of bosons through a macroscopic wave function (order parameter) ψ, with V being an external potential, and g the Gross-Pitaevskii constant given by

$$g = \frac{4\pi\hbar^2 a_s}{m}.$$

The constant a_s denotes the $s-$wave scattering length that characterizes the interaction between atoms in the low energy limit. The size and even the sign of a_s can be controlled using the phenomenon of Feshbach resonance [93]. Thus, considering $g > 0$ corresponds to a repulsive interaction term (*defocusing NLS*), whereas $g < 0$ provides an attractive interaction (*focusing NLS*).

There is a conserved energy associated with (1.1) that reads

$$E = \int_{\mathbf{R}^3} \frac{\hbar^2}{2m}|\nabla\psi|^2 + V|\psi|^2 + \frac{g}{2}|\psi|^4\,dx \tag{1.2}$$

Notice that (1.1) may also be interpreted as the Hamilton equation associated with the functional (1.2), with symplectic structure given by (a multiple of) the imaginary unit i,

$$2i\hbar\partial_t\psi = \frac{\delta E}{\delta\bar{\psi}}. \tag{1.3}$$

In this way, Eq. (1.1) and, more generally, NLS-type equations may be interpreted as infinite dimensional Hamiltonian dynamical systems [28]. More details about the Lagrangian and Hamiltonian formalism associated with NLS-type equations may be found in [98, Chapter 2].

There is a deep link between the GP Eq. (1.1) and fluid dynamics, as it is possible to interpret the condensate as a gas characterized by density and momentum distributions. This relation may be formalized through the so-called Madelung transformation [81], expressing the wave function in terms of its amplitude and phase, namely

$$\psi = \sqrt{n}e^{iS}, \tag{1.4}$$

where $n = n(t, x) = |\psi(t, x)|^2$ is the number of particles density, and $S = S(t, x)$ is the phase. The mass density $\rho = \rho(t, x)$ is defined by

$$\rho = mn, \tag{1.5}$$

while the fluid velocity $v = v(t, x)$ may be characterized by

$$v = \frac{\hbar}{m}\nabla S. \tag{1.6}$$

Notice that the definition above implies that the velocity field is irrotational, and the (scaled) phase density $\frac{\hbar}{m}S$ plays the role of the velocity potential.

The condensate momentum density $J = J(t, x)$ is given by

$$J = \rho v. \tag{1.7}$$

The hydrodynamic formulation of the GP model can be recast by plugging the Madelung transformation (1.4) into Eq. (1.1). By separating the real and imaginary parts, we obtain

$$\begin{cases} \hbar\partial_t\sqrt{n} + \dfrac{\hbar^2}{m}\nabla\sqrt{n}\cdot\nabla S + \dfrac{\hbar^2}{2m}\sqrt{n}\Delta S = 0 \\ \hbar\sqrt{n}\partial_t S - \dfrac{\hbar^2}{2m}\Delta\sqrt{n} + \dfrac{\hbar^2}{2m}\sqrt{n}|\nabla S|^2 + V\sqrt{n} + gn^{\frac{3}{2}} = 0. \end{cases} \tag{1.8}$$

We now multiply the first and second equation by $\frac{2m\sqrt{n}}{\hbar}$ and $\frac{1}{\sqrt{n}}$, respectively. By recalling the definitions (1.5), (1.6), the following system is obtained

$$\begin{cases} \partial_t n + \operatorname{div}(nv) = 0 \\ \hbar\partial_t S + \dfrac{\hbar^2}{2m}|\nabla S|^2 + V + gn - \dfrac{\hbar^2}{2m}\dfrac{\Delta\sqrt{n}}{\sqrt{n}} = 0, \end{cases} \tag{1.9}$$

which consists of a continuity equation for the particle density n and a Bernoulli-type equation for the (scaled) velocity potential. Notice that the quantum term $\frac{\hbar^2}{2m}\frac{\Delta\sqrt{n}}{\sqrt{n}}$ appears as a self-consistent potential, depending on the derivatives of n and may be interpreted as a quantum correction to the classical Bernoulli equation. If we apply the gradient to the second equation above, we obtain an Euler-type equation for the velocity field v that reads

$$m(\partial_t v + v\cdot\nabla v) + \nabla\left(V + gn - \frac{\hbar^2}{2m}\frac{\Delta\sqrt{n}}{\sqrt{n}}\right) = 0. \tag{1.10}$$

It is straightforward to see that the irrotationality constraint on the velocity field is preserved along the flow of the equation above.

After multiplication by ρ, the previous equation may be written as follows,

$$\rho(\partial_t v + v \cdot \nabla v) = -\rho \frac{1}{m} \nabla V - \nabla p + \operatorname{div} \mathcal{T}_Q, \tag{1.11}$$

where p is a (classical) pressure field given by

$$p = p(\rho) = \frac{g}{2m^2} \rho^2, \tag{1.12}$$

and $\mathcal{T}_Q$ is the quantum stress tensor, defined through the following identities,

$$\begin{aligned} \operatorname{div} \mathcal{T}_Q &= \frac{\hbar^2}{2m^2} \rho \nabla \left(\frac{\Delta\sqrt{\rho}}{\sqrt{\rho}} \right) = \operatorname{div} \left(\frac{\hbar^2}{4m^2} \rho \nabla^2 \log \rho \right) \\ &= \operatorname{div} \left(\frac{\hbar^2}{4m^2} \nabla^2 \rho - \frac{\hbar^2}{m^2} \nabla\sqrt{\rho} \otimes \nabla\sqrt{\rho} \right). \end{aligned} \tag{1.13}$$

By exploiting the continuity equation in (1.9), we write the local balance law for the momentum density ρv as follows

$$\partial_t(\rho v) + \operatorname{div}(\rho v \otimes v) + \nabla p(\rho) + \frac{1}{m} \rho \nabla V = \frac{\hbar^2}{2m^2} \rho \nabla \left(\frac{\Delta\sqrt{\rho}}{\sqrt{\rho}} \right) \tag{1.14}$$

Thus, if $V \equiv 0$ and by using (1.13), the total momentum defined by

$$\mathcal{P} = \int_{\mathbf{R}^3} \rho v \, dx,$$

is formally conserved for all times.

The system of equations given by the continuity equation in (1.9) and the momentum Eq. (1.14) defines the *quantum hydrodynamics* (QHD) system,

$$\begin{cases} \partial_t \rho + \operatorname{div}(\rho v) = 0 \\ \partial_t(\rho v) + \operatorname{div}(\rho v \otimes v) + \nabla p(\rho) + \frac{1}{m} \rho \nabla V = \frac{\hbar^2}{2m^2} \rho \nabla \left(\frac{\Delta\sqrt{\rho}}{\sqrt{\rho}} \right). \end{cases} \tag{1.15}$$

The QHD system (1.15) consists of a compressible Euler-type system with a (barotropic) pressure $p = p(\rho)$ given by (1.12), where the momentum equation is augmented by the quantum pressure term defined in (1.13). Thus, given $\psi = |\psi| e^{iS}$ solution to (1.1), a solution to (1.15) is formally given by $(\rho, v) = (m|\psi|^2, \frac{\hbar}{m} \nabla S)$, where $\psi = |\psi| e^{iS}$ is a solution to the GP equation (1.1).

To appreciate how (1.15) differs from the classical compressible Euler system, we compare the contributions given by the classical pressure p and the quantum stress tensor $\mathcal{T}_Q$ by formally estimating their orders of magnitude at different length scales. If $\bar{\rho}$ is a reference mass density and L the typical variation length scale for ρ, we have

$$p \sim \frac{g}{m^2}\bar{\rho}^2, \qquad \mathcal{T}_Q \sim \frac{\hbar^2}{m^2 L^2}\bar{\rho}.$$

The two contributions become comparable at the length scale

$$\xi = \frac{\hbar}{\sqrt{g\bar{\rho}}}, \tag{1.16}$$

that is usually called the *healing length.* At length scales smaller than ξ, the quantum stress tensor becomes dominant, while for large length scales $L \gg \xi$, its contribution becomes negligible, and the QHD system formally reduces to the system of (classical) barotropic Euler equations.

There is a total energy associated to (1.15), given by

$$E = \int_{\mathbf{R}^3} \frac{\hbar^2}{2m^2}|\nabla\sqrt{\rho}|^2 + \frac{1}{2}\rho|v|^2 + \frac{1}{m}\rho V + \frac{g}{2m^2}\rho^2\,dx, \tag{1.17}$$

and formally conserved by solutions to the QHD system. Indeed, no dissipative effects are encoded in system (1.15). Notice that the total energy in (1.17) can also be obtained by writing the energy in (1.2) in terms of the hydrodynamic quantities and using the Madelung transformations (1.4).

Let us further remark that the QHD equations may also be (formally) recast as the Hamilton equations for (ρ, v) when we consider the Hamiltonian given by the energy functional (1.17), the following symplectic structure

$$\mathcal{J} = \begin{pmatrix} 0 & -\operatorname{div} \\ -\nabla & 0 \end{pmatrix},$$

and we assume the velocity field to be irrotational, i.e., $\nabla \wedge v = 0$. This last assumption is consistent with the definition (1.6). In this framework, Hamilton's equations read

$$\begin{cases} \partial_t \rho = -\operatorname{div}\left(\dfrac{\delta E}{\delta v}\right) \\ \partial_t v = -\nabla\left(\dfrac{\delta E}{\delta \rho}\right). \end{cases} \tag{1.18}$$

Notice that the two equations above are nothing but the continuity equation for ρ and Eq. (1.10) for v, respectively, where we also used the following vector identity

$$v \cdot \nabla v = \nabla \left(\frac{1}{2} |v|^2 \right) - v \wedge (\nabla \wedge v),$$

and the irrotationality of v.

Usually, the quantity $\frac{\delta E}{\delta \rho}$ is identified as the chemical potential associated to the fluid dynamical system. In Sect. 1.4, we will see how we exploit a generalized notion of chemical potential to identify a regularity class of weak solutions for (1.15). More details about the Hamiltonian formalism for the QHD system (and for the more general class of Euler-Korteweg systems) may be found in the interesting review [25, Chapter 1.6].

1.1.2 Quantized Vortices

As previously mentioned, one of the most striking features of quantum fluids is the existence of quantized vortices. These coherent structures play a prominent role in many physical applications, especially in quantum turbulence studies [24, 100]. In this paragraph, we highlight some of their properties.

To simplify our discussion, let us consider for the moment a two-dimensional domain and let ψ be a wave function. By the Madelung transformation (1.4), we write $\psi = \sqrt{n} e^{iS}$. If C is a closed path in the region $\{|\psi| > 0\}$, then we (formally) compute the integrated change in phase as

$$S_{diff} = \oint_C \nabla S \cdot d\ell. \tag{1.19}$$

Being C a closed path, and since we require ψ to be single-valued, we must have that

$$S_{diff} = 2\pi q, \qquad q \in \mathbf{Z}.$$

If $q \neq 0$, there must be a *phase defect* somewhere in the region enclosed by C, namely a point x_0 where the phase wraps by $2\pi q$. At that point, the phase takes every value. Thus, the only way for ψ to remain single-valued is to vanish at that point, $\psi(x_0) = 0$, and consequently $\rho(x_0) = 0$.

By plugging (1.6) into (1.19) we find

$$\Gamma = \oint_C v \cdot d\ell = q\kappa, \qquad q \in \mathbf{Z}, \tag{1.20}$$

where $\kappa = \frac{2\pi\hbar}{m}$ is the so-called *quantum of circulation*. By Stokes' Theorem, formula (1.20) formally implies that the vorticity satisfies $\nabla \wedge v = q\kappa\delta_{x_0}$ in the interior of the curve C.

This fact, together with the definition (1.6), implies that the velocity is irrotational except at $x = x_0$, where ψ vanishes. Thus, the only vorticity carried by the fluid is concentrated in the vacuum region and is quantized. In particular, the following identity

$$\rho \nabla \wedge v = 0, \tag{1.21}$$

is satisfied in the whole space.

1.1.3 Towards a Rigorous Analysis of QHD Systems

These lecture notes aim to set up a rigorous mathematical framework to study QHD models, such as (1.15). We focus on the associated Cauchy problem and investigate the existence of global-in-time, finite energy weak solutions. Our main strategy exploits the underlying NLS dynamics, such as (1.1), to construct such solutions. However, most of the arguments presented above are formal, and several mathematical difficulties are encountered when providing a rigorous justification.

First, the approach exposed above fails in the presence of non-trivial vacuum regions, namely in the set $\{\psi = 0\} = \{\rho = 0\}$. The Madelung transformation (1.4) becomes singular in vacuum, where the phase cannot be defined. Moreover, even the derivation of hydrodynamic equations presented above faces a similar issue, as passing from (1.8) to (1.9) would require the mass density to be positive. Finally, the low regularity framework of finite energy states can not provide sufficient information on the vacuum region. Consequently, even some approximation arguments close to vacuum become delicate.

In our approach, we overcome the need to define the velocity field and the phase density in (and close to) the vacuum region. Instead, we exploit a more general notion of polar decomposition for the wave function that allows to unambiguously define the hydrodynamic state even in vacuum. This approach is developed in Sect. 1.5.

The polar factorization allows to construct global-in-time, finite energy weak solutions to the QHD system without any smallness and/or high regularity assumptions and with possible vacuum regions where quantized vortices may be located. The solutions constructed in this way further inherit qualitative properties from the NLS dynamics, such as dispersive estimates and information about their asymptotic behavior.

On the other hand, it is not guaranteed that every physically suitable solution to QHD is provided through a wave function solution to an NLS evolution. Thus, we develop an intrinsically hydrodynamic approach to these systems to study stability properties for general solutions. More precisely, we infer new a priori bounds for general solutions, which are not necessarily constructed through NLS equations. As an application, such estimates yield a compactness result for a large class of weak solutions.

This stability result can also be seen as a preliminary step to extend the present theory to the more general class of Euler-Korteweg systems. Even though such models describe completely different phenomena in continuum mechanics (for instance, diffuse liquid-vapor interfaces [25, 74]), from the mathematical point of view, they share many interesting peculiarities.

The Euler-Korteweg system reads

$$\begin{cases} \partial_t \rho + \operatorname{div}(\rho v) = 0 \\ \partial_t(\rho v) + \operatorname{div}(\rho v \otimes v) + \nabla p(\rho) = \rho \nabla \left(\operatorname{div}(k(\rho)\nabla\rho) - \frac{1}{2}k'(\rho)|\nabla\rho|^2 \right), \end{cases} \tag{1.22}$$

where $k = k(\rho) \geq 0$ is the capillarity coefficient. The QHD system can be recast by choosing $k(\rho) \sim \frac{1}{4\rho}$. However, the Cauchy problem associated with (1.22) cannot be studied by adapting the strategy presented in these notes. Indeed, the corresponding wave function dynamics would be given by

$$i\partial_t\psi = -\frac{1}{2}\Delta\psi - \operatorname{div}\left(\widetilde{k}(|\psi|^2)\nabla|\psi|^2\right)\psi + \frac{1}{2}\widetilde{k}'(|\psi|^2)|\nabla|\psi|^2|^2\psi + f'(|\psi|^2)\psi, \tag{1.23}$$

where we defined $\widetilde{k}(\rho) = k(\rho) - \frac{1}{4\rho}$. Equation (1.23) is quasilinear, and presently there is no satisfactory theory for arbitrarily large initial data.

These lecture notes are structured as follows: Sect. 1.3 reviews some well-known results about NLS-type equations. Section 1.4 introduces the main hydrodynamical objects we will use to analyze QHD systems. We also collect the rigorous definitions of weak solutions we use throughout the review. Section 1.5 presents the polar factorization method that thoroughly justifies the Madelung transformation. Its inverse map, given by the wave function lifting, is also presented here. In Sect. 1.6, we list several existence results for weak solutions. Section 1.7 is dedicated to the analysis of weak solutions to a QHD system with a linear collision term, constructed by means of a fractional step argument. In Sect. 1.8, we collect some a priori dispersive estimates. Finally, in Sect. 1.9, we exploit them to show a compactness result for sequences of weak solutions to the one-dimensional QHD system.

1.2 Notations

This section introduces some of the notations used in these lecture notes. Given a number $a \in \mathbf{R}$, we denote $a_+ = \max\{0, a\}$. In particular, for $d \in \mathbf{N}$, we usually write $\frac{1}{(d-2)_+}$, that will denote the number $\frac{1}{d-2}$ for $d \geq 3$, while $\frac{1}{(d-2)_+} = \infty$ for $d = 1, 2$.

Given a measurable set $E \subset \mathbf{R}^m$, the symbol $\mathbf{1}_E$ will stand for the characteristic function of the set E. The identity matrix will be denoted by $\mathbb{I}$. Given a vector $v \in \mathbf{R}^d$, the symbol $v \otimes v$ will denote the rank one matrix whose entries are provided by $(v \otimes v)^{jk} = v^j v^k$. Given two matrices, the symbol $A : B$ will denote the Frobenius product given by summing the product component by component $A : B = \mathrm{Tr}(AB^T) = \sum_{jk} A^{jk} B^{jk}$. For any $X, Y \in \mathbf{R}$, $X \lesssim Y$ denotes $X \leq CY$, for some constant $C > 0$. The Japanese bracket $\langle \cdot \rangle$ denotes the quantity $\langle \theta \rangle = (1+|\theta|^2)^{1/2}$, for some $\theta \in \mathbf{R}^m$. Let $\mathcal{S}(\mathbf{R}^d)$ be the space of Schwartz functions and let $\mathcal{S}'(\mathbf{R}^d)$ be its dual, namely the space of tempered distributions. For any $f \in \mathcal{S}'(\mathbf{R}^d)$ we denote its Fourier transform by

$$\widehat{f}(\xi) = \mathcal{F}(f)(\xi) = \frac{1}{(2\pi)^{d/2}} \int e^{-ix\cdot\xi} f(x)\, dx.$$

The Lebesgue space with exponent $p \in [1, \infty]$ is denoted by $L^p(\mathbf{R}^d)$, the exponent p' satisfying $\frac{1}{p} + \frac{1}{p'} = 1$ will denote the dual Lebesgue exponent of p. For $s \geq 0$, $H^s(\mathbf{R}^d)$ is the Sobolev space whose norm is defined by $\|f\|_{H^s} := \|\langle \cdot \rangle^s \widehat{f}\|_{L^2}$. We often make use of mixed space-time Lebesgue-Sobolev spaces, for instance, $L^q(I; W^{k,r}(\mathbf{R}^d))$, where I is an arbitrary interval and q, r two Lebesgue exponents. If $I = [0, T]$, we usually denote $L^q(I; W^{k,r}(\mathbf{R}^d)) = L^q(0, T; W^{k,r}(\mathbf{R}^d))$. To shorten the notations, we also write $L_t^q W_x^{k,r}(I \times \mathbf{R}^d)$ or even simply $L_t^q W_x^{k,r}$ when there is no ambiguity. For any interval $I \subset \mathbf{R}$, $C(I; H^s(\mathbf{R}^d))$ denotes the space of continuous H^s-valued functions and C_c^∞ the space of infinitely differentiable, compactly supported functions.

1.3 A Short Review on Nonlinear Schrödinger Equations

This section condenses some fundamental concepts about nonlinear Schrödinger equations that will be exploited later in our analysis. In particular, we focus on the following Cauchy problem,

$$\begin{cases} i\partial_t \psi = -\frac{1}{2}\Delta\psi + \mathcal{N}(\psi), & (t, x) \in I \times \mathbf{R}^d \\ \psi(0) = \psi_0 \in H^1(\mathbf{R}^d), \end{cases} \tag{1.24}$$

where $\mathcal{N}(\cdot)$ is a suitable nonlinear operator, acting on the unknown (wave function) $\psi : I \times \mathbf{R}^d \to \mathbf{C}$. Here, I is an arbitrary time interval such that $0 \in I \subset \mathbf{R}$ and d is the space dimension, which may be assumed arbitrary (even though most of the analysis in the subsequent sections will focus on the physically relevant cases $d = 1, 2, 3$.).

To illustrate the various terms that may appear in the models under our consideration, in what follows, we list some typical nonlinearities that may be considered in $\mathcal{N}(\cdot)$, alongside with some physically relevant examples.

Assumption 1.1 (On the Structure of $\mathcal{N}(\cdot)$) We consider three main nonlinear terms.

- **Local nonlinearities**: $\mathcal{N}(\psi) = f'(|\psi|^2)\psi$, where $f : \mathbf{R}_+ \to \mathbf{R}$ satisfies
  - $f \in C^1([0,\infty)) \cap C^2((0,\infty))$, $f(0) = 0$;
  - $|f'(\rho)|, |\rho f''(\rho)| \lesssim 1 + \rho^\alpha$, with $0 < \alpha < \infty$ for $d = 1, 2$ and $0 < \alpha < \frac{2}{d-2}$ for $d \geq 3$;
  - $f(\rho) \gtrsim -\rho - \rho^\beta$, with $1 < \beta < 1 + \frac{2}{d}$.

  Example: by considering $\mathcal{N}(\psi) = g|\psi|^2\psi$ results in the GP equation (1.1).
- **External potentials**: $\mathcal{N}(\psi) = V\psi$, where $V : \mathbf{R}^d \to \mathbf{R}$ is given [1] and $V = V_1 + V_2$, such that
  - $V_1 = V_1(x)$ satisfies $V_1, \nabla V_1 \in L^\delta(\mathbf{R}^d) + L^\infty(\mathbf{R}^d)$, for some $\delta \geq 1$, $\delta > d/2$;
  - $V_2 = V_2(t,x)$ satisfies $V_2 \in L^p(\mathbf{R}; W^{1,s}(\mathbf{R}^d))$ with $\frac{2}{p} + \frac{d}{s} = 2$, $(p,s) \neq (\infty, \frac{d}{2})$, $d \geq 3$ and moreover $\partial_t V \in L^1_{loc}(\mathbf{R}; L^\eta(\mathbf{R}^d) + L^\infty(\mathbf{R}^d))$, with $\max\{1, \frac{d}{2}\} \leq \eta < \infty$, $\eta > 1$ for $d = 2$.
- **Non-local nonlinear interactions**: $\mathcal{N}(\psi) = \widetilde{V}\psi$, where $\widetilde{V} = W * |\psi|^2$ and $W : \mathbf{R}^d \to \mathbf{R}$ is even and satisfies
  - $W \in L^\sigma(\mathbf{R}^d) + L^\infty(\mathbf{R}^d)$ for some $\sigma \geq 1$, $\sigma > d/4$;
  - its negative part $W^- := \min\{0, W\}$ satisfies $W^- \in L^\theta(\mathbf{R}^d) + L^\infty(\mathbf{R}^d)$ for some $\theta \geq \max\{1, d/2\}$ ($\theta > 1$ for $d = 2$).

  Example: $\widetilde{V}$ solution to $\Delta\widetilde{V} = |\psi|^2$ gives the *Schrödinger-Poisson* equation.

Remark 1.1 The above assumptions on the nonlinear term $\mathcal{N}(\cdot)$ are also motivated by the various terms we are going to consider for the general QHD system in Sect. 1.6, see the Assumptions 1.2 therein.

Let us notice that all nonlinearities listed above are Hamiltonian. More precisely, under the above assumptions on $\mathcal{N}(\cdot)$, it is possible to determine a functional $G = G[\psi]$, such that $\mathcal{N}(\psi) = \frac{\delta G}{\delta \psi}[\psi]$. Consequently, if the NLS equation (1.24) is time autonomous (i.e., the external potential V is time-independent), then the total energy functional

$$E = E[\psi] = \int_{\mathbf{R}^d} \frac{1}{2}|\nabla\psi|^2\,dx + 2G[\psi], \tag{1.25}$$

is formally conserved along the flow determined by (1.24).

[1] In this case, $\mathcal{N}(\psi)$ is actually linear.

For this reason, we usually call *finite energy solutions* those with H^1 regularity.

We now briefly recall the notion of solutions we will use throughout the paper. For a more comprehensive discussion about different notions of solutions to NLS equations, we refer to [34, Section 3.1].

Definition 1.1 Let $0 \in I \subset \mathbf{R}$ be an arbitrary time interval, $\mathcal{N} \in C(H^1(\mathbf{R}^d), H^{-1}(\mathbf{R}^d))$, $\psi_0 \in H^1(\mathbf{R}^d)$, we say that ψ is a strong H^1-solution to (1.24) on $I \times \mathbf{R}^d$, if the following conditions hold:

- $\psi \in C(I; H^1(\mathbf{R}^d)) \cap C^1(I; H^{-1}(\mathbf{R}^d))$;
- the identity $i\partial_t \psi = -\frac{1}{2}\Delta\psi + \mathcal{N}(\psi)$ is satisfied in $H^{-1}(\mathbf{R}^d)$, for all $t \in I$;
- we have $\psi(0) = \psi_0$.

Remark 1.2 Here and hereafter, we are going to consider only positive time intervals, say $I = [0, T]$. However, let us notice that, if $\mathcal{N}$ satisfies Assumption 1.1, then the NLS equation (1.24) is time reversible. More precisely, $\psi = \psi(t, x)$ is a solution if and only if $\overline{\psi(-t, x)}$ is also a solution. This straightforwardly implies that most of the results presented in these notes can be stated for negative times, too.

The following proposition shows that the Cauchy problem (1.24) can be equivalently reformulated as a nonlinear integral equation. We refer to [34, Proposition 3.1.3.] for a detailed proof.

Proposition 1.1 (Duhamel's Formula) *Let I be an interval with $0 \in I$, $\mathcal{N} \in C(H^1(\mathbf{R}^d); H^{-1}(\mathbf{R}^d))$, $\psi_0 \in H^1(\mathbf{R}^d)$. A function $\psi \in C(I; H^1(\mathbf{R}^d))$ is a strong H^1-solution to* (1.24) *on I if and only if for any $t \in I$, ψ satisfies*

$$\psi(t) = e^{\frac{i}{2}t\Delta}\psi_0 - i\int_0^t e^{\frac{i}{2}(t-s)\Delta}\mathcal{N}(\psi)(s)\,ds. \tag{1.26}$$

The nonlinear integral Eq. (1.26) can be analyzed by exploiting the smoothing properties associated with the linear propagator $e^{\frac{i}{2}t\Delta}$, that we are going to recall in what follows. By writing the fundamental solution of the free Schrödinger equation

$$e^{\frac{i}{2}t\Delta}f(x) = (2\pi i t)^{-\frac{d}{2}}\int_{\mathbf{R}^d} e^{i\frac{|x-y|^2}{2t}}f(y)\,dy,$$

it is straightforward to derive the following dispersive estimate

$$\|e^{\frac{i}{2}t\Delta}f\|_{L^\infty(\mathbf{R}^d)} \lesssim |t|^{-\frac{d}{2}}\|f\|_{L^1(\mathbf{R}^d)}.$$

We can now interpolate the previous estimate with the usual $L^2 - L^2$ estimate given by the unitarity of the operator $e^{\frac{i}{2}t\Delta}$, namely $\|e^{\frac{i}{2}t\Delta}f\|_{L^2} = \|f\|_{L^2}$, to obtain

$$\|e^{\frac{i}{2}t\Delta}f\|_{L^p(\mathbf{R}^d)} \lesssim |t|^{-d\left(\frac{1}{2}-\frac{1}{p}\right)}\|f\|_{L^{p'}(\mathbf{R}^d)}, \quad 2 \le p \le \infty. \tag{1.27}$$

These estimates further imply a whole family of dispersive space-time estimates, commonly referred to in the literature as Strichartz estimates [47, 70, 97, 101, 105].

Definition 1.2 The pair of Lebesgue exponents (q, r) is called admissible if

$$2 \leq q, r \leq \infty, \quad \frac{2}{q} + \frac{d}{r} = \frac{d}{2}, \quad (q, r, d) \neq (2, \infty, 2).$$

Theorem 1.1 ([70]) *Let (q_1, r_1) and (q_2, r_2) be two arbitrary admissible pairs and let I be any time interval (possibly unbounded), then the following estimates hold*

$$\|e^{\frac{i}{2}t\Delta} f\|_{L_t^q L_x^r(I\times\mathbf{R}^d)} \lesssim \|f\|_{L^2(\mathbf{R}^d)};$$

$$\|\int_0^t e^{\frac{i}{2}(t-s)\Delta} F(s)\, ds\|_{L_t^q L_x^r(I\times\mathbf{R}^d)} \lesssim \|F\|_{L_t^{q_1'} L_x^{r_1'}(I\times\mathbf{R}^d)}.$$

Besides the Strichartz estimates, additional estimates for the linear propagator result from its dispersive behavior, one of the most relevant being the well-known Kato smoothing effect [68]. There is extensive literature on local smoothing estimates; our analysis uses the result as stated in [37].

Theorem 1.2 ([37]) *Let $\psi \in C([0, T]; L^2(\mathbf{R}^d))$ be an L^2−strong solution to*

$$\begin{cases} i\partial_t \psi = -\dfrac{1}{2}\Delta\psi + F \\ \psi(0) = \psi_0, \end{cases}$$

with $\psi_0 \in L^2(\mathbf{R}^d)$ and $F \in L^1([0, T]; L^2(\mathbf{R}^d))$. Then, the following smoothing estimate holds

$$\|\psi\|_{L^2([0,T];H^{1/2}_{loc}(\mathbf{R}^d))} \lesssim \|\psi_0\|_{L^2(\mathbf{R}^d)} + \|F\|_{L^1}.$$

By exploiting the Strichartz estimates stated in Theorem 1.1, it is possible to prove the global well-posedness in H^1 for the Cauchy problem (1.24). Since we are interested in global-in-time, finite energy solutions, below, we only recall the definition of global well-posedness in H^1. For a more comprehensive discussion about different notions of well-posedness for nonlinear dispersive equations, we refer to [99, Section 3.2].

Definition 1.3 We say that the Cauchy problem (1.24) is globally well-posed in $H^1(\mathbf{R}^d)$ if, for any initial datum $\psi_0 \in H^1$, there exists a unique global in time strong H^1− solution $\psi \in C(\mathbf{R}; H^1(\mathbf{R}^d))$ to (1.24). Moreover, the solution depends continuously on the initial data, namely for any $0 < T < \infty$ and for any sequence $\{\psi_{0,n}\} \subset H^1(\mathbf{R}^d)$ such that $\psi_{0,n} \to \psi_0$ in $H^1(\mathbf{R}^d)$, then $\psi_n \to \psi$ in $C([0, T]; H^1(\mathbf{R}^d))$, where ψ_n is the solution to (1.24), with $\psi_n(0) = \psi_{0,n}$.

Theorem 1.3 *Under the Assumptions 1.1 on $\mathcal{N}(\cdot)$, the Cauchy problem* (1.24) *is globally well-posed in $H^1(\mathbf{R}^d)$, in the sense of Definition 1.3. Moreover, the total mass functional, defined by*

$$M(t) = \int_{\mathbb{R}^d} |\psi(t, x)|^2 \, dx$$

is conserved for all times, whereas the total energy defined in (1.25) *satisfies the balance equation*

$$E(t) + \int_0^t \int_{\mathbf{R}^d} |\psi|^2 \partial_t V \, dx dt' = E(0), \quad t \in \mathbf{R}. \tag{1.28}$$

For any $0 < T < \infty$ and any admissible pair (q, r), we have

$$\|\psi\|_{L^q([0,T]; W^{1,r}(\mathbf{R}^d))} \leq C(T, \|\psi_0\|_{H^1(\mathbf{R}^d)}). \tag{1.29}$$

Finally, if we additionally assume $\psi_0 \in H^2(\mathbf{R}^d)$, then the solution satisfies $\psi \in C(\mathbf{R}; H^2(\mathbf{R}^d)) \cap C^1(\mathbf{R}; L^2(\mathbf{R}^d))$ and we have

$$\|\psi\|_{L^\infty(0,T; H^2(\mathbf{R}^d))} + \|\partial_t \psi\|_{L^\infty(0,T; L^2(\mathbf{R}^d))} \leq C(T, \|\psi_0\|_{H^2(\mathbf{R}^d)}), \tag{1.30}$$

for any $0 < T < \infty$ and (q, r) admissible pair.

The proof of Theorem 1.3 is now standard, see for instance [34, Corollary 6.1.2], [84, Theorem XI] and [69], for a treatment of all nonlinear terms considered by Assumption 1.1.

1.4 Quantum Fluid Mechanics and Rigorous Notions of Weak Solutions

In this section we will introduce the rigorous definitions for the weak solutions under discussion in these lecture notes. As already anticipated in the Introduction, the notion of weak solutions will encode some of the main features relevant to the physical phenomena we describe through these models. To simplify the presentation, we focus on the following prototypical example of a QHD system, given (in adimensional form) by

$$\begin{cases} \partial_t \rho + \operatorname{div}(\rho v) = 0 \\ \partial_t (\rho v) + \operatorname{div}(\rho v \otimes v) + \nabla p(\rho) = \dfrac{1}{2} \rho \nabla \left(\dfrac{\Delta \sqrt{\rho}}{\sqrt{\rho}} \right), \end{cases} \tag{1.31}$$

where again ρ denotes the particle density, v the velocity field and $p = p(\rho)$ is a pressure term. The definitions presented below can be easily adapted to related systems with additional terms describing other interactions and force fields.

The internal energy $f(\rho)$ associated to the pressure $p(\rho)$ is defined by

$$f(\rho) = \rho \int_{\rho^*}^{\rho} \frac{p(s)}{s^2}\,ds - \frac{p(\rho^*)}{\rho^*}(\rho - \rho^*) = \rho \int_{\rho^*}^{\rho} \frac{p(s) - p(\rho^*)}{s^2}\,ds, \tag{1.32}$$

where the constant $\rho^* \geq 0$ is a given reference density. Formula (1.32) also yields that the internal energy and the pressure are related through the identity

$$p(\rho) - p(\rho^*) = \rho f'(\rho) - f(\rho).$$

Notice that $f(\rho)$ can be defined up to adding an arbitrary affine function. The definition given in (1.32) satisfies $f(\rho^*) = f'(\rho^*) = 0$.

The total energy associated to system (1.31) is given by

$$E = \int_{\mathbf{R}^d} \frac{1}{2}|\nabla\sqrt{\rho}|^2 + \frac{1}{2}\rho|v|^2 + f(\rho)\,dx \tag{1.33}$$

and is formally conserved along the flow of solutions to (1.31).

Notice that the finiteness of the total energy, and hence the integrability of the internal energy density, implies that $\rho(x) \to \rho^*$ as $|x| \to \infty$. This review is mainly dedicated to the case when $\rho^* = 0$, namely ρ is integrable, and the system also has finite total mass. However, we stress that also the case $\rho^* > 0$ is very interesting for applications, especially related to the study of special solutions, including the analysis of quantized vortex dynamics. Although mathematically the analysis is more involved and requires a nontrivial amount of additional work [18, 58], these problems can also be addressed with a natural extension of the framework developed here.

When studying weak solutions to (1.31), the finite energy framework readily presents two main mathematical difficulties. First, the only available bound for the velocity field v is in the space $L^2(\rho dx)$, which implies that v cannot be controlled (or even uniquely defined) in the vacuum region. On the contrary, the momentum density $J = \rho v$ is always well defined. Indeed the energy bound (1.33) implies $|J|^2/\rho \in L^1$, consequently $J \in L^1_{loc}(\mathbf{R}^d)$ and in particular $J = 0$, a.e. in the vacuum region $\{\rho = 0\}$. However, the convective term written in terms of the physical variables (ρ, J) as $\operatorname{div}\left(\frac{J\otimes J}{\rho}\right)$ exhibits well-known difficulties in the low regularity framework. On the other hand, the quantity

$$\Lambda := \sqrt{\rho}v = J/\sqrt{\rho} \tag{1.34}$$

is always well defined since $\Lambda \in L^2$ by the energy bound (1.33) and we can rewrite the convective term in the more convenient form as $\operatorname{div}(\Lambda \otimes \Lambda)$.

The second mathematical difficulty we encounter in studying finite energy weak solutions for (1.31) is the quantum force appearing on the right-hand side of the momentum equation. Indeed, our regularity framework does not seem to provide a reasonable a priori definition of the distributional solution because of the high degree of nonlinearity of the quantum term. However, formally, the following identities hold

$$\frac{1}{2}\rho\nabla\left(\frac{\Delta\sqrt{\rho}}{\sqrt{\rho}}\right) = \operatorname{div}\left(\frac{1}{4}\rho\nabla^2\log\rho\right) = \operatorname{div}(\frac{1}{4}\nabla^2\rho - \nabla\sqrt{\rho}\otimes\nabla\sqrt{\rho}). \tag{1.35}$$

The last expression on the right-hand side allows us to provide a distributional definition of the quantum term in the finite energy framework. Moreover, since the identity (1.35) implies that the quantum term can be written as the divergence of a stress tensor, the total momentum is (formally) conserved for all times.

The previous comments suggest that it is more convenient to define the weak solutions to (1.31) through the quantities $\sqrt{\rho}$ and Λ. Indeed, $(\sqrt{\rho}, \Lambda)$ are well defined given the natural bounds provided by the total energy. Moreover, they define the physical observables (ρ, J) through the quadratic relations $\rho = (\sqrt{\rho})^2$, $J = \sqrt{\rho}\Lambda$.

For this reason, in what follows, we refer to $(\sqrt{\rho}, \Lambda)$ as the *hydrodynamic state* variables, whereas the usual macroscopic observables (ρ, J) denote the solution to the system (1.31).

Remark 1.3 (About the State Equation for the Pressure) For what concerns the discussion in the present section, the only assumption we need on the pressure is that $p(\rho) \in L^1_{loc}(\mathbf{R}^d)$, for any $\sqrt{\rho} \in H^1_{loc}(\mathbf{R}^d)$. Notice however that the existence results showed in Sect. 1.6 require further structure on the state law $\rho \mapsto p(\rho)$. We refer to Assumption 1.2 below for a more precise characterization, in particular see the hypotheses required on the internal energy $f(\rho)$. We stress that our theory allows a quite general class of state equations, in particular van der Waals type pressure terms may be considered.

Resuming, we can write the QHD system (1.31) in terms of the hydrodynamic state $(\sqrt{\rho}, \Lambda)$ as follows,

$$\begin{cases} \partial_t\rho + \operatorname{div} J = 0 \\ \partial_t J + \operatorname{div}(\Lambda\otimes\Lambda) + \nabla p(\rho) = \frac{1}{4}\nabla\Delta\rho - \operatorname{div}(\nabla\sqrt{\rho}\otimes\nabla\sqrt{\rho}), \end{cases}$$

where we defined $\rho = (\sqrt{\rho})^2$ and $J = \sqrt{\rho}\Lambda$, see also (1.34). This formulation motivates our definition of weak solutions.

Definition 1.4 (Weak Solutions) Let $0 < T \leq \infty$, $\rho_0 \in L^1_{loc}(\mathbf{R}^d)$, $J_0 \in L^1_{loc}(\mathbf{R}^d)$. We say that (ρ, J) is a weak solution to (1.31) in $[0, T) \times \mathbf{R}^d$ if there exists a pair of functions $(\sqrt{\rho}, \Lambda)$, with $\sqrt{\rho} \in L^2_{loc}(0, T; H^1_{loc}(\mathbf{R}^d))$, $\Lambda \in L^2_{loc}(0, T; L^2_{loc}(\mathbf{R}^d))$, such that

- $\rho = (\sqrt{\rho})^2$, $J = \sqrt{\rho}\Lambda$;
- (continuity equation) for any $\eta \in C_c^\infty([0, T) \times \mathbf{R}^d)$, we have

$$\int_0^T \int_{\mathbf{R}^d} \rho \partial_t \eta + J \cdot \nabla \eta \, dx dt + \int_{\mathbf{R}^d} \rho_0 \eta(0, \cdot) \, dx = 0; \tag{1.36}$$

- (momentum equation) for any test function $\zeta \in C_c^\infty([0, T) \times \mathbf{R}^d; \mathbf{R}^d)$, we have

$$\begin{aligned} \int_0^T \int_{\mathbf{R}^d} & J \cdot \partial_t \zeta + \Lambda \otimes \Lambda : \nabla \zeta + p(\rho) \operatorname{div} \zeta + \nabla \sqrt{\rho} \otimes \nabla \sqrt{\rho} : \nabla \zeta \\ & - \rho \frac{1}{4} \Delta \operatorname{div} \zeta \, dx dt + \int_{\mathbf{R}^d} J_0 \cdot \zeta(0, \cdot) \, dx = 0; \end{aligned} \tag{1.37}$$

- (generalized irrotationality condition) for a.e. $t \in [0, T)$ the identity

$$\nabla \wedge J = 2 \nabla \sqrt{\rho} \wedge \Lambda, \tag{1.38}$$

is satisfied in the sense of distributions.

Remark 1.4 As said before, for QHD systems like (1.31), it is more convenient to deal with Λ rather than the velocity field v. This also happens for other fluid dynamical systems, where some macroscopic functions are not well defined in the vacuum region, like, for instance, the Navier-Stokes system with degenerate viscosity [31, 77, 79]. In that context, Λ is used not only in the convective term but also to provide the weak formulation of the viscous stress tensor; see, for example, formula (2.11) in [79]. The same problem is present for the more general class of Navier-Stokes-Korteweg systems, see for example [12, 17, 31].

Let us also mention that in some cases, it is possible to show that Λ is "physically admissible", in the sense that the constructed solutions satisfy $\Lambda = 0$ a.e. on $\{\rho = 0\}$, see, for instance, the existence results we are going to present in Sect. 1.6, or [9] for the analysis of a viscous flow. This is not always true in general; see, for instance, [11, 12]. The property that Λ vanishes almost everywhere in the vacuum region allows to define an associated velocity field, but not in a unique way. As said, the solutions under consideration in these lecture notes will always be physically admissible in this sense. However, in our presentation, we will avoid the definition of the velocity field and construct the theory by using the hydrodynamic state $(\sqrt{\rho}, \Lambda)$.

Remark 1.5 (Generalized Irrotationality Condition) To understand the meaning of condition (1.38), let (ρ, J) be a smooth solution and let us assume for the moment we may define the velocity field v so that $J = \rho v$. Then, the generalized irrotationality condition (1.38) implies

$$\rho \nabla \wedge v = 0. \tag{1.39}$$

Notice this condition was also derived in the Introduction, see (1.21), to express that a quantum flow is irrotational outside the vacuum. In contrast, quantized vortices may appear in the vacuum region. In this sense, the generalized irrotationality condition can be seen as the weak formulation of (1.39), expressed in terms of the hydrodynamic state $(\sqrt{\rho}, \Lambda)$.

As we will show in Sect. 1.6, our theory allows to define finite energy weak solutions with vacuum regions and non-trivial vorticity.

Moreover, let us further remark that condition (1.38) is somehow analogous to the weak formulation of the viscous stress tensor for the compressible Navier-Stokes equations with degenerate viscosity, see for instance formula (2.11) in [79] and the discussion in the previous Remark.

Finally, we also mention that there are also other physical phenomena described by similar models where the generalized irrotationality condition does not play a prominent role. This is the case of the Korteweg theory of capillarity [41, 74], described by the so-called Euler-Korteweg system. This review will not discuss such solutions, and we refer to [26] for a local well-posedness result.

Definition 1.5 (Finite Energy Weak Solutions) Let (ρ_0, J_0) be such that

$$E_0 = \int_{\mathbf{R}^d} \frac{1}{2}\frac{|J_0|^2}{\rho_0} + \frac{1}{2}|\nabla\sqrt{\rho_0}|^2 + f(\rho_0)\, dx < \infty.$$

Let $0 < T < \infty$ and let (ρ, J) be a weak solution to (1.31) in $[0, T) \times \mathbf{R}^d$ as in Definition 1.4. Let the total energy $E(t)$ be given by

$$E(t) = \int_{\mathbf{R}^d} \frac{1}{2}|\nabla\sqrt{\rho}|^2 + \frac{1}{2}|\Lambda|^2 + f(\rho)\, dx, \tag{1.40}$$

where f is given by (1.32) with $\rho^* = 0$. We say that (ρ, J) is a finite energy weak solution to (1.31) if additionally $E(t) \leq E(0)$, for a.e. $t \in [0, T]$, where $E(t)$ is defined in (1.40).

Let

$$M(t) = \int \rho(t, x)\, dx. \tag{1.41}$$

We say that the weak solution (ρ, J) have finite mass if $M(0) < \infty$ and $M(t) \leq M(0)$.

Remark 1.6 Since the present review deals only with the case when the mass density is integrable at space infinity, namely $\rho^* = 0$ in (1.32), here we shall use the terminology finite energy weak solutions to denote weak solutions with both finite energy and finite mass. Let us remark that, in the case $\rho^* > 0$, it is possible to consider solutions with finite energy but infinite mass [18, 58].

Finally, (ρ, J) is a global-in-time finite energy solution if the definitions above are satisfied for any $0 < T < \infty$.

Remark 1.7 It is straightforward to see that for a finite mass weak solution (ρ, J) to (1.31) we actually have that $M(t) = M(0)$ for a.e. $t \in [0, T)$, see for instance Theorem 1.3.4 in [38].

Analogously, for weak solutions that have finite mass and energy, we also have the total momentum, defined by

$$P(t) = \int J(t, x)\, dx \tag{1.42}$$

is finite and conserved for a.e. $t \in [0, T)$, i.e. $P(t) = P(0)$.

1.4.1 *The Generalized Chemical Potential*

The functionals (1.41), (1.42), (1.33) defined above are related to the physical quantities and are formally conserved along the flow of (1.31). In our analysis, we introduce a further functional, first appeared in [14] in the context of quantum fluid dynamics, which involves a new notion of the generalized chemical potential.

Before providing its rigorous definition, let us give a heuristic presentation of the objects we deal with. Given the energy functional in (1.33), we can formally define the chemical potential μ to be its first variation with respect to the mass density, i.e.

$$\mu = \frac{\delta E}{\delta \rho} = -\frac{1}{2}\frac{\Delta\sqrt{\rho}}{\sqrt{\rho}} + \frac{1}{2}|v|^2 + f'(\rho). \tag{1.43}$$

By recalling the (formal) Hamiltonian formulation of QHD, see, for instance, (1.18), we see that the chemical potential appears in the equation for the velocity field, namely

$$\partial_t v + \nabla\mu = 0.$$

Given the chemical potential introduced in (1.43), we formally define the following functional

$$\begin{aligned} I(t) &= \int_{\mathbf{R}^d} \frac{1}{2}\rho\mu^2 + \frac{1}{8}\frac{(\operatorname{div} J)^2}{\rho}\, dx \\ &= \int \frac{1}{2}\rho\mu^2 + \frac{1}{2}(\partial_t\sqrt{\rho})^2\, dx, \end{aligned} \tag{1.44}$$

where we used the continuity equation to write the second contribution in $I(t)$. As we will show in Sect. 1.8, the functional $I(t)$ is formally bounded along the flow of solutions to (1.31) in the one-dimensional case. The a priori estimates inferred from its boundedness will yield a stability result for weak solutions to (1.31) in the one-dimensional case; see Theorem 1.10. However, the chemical potential μ introduced in (1.43) cannot be used to conduct a satisfactory analysis in the finite energy weak solutions framework. On the other hand, in this general context, it is always possible to define the distribution

$$\begin{aligned}\xi = \rho\mu &= -\frac{1}{4}\Delta\rho + \frac{1}{2}|\nabla\sqrt{\rho}|^2 + \frac{1}{2}|\Lambda|^2 + \rho f'(\rho)\\ &= -\frac{1}{4}\Delta\rho + e + p(\rho),\end{aligned} \tag{1.45}$$

where e is the total energy density,

$$e = \frac{1}{2}|\nabla\sqrt{\rho}|^2 + \frac{1}{2}|\Lambda|^2 + f(\rho), \tag{1.46}$$

see (1.40). Thus, the finiteness of the functional in (1.44) formally yields $\xi/\sqrt{\rho} \in L^2(\mathbf{R}^d)$. Analogously to our previous discussion about the velocity field v, the momentum density $J = \rho v$ and the hydrodynamic quantity $\Lambda = J/\sqrt{\rho}$, we see that also for the chemical potential, it will be convenient to deal with the quantity given by $\lambda = \sqrt{\rho}\mu = \xi/\sqrt{\rho}$. In this way the finiteness of $I(t)$ implies $\lambda(t) \in L^2(\mathbf{R}^d)$. This means that we may define the functional $I(t)$ as

$$I(t) = \int_{\mathbf{R}^d} \frac{1}{2}\lambda^2 + \frac{1}{8}\frac{(\operatorname{div} J)^2}{\rho}\, dx, \tag{1.47}$$

where $\lambda \in L^2(\mathbf{R}^d)$, and it satisfies the following identity

$$\sqrt{\rho}\lambda = -\frac{1}{4}\Delta\rho + e + p(\rho). \tag{1.48}$$

The identification of λ through (1.48) leaves an ambiguity in explicitly defining λ in the vacuum region. However, it turns out that it is reasonable to assume that λ vanishes there. This will be further discussed in Sect. 1.5.2.1 later.

Definition 1.6 Let $(\sqrt{\rho}, \Lambda) \in H^1(\mathbf{R}^d) \times L^2(\mathbf{R}^d)$ be a finite energy hydrodynamic state such that $\Lambda = 0$ a.e. on $\{\rho = 0\}$ and $\Delta\rho \in L^1_{loc}(\mathbf{R})$. We define the generalized chemical potential λ to be the measurable function given by

$$\lambda = \begin{cases} \frac{\xi}{\sqrt{\rho}} & \text{in } \{\rho > 0\} \\ 0 & \text{elsewhere} \end{cases} \tag{1.49}$$

In the one-dimensional case, the generalized chemical potential can also be characterized more precisely by the hydrodynamical quantities $\sqrt{\rho}$ and Λ, see [14]. In this review, we will not use this fine characterization, so for more details, we address the interested reader to [14] and especially Section 7 therein.

The generalized chemical potential allows us also to introduce another class of hydrodynamic states.

Definition 1.7 (GCP States) We say that the pair $(\sqrt{\rho}, \Lambda)$ is a hydrodynamic state with bounded generalized chemical potential (or GCP state, in short) if:

- $\sqrt{\rho} \in H^1(\mathbf{R}^d)$ and $\Lambda \in L^2(\mathbf{R}^d)$;
- $\Lambda = 0$ a.e. on $\{\rho = 0\}$;
- $\Delta\rho, \operatorname{div} J \in L^1_{loc}(\mathbf{R}^d)$, where $(\rho, J) = ((\sqrt{\rho})^2, \sqrt{\rho}\Lambda)$;
- the following bounds are satisfied

$$\|\mathbf{1}_{\{\rho>0\}} \operatorname{div} J/\sqrt{\rho}\|_{L^2(\mathbf{R}^d)} + \|\lambda\|_{L^2(\mathbf{R}^d)} \le C,$$

where λ is defined as in (1.49).

Notice that the last assumption in the definition above implies the finiteness of the functional defined in (1.47). Being related to the boundedness of $I(t)$, the notion of GCP states naturally yields a regularity class for weak solutions. As we will show later, this indeed identifies a class of solutions that are sequentially stable, see Theorem 1.10 below.

Definition 1.8 (GCP Solutions) Let (ρ, J) be a finite energy weak solution to (1.31) on $[0, T] \times \mathbf{R}^d$. We say that (ρ, J) is a GCP solution if and only if

- there exists $\lambda \in L^\infty(0, T; L^2(\mathbf{R}^d))$, such that

$$\sqrt{\rho}\lambda = -\frac{1}{4}\Delta\rho + e + p(\rho),$$

where the energy density e is defined by

$$e = \frac{1}{2}|\nabla\sqrt{\rho}|^2 + \frac{1}{2}|\Lambda|^2 + f(\rho);$$

- for some $0 < M_1, M_2 < \infty$, the following bounds are satisfied

$$\begin{aligned}\|\sqrt{\rho}\|_{L^\infty(0,T;H^1(\mathbf{R}^d))} + \|\Lambda\|_{L^\infty(0,T;L^2(\mathbf{R}^d))} &\le M_1,\\ \|\partial_t\sqrt{\rho}\|_{L^\infty(0,T;L^2(\mathbf{R}^d))} + \|\lambda\|_{L^\infty(0,T;L^2(\mathbf{R}^d))} &\le M_2.\end{aligned}$$

Let us remark that the above definition of GCP solutions does not require the precise characterization of the generalized chemical potential as in (1.49). We only need the existence of some $\lambda \in L^\infty_t L^2_x$ such that identity (1.48) is satisfied.

1.5 The Polar Factorization and the Wave Function Lifting

This section establishes the rigorous correspondence between wave function and quantum fluid mechanics. Given a wave function, it is straightforward to define the observable quantities given by the mass and momentum densities, namely $\rho = |\psi|^2$ and $J = \operatorname{Im}(\bar{\psi}\nabla\psi)$. On the other hand, as we highlighted in Sect. 1.4 , it is suitable to consider the hydrodynamic state $(\sqrt{\rho}, \Lambda)$ as providing the minimal information about the hydrodynamic system. Thus, our first goal is to determine the pair $(\sqrt{\rho}, \Lambda)$ from a given wave function. This will be achieved in the following Sect. 1.5.1 using the polar factorization, which will provide the correct setup to rigorously justify the Madelung transformation without the ambiguity of defining the phase in the vacuum region.

Contrarily, one could ask under which assumptions it is possible to go in the opposite direction, namely given a hydrodynamic state $(\sqrt{\rho}, \Lambda)$ we want to determine an associated wave function. If we think of the Madelung transform (or, more precisely, the polar factorization) as a map from the space of wave functions to the space of hydrodynamic states, then we want to see whether it is possible to invert this map. This is somehow reminiscent of the so-called Pauli problem [86], which can be stated as follows. Is it possible to determine a quantum state by performing a (finite) number of measurements? The original problem was stated given the knowledge of the position and momentum probability distributions, $|\psi(x)|^2$ and $|\widehat{\psi}(\xi)|^2$, respectively, but a similar question can also be asked by giving the position and current densities ρ and J as data.

As remarked in [103], it is straightforward to see that, given ρ and J, the quantum state ψ is not uniquely determined. Indeed, it is sufficient to consider two real-valued wave functions $\psi_\pm$ satisfying

$$\psi_\pm(x) = \pm\psi_\pm(-x), \quad \psi_\pm(0) = 0,$$

then we see that the measurement of ρ and J can not distinguish the two different states. This can be generalized by considering arbitrary phase shifts on the different connected components of the region where the wave function does not vanish (the non-vacuum set). See also Lemma 1.3 below. This fact, combined with the possible change in time of the number of connected components of the non-vacuum set, has been exploited recently in [82] to provide a non-uniqueness theorem for weak solutions to a pressureless QHD system, see also [102] for a previous formal discussion.

We address these questions in the Sect. 1.5.3. First, we provide some results related to the wave function lifting, namely, we determine under which conditions on $(\sqrt{\rho}, \Lambda)$ it is possible to associate a wave function to it. Moreover, by focusing on the one-dimensional case, we discuss some stability and instability properties of this lifting; in particular, we will also provide conditions that allow us to prove the H^2 regularity of the wave function.

Finally, let us also mention that it is possible to give a geometrical interpretation of the wave function lifing, in the spirit of Arnold's geometrical description of incompressible fluid dynamics, see [72].

The results presented here will be then used in the next Sect. 1.6 to prove some existence theorems for solutions to QHD systems. Nevertheless, they have an independent interest in establishing a correspondence between quantum fluid and wave function dynamics.

1.5.1 Polar Factorization

This section introduces the polar factorization method to define the hydrodynamical state $(\sqrt{\rho}, \Lambda)$ from a given wave function ψ. Unlike the Madelung transformation, which defines the velocity field as the phase gradient, here we exploit the polar factor associated with ψ. More precisely, given an arbitrary function $\psi : \mathbf{R}^d \to \mathbf{C}$, $\psi \in L^2(\mathbf{R}^d)$, we define the set of its polar factors as

$$P(\psi) = \left\{ \phi \in L^\infty(\mathbf{R}^d) \ : \ \psi = |\psi|\phi \text{ a.e. }, \|\phi\|_{L^\infty(\mathbf{R}^d)} \leq 1 \right\}.$$

Any element $\phi \in P(\psi)$ is called a *polar factor* associated to ψ. It is straightforward to see that $P(\psi)$ may contain more than one element, depending on the values of $\phi \in P(\psi)$ attained in the vacuum region $\{\psi = 0\}$. Since we consider functions of mere Lebesgue or Sobolev regularity, let us recall that by $\{\psi = 0\}$, we mean the set

$$\Sigma(\psi) = \{x \mid \exists \lim_{r \to 0^+} \frac{1}{|B_r(x)|} \int_{B_r(x)} |\psi(y)| \, dy = 0\}.$$

The polar factors satisfy the following properties:

- ϕ is uniquely defined a.e. with respect to the measure $\rho \, dx$;
- $|\phi| = 1$ a.e. in $\rho \, dx$,

where $\rho = |\psi|^2$.

The definition of polar factors is local. Hence, given a compact set $K \subset \mathbf{R}^d$, we may also define

$$P(\psi, K) := \left\{ \phi \in L^\infty(K) \ : \ \psi = |\psi|\phi \text{ a.e. } x \in K, \ \|\phi\|_{L^\infty(K)} \leq 1 \right\}.$$

The set $P(\psi, K)$ can also be characterized in a variational way. Let us define the functional

$$\Phi_K[\phi] = \mathrm{Re} \int_K \bar{\psi}(x)\phi(x) \, dx,$$

where ϕ is taken in the set

$$S_K := \left\{ \phi \in L^\infty(K) \ : \|\phi\|_{L^\infty(K)} \leq 1 \right\}.$$

Then, $P(\psi, K)$ is the set of maximizers of Φ_K over S_K.

Lemma 1.1 *Let $\psi \in L^1_{loc}(\mathbf{R}^d)$ and let $K \subset \mathbf{R}^d$ be a compact set. Then the following are equivalent:*

- *$\phi \in P(\psi, K)$;*
- *ϕ is a maximizer of the functional $\Phi_K[\phi]$ over the set S_K.*

The above lemma, even though straightforward, has two relevant implications. First, it relates the local definition of the polar factors with a (global) variational problem. More importantly, it relates our approach to Brenier's polar factorization [29], performed in a (very) simplified setting.

Proof First of all, let us notice that, by definition, we have

$$\Phi_K[\phi] \leq \int_K |\psi|\, dx, \quad \text{for any } \phi \in S_K.$$

It is straightforward that $P(\psi, K) \subset S_K$ and moreover for $\varphi \in P(\psi, K)$ we have that $\Phi_K[\varphi] = \int_K |\psi|\, dx$, so that φ is also a minimizer.

On the other hand, let $\phi \in S_K$ be a minimizer for Φ_K. Then $\mathrm{Re}(\bar{\psi}(x)\phi(x)) = |\psi(x)|$ a.e. $x \in K$, because otherwise we had $\mathrm{Re}(\bar{\psi}(x)\phi(x)) < |\psi(x)|$ on a set $E \subset K$ of positive measure and consequently

$$\widetilde{\varphi}(x) = \begin{cases} \phi(x) & x \in K \setminus E \\ \frac{\psi(x)}{|\psi(x)|} & x \in E \end{cases}$$

would satisfy $\Phi_K[\phi] < \Phi_K[\widetilde{\varphi}]$. To prove that $\phi \in P(\psi, K)$ it then suffices to prove that $\mathrm{Im}(\bar{\psi}\phi) = 0$. Let us assume this is not the case, namely there exists a measurable set $E \subset K$ of positive measure such that $\mathrm{Im}(\bar{\psi}(x)\phi(x)) \neq 0$ in this set. In this case, it is possible to define on E the following function

$$\chi = \frac{\mathrm{Im}(\bar{\psi}\phi)}{|\psi|}$$

such that, again, on the set E, we have

$$\bar{\psi}(x)\phi(x) = |\psi(x)|(1 + i\chi(x)).$$

As a consequence we would obtain on the set E that $|\phi(x)|^2 = 1 + |\chi(x)|^2 > 1$, contradicting the fact that $\phi \in S_K$. Hence ϕ satisfies $\bar{\psi}\phi = |\psi|$, and consequently $\phi \in P(\psi, K)$. □

We are now going to show that, in the framework of finite energy wave functions, it is possible to exploit the polar factor to define the finite energy hydrodynamic state $(\sqrt{\rho}, \Lambda)$, which in turn determine the particle and momentum densities by $\rho = (\sqrt{\rho})^2$ and $J = \sqrt{\rho}\Lambda$, respectively, see the discussion in Sect. 1.4.

As already said, a wave function $\psi \in H^1(\mathbf{R}^d)$ straightforwardly defines the physical observables through the Madelung transformation $\rho = |\psi|^2$ and $J = \mathrm{Im}(\bar{\psi}\nabla\psi)$. Since we have $J = \sqrt{\rho}\Lambda$, this suggests that the natural definition of Λ in terms of a wave function is given by $\Lambda = \mathrm{Im}(\bar{\phi}\nabla\psi)$, where $\phi \in P(\psi)$.

Moreover, this definition for Λ does not depend on the particular choice of $\phi \in P(\psi)$. This is a consequence of the following fundamental result for Sobolev functions, which may be found in [80, Theorem 6.19.].

Theorem 1.4 (Theorem 6.19, [80]) *Let $A \subset \mathbf{R}$ be a Borel set with zero Lebesgue measure and let $f : \Omega \to \mathbf{R}$ be in $W^{1,1}_{loc}(\Omega)$. Let*

$$B = f^{-1}(A) = \{x \in \Omega \,:\, f(x) \in A\} \subset \Omega.$$

Then $\nabla f(x) = 0$ for a.e. $x \in B$.

The theorem above implies that in the vacuum region $\{\rho = 0\}$, where the polar factor is not uniquely defined, we have $\nabla \psi = 0$ a.e. As a consequence, we also have that $\Lambda = \mathrm{Im}(\bar{\phi} \nabla \psi) = 0$ a.e. on $\{\rho = 0\}$, see also Remark 1.8 below.

Definition 1.9 Given $\psi \in H^1(\mathbf{R}^d)$, $(\sqrt{\rho}, \Lambda) \in H^1(\mathbf{R}^d) \times L^2(\mathbf{R}^d)$, we say that $(\sqrt{\rho}, \Lambda)$ is the hydrodynamic state associated to ψ if $\sqrt{\rho} = |\psi|$, $\Lambda = \mathrm{Im}(\bar{\phi} \nabla \psi)$, where $\phi \in P(\psi)$.

The following proposition provides the rigorous justification of hydrodynamic quantities in terms of wave function and polar factor. Moreover, it shows some stability properties related to the polar factorization.

Proposition 1.2 *Let $\psi \in H^1(\mathbf{R}^d)$ and $\phi \in P(\psi)$ be a polar factor associated to ψ. Then, the hydrodynamic state $(\sqrt{\rho}, \Lambda)$ associated to ψ, defined by $\sqrt{\rho} := |\psi|$ and $\Lambda := \mathrm{Im}(\bar{\phi} \nabla \psi)$, respectively, satisfies the following properties:*

- $\sqrt{\rho} \in H^1(\mathbf{R}^d)$, *and we have*

$$\nabla \sqrt{\rho} = \mathrm{Re}(\bar{\phi} \nabla \psi) \quad \text{a.e. } x \in \mathbf{R}^d; \tag{1.50}$$

-

$$\mathrm{Re}\,(\nabla \psi \otimes \nabla \psi) = \nabla \sqrt{\rho} \otimes \nabla \sqrt{\rho} + \Lambda \otimes \Lambda, \quad \text{a.e. } x \in \mathbf{R}^d; \tag{1.51}$$

- *by defining $J := \sqrt{\rho} \Lambda$, then $\nabla \wedge J \in L^1(\mathbf{R}^d)$ and the following identity*

$$\nabla \wedge J = 2 \nabla \sqrt{\rho} \wedge \Lambda, \tag{1.52}$$

  holds a.e. in $\mathbf{R}^d$.

Moreover, if $\{\psi_n\} \subset H^1(\mathbf{R}^d)$ satisfies $\psi_n \to \psi$ in $H^1(\mathbf{R}^d)$ and $(\sqrt{\rho_n}, \Lambda_n)$ is the hydrodynamic state associated to ψ_n, then we have

$$\begin{aligned} \sqrt{\rho_n} &\to \sqrt{\rho}, \quad H^1(\mathbf{R}^d), \\ \Lambda_n &\to \Lambda, \quad L^2(\mathbf{R}^d). \end{aligned} \tag{1.53}$$

Proof Let us consider $\sqrt{\rho} = |\psi|$, it is straightforward to see that $\left|\nabla\sqrt{\rho}\right| \leq |\nabla\psi|$ a.e. so that we have $\sqrt{\rho} \in H^1(\mathbf{R}^d)$ and also $\nabla\sqrt{\rho} = 0$ a.e. in the vacuum region. Moreover,

$$2\sqrt{\rho}\nabla\sqrt{\rho} = \nabla\rho = 2\,\mathrm{Re}(\bar{\psi}\nabla\psi)$$

and consequently

$$\sqrt{\rho}\left(\nabla\sqrt{\rho} - \mathrm{Re}(\bar{\phi}\nabla\psi)\right) = 0.$$

The last identity implies that $\nabla\sqrt{\rho} = \mathrm{Re}(\bar{\phi}\nabla\psi)$ a.e. in $\sqrt{\rho}\,dx$. On the other hand, we already observed that both $\nabla\sqrt{\rho}$ and $\mathrm{Re}(\bar{\phi}\nabla\psi)$ vanish in the vacuum region, so that the identity (1.50) holds a.e. Let us now consider the quadratic term on the left-hand side of (1.51), again by using Theorem 1.4 and the fact that $|\phi| = 1$ a.e. in $\sqrt{\rho}\,dx$, we obtain

$$\mathrm{Re}(\nabla\bar{\psi} \otimes \nabla\psi) = \mathrm{Re}\left((\phi\nabla\bar{\psi}) \otimes (\bar{\phi}\nabla\psi)\right) = \nabla\sqrt{\rho} \otimes \nabla\sqrt{\rho} + \Lambda \otimes \Lambda,$$

where in the last identity we used the definition of Λ and the formula (1.50).

We now prove identity (1.52). By definition we have $J = \mathrm{Im}(\bar{\psi}\nabla\psi)$, so that $\nabla \wedge J = \mathrm{Im}(\nabla\bar{\psi} \wedge \nabla\psi)$. By using a similar argument as for the identity (1.51), we can argue as follows,

$$\nabla \wedge J = \mathrm{Im}(\nabla\bar{\psi} \wedge \nabla\psi) = \mathrm{Im}\left((\phi\nabla\bar{\psi}) \wedge (\bar{\phi}\nabla\psi)\right) = 2\nabla\sqrt{\rho} \wedge \Lambda.$$

Let us now prove the H^1 stability property. We consider a sequence $\{\psi_n\} \subset H^1(\mathbf{R}^d)$ such that $\|\psi_n - \psi\|_{H^1} \to 0$ as $n \to \infty$. Let $(\sqrt{\rho_n}, \Lambda_n)$ be the hydrodynamic quantities associated to ψ_n. First of all, by using that $|\sqrt{\rho_n} - \sqrt{\rho}| \leq |\psi_n - \psi|$ a.e. we conclude

$$\sqrt{\rho_n} \to \sqrt{\rho}, \qquad \text{in } L^2(\mathbf{R}^d). \tag{1.54}$$

Moreover, since the polar factors $\phi_n \in P(\psi_n)$ are uniformly bounded in $L^\infty(\mathbf{R}^d)$ by definition, there exists a subsequence such that $\phi_{n_k} \overset{*}{\rightharpoonup} \phi$ in L^∞, for some $\phi \in P(\psi)$. By using again the identity (1.50) and the definition of Λ, we infer

$$\begin{aligned} \nabla\sqrt{\rho_{n_k}} &\rightharpoonup \mathrm{Re}(\bar{\phi}\nabla\psi) = \nabla\sqrt{\rho}, \quad L^2 \\ \Lambda_{n_k} &\rightharpoonup \mathrm{Im}(\bar{\phi}\nabla\psi) = \Lambda, \quad L^2, \end{aligned} \tag{1.55}$$

where $(\sqrt{\rho}, \Lambda)$ are the hydrodynamic quantities associated to ψ. Now, we claim that the convergence in (1.55) is actually strong. Indeed, by using the lower semicontinuity of the norms with respect to the weak convergence and by exploiting the identity (1.51), we have that

$$\|\nabla\sqrt{\rho}\|_{L^2}^2 + \|\Lambda\|_{L^2}^2 \le \liminf_{k\to\infty}\left(\|\nabla\sqrt{\rho}_{n_k}\|_{L^2}^2 + \|\Lambda_{n_k}\|_{L^2}^2\right)$$
$$= \liminf_{k\to\infty}\|\nabla\psi_{n_k}\|_{L^2}^2 = \|\nabla\psi\|_{L^2}^2 = \|\nabla\sqrt{\rho}\|_{L^2}^2 + \|\Lambda\|_{L^2}^2,$$

hence

$$\lim_{k\to\infty}\|\nabla\sqrt{\rho}_{n_k}\|_{L^2} = \|\nabla\sqrt{\rho}\|_{L^2}, \quad \lim_{k\to\infty}\|\Lambda_{n_k}\|_{L^2} = \|\Lambda\|_{L^2}.$$

By combining the convergence of the norms with (1.55) we then obtain the strong convergence, $\nabla\sqrt{\rho}_{n_k} \to \nabla\sqrt{\rho}$ and $\Lambda_{n_k} \to \Lambda$ in $L^2(\mathbf{R}^d)$. Finally, we claim that it is actually the whole sequence that is converging strongly. Indeed, let us assume the claim is not true. Then there is for instance a subsequence $\{\nabla\sqrt{\rho}_{n_\ell}\}$ which is not a Cauchy sequence, namely there exists $\varepsilon_\star > 0$ such that

$$\|\nabla\sqrt{\rho}_{n_{\ell_1}} - \nabla\sqrt{\rho}_{n_{\ell_2}}\|_{L^2} \ge \varepsilon_\star,$$

for any $\ell_1, \ell_2 \in \mathbf{N}$. On the other hand, the sequences $\{\nabla\sqrt{\rho}_{n_\ell}\}, \{\Lambda_{n_\ell}\}$ are still uniformly bounded in L^2, so that there exist two subsequences converging weakly and for which we can repeat the previous argument, also showing the convergence of norms. This eventually leads to a contradiction; hence, the whole sequence of hydrodynamic states $\{(\sqrt{\rho}_n, \Lambda_n)\}$ converges strongly. □

Remark 1.8 From Definition 1.9 and Theorem 1.4, we see that the hydrodynamic quantity $\Lambda = \mathrm{Im}(\bar{\phi}\nabla\psi)$ satisfies $\Lambda = 0$ a.e. on $\{\rho = 0\}$. This implies that it is possible to define a velocity field; for instance, we can define

$$v(x) = \begin{cases} \frac{\Lambda(x)}{\sqrt{\rho(x)}} & \text{if } \sqrt{\rho}(x) \neq 0 \\ 0 & \text{if } \sqrt{\rho}(x) = 0. \end{cases}$$

This definition does not overcome the problem in the vacuum region, where the velocity field can take arbitrary values. Moreover, the only available information would still be that $\sqrt{\rho}v \in L^2(\mathbf{R}^d)$. On the other hand, from the generalized irrotationality condition (1.52), we can infer that in the distributional sense we have $\rho\nabla \wedge v = \nabla \wedge J - 2\nabla\sqrt{\rho} \wedge \Lambda = 0$, see also Remark 1.5.

Corollary 1.1 *Let* $\psi : I \times \mathbf{R}^d \to \mathbf{C}$, $I \subset \mathbf{R}$, *be such that* $\psi \in C(I; H^1(\mathbf{R}^d))$. *Then, the time-dependent hydrodynamic state* $(\sqrt{\rho}, \Lambda)$ *associated to* ψ *satisfies*

$$(\sqrt{\rho}, \Lambda) \in C(I; H^1(\mathbf{R}^d) \times L^2(\mathbf{R}^d)).$$

Proof This is a straightforward consequence of the H^1-stability property of the polar factorization. □

In the next lemma, we prove an approximation result, which will be used later in Sect. 1.7.

Lemma 1.2 *Let $\varepsilon, \tau > 0$ be two arbitrary (small) real numbers and let $\alpha \geq 0$. Let us consider a wave function $\Phi \in H^1(\mathbf{R}^d)$ and let $(\sqrt{\rho}, \Lambda) \in H^1(\mathbf{R}^d) \times L^2(\mathbf{R}^d)$ be its associated hydrodynamic state. Then there exists $\Psi_\star \in H^1(\mathbf{R}^d)$ such that if $(\sqrt{\rho}_\star, \Lambda_\star)$ is its associated hydrodynamic state, we have*

$$\begin{cases} \sqrt{\rho}_\star = \sqrt{\rho} + r_\varepsilon \\ \Lambda_\star = (1 - \alpha\tau)\Lambda + \Gamma_\varepsilon \\ \Psi_\star = \Phi + R^1_{\varepsilon,\tau} \\ \nabla\Psi_\star = \nabla\Phi - i\alpha\tau\widehat{\varphi}\Lambda + R^2_{\varepsilon,\tau}, \end{cases} \tag{1.56}$$

where $\widehat{\varphi}$ satisfies $\|\widehat{\varphi}\|_{L^\infty} \leq 1$ and the remainder terms in (1.56) are bounded by

$$\begin{aligned} \|r_\varepsilon\|_{H^1} + \|\Gamma_\varepsilon\|_{L^2} &\leq \varepsilon \\ \|R^1_{\varepsilon,\tau}\|_{L^2} + \|R^2_{\varepsilon,\tau}\|_{L^2} &\lesssim \varepsilon + \alpha\tau\|\Phi\|_{H^1}. \end{aligned}$$

Proof Let $\Phi \in H^1(\mathbf{R}^d)$, we consider $\{\psi_n\} \subset C_0^\infty(\mathbf{R}^d)$ a sequence of smooth, compactly supported functions, converging strongly to Φ in H^1. Then, we may write

$$\psi_n = e^{i\theta_n}\sqrt{\rho}_n,$$

where θ_n is a piecewise smooth function, defined on the bounded set $\Omega_n := \{x \in \mathbf{R}^d : |\psi_n(x)| > 0\}$. From the H^1-stability property in (1.53), we have

$$\sqrt{\rho_n} \to \sqrt{\rho} \text{ in } H^1, \quad \Lambda_n \to \Lambda \text{ in } L^2,$$

where $(\sqrt{\rho}_n, \Lambda_n)$ is the hydrodynamic state associated to ψ_n. Let $n^* = n^*(\varepsilon) \in \mathbf{N}$ be such that

$$\|\psi_{n^*} - \Phi\|_{H^1} + \|\Lambda_{n^*} - \Lambda\|_{L^2} \leq \varepsilon. \tag{1.57}$$

We may thus define

$$\Psi_\star := e^{i(1-\alpha\tau)\theta_{n^*}}\sqrt{\rho}_{n^*},$$

so that its associated hydrodynamic state is given by $(\sqrt{\rho}_\star, \Lambda_\star) = (\sqrt{\rho}_{n^*}, (1 - \alpha\tau)\Lambda_{n^*})$. The first two identities in (1.56) are satisfied with $r_\varepsilon = \sqrt{\rho}_{n^*} - \sqrt{\rho}$ and $\Gamma_\varepsilon = (1 - \alpha\tau)(\Lambda_{n^*} - \Lambda)$, whose norms are controlled by (1.57). We now compute the gradient of $\Psi_\star$. We have

$$
\begin{aligned}
\nabla\Psi_\star =& e^{i(1-\alpha\tau)\theta_{n^*}}\left(\nabla\sqrt{\rho}_{n^*} + i(1-\alpha\tau)\sqrt{\rho}_{n^*}\nabla\theta_{n^*}\right) \\
=& \nabla\psi_{n^*} - i\alpha\tau e^{i(1-\alpha\tau)\theta_{n^*}}\Lambda_{n^*} + \left(e^{i(1-\alpha\tau)\theta_{n^*}} - e^{i\theta_{n^*}}\right)\left(\nabla\sqrt{\rho}_{n^*} + i\Lambda_{n^*}\right) \\
=& \nabla\Phi - i\alpha\tau e^{i(1-\alpha\tau)\theta_{n^*}}\Lambda + \left(e^{i(1-\alpha\tau)\theta_{n^*}} - e^{i\theta_{n^*}}\right)\left(\nabla\sqrt{\rho}_{n^*} + i\Lambda_{n^*}\right) \\
&+ (\nabla\psi_{n^*} - \nabla\Phi) - i\alpha\tau e^{i(1-\alpha\tau)\theta_{n^*}}(\Lambda_{n^*} - \Lambda),
\end{aligned}
$$

so that the third identity in (1.56) is satisfied with $\widehat{\varphi} = e^{i(1-\alpha\tau)\theta_{n^*}}$ and

$$
\begin{aligned}
R^2_{\varepsilon,\tau} =& \left(e^{i(1-\alpha\tau)\theta_{n^*}} - e^{i\theta_{n^*}}\right)\left(\nabla\sqrt{\rho}_{n^*} + i\Lambda_{n^*}\right) + (\nabla\psi_{n^*} - \nabla\Phi) \\
&- i\alpha\tau e^{i(1-\alpha\tau)\theta_{n^*}}(\Lambda_{n^*} - \Lambda).
\end{aligned}
$$

In order to estimate the L^2 norm of $R^2_{\varepsilon,\tau}$, we use that

$$
e^{i(1-\alpha\tau)\theta_{n^*}} - e^{i\theta_{n^*}} = \int_0^1 \frac{d}{ds} e^{i(1-s\alpha\tau)\theta_{n^*}}\,ds = -\alpha\tau\theta_{n^*}\int_0^1 e^{i(1-s\alpha\tau)\theta_{n^*}}\,ds,
$$

and hence

$$
\| e^{i(1-\alpha\tau)\theta_{n^*}} - e^{i\theta_{n^*}} \|_{L^\infty(\mathbf{R}^d)} \le 2\pi\alpha\tau. \tag{1.58}
$$

By combining (1.57) and (1.58), we then obtain the control on the L^2 norm of $R^2_{\varepsilon,\tau}$. The estimate for $R^1_{\varepsilon,\tau} = \Psi_\star - \Phi$ is done similarly. □

Let us mention that the polar factorization approach has been proved useful also in the presence of magnetic fields [8, 16]. More importantly, it can also be adapted in the case when the mass density satisfies a non-trivial far-field behavior, namely when $\rho^* > 0$ in (1.32). This case is particularly relevant for studying special solutions, like traveling waves and especially vortices. The analysis is more involved in the latter case since vortices carry infinite energy. For more details about this case, we refer to [58] and the forthcoming paper [18].

1.5.2 Regularity Properties of Hydrodynamic States

1.5.2.1 The Generalized Chemical Potential for Schrödinger-generated States

We conclude this section on the polar factorization method with a brief discussion on the generalized chemical potential, already introduced in Sect. 1.4.1. We also recall that the notion of generalized chemical potential and the entropy functional

$I(t)$ defined in (1.47) will be used in Sect. 1.9 to prove a stability result for weak solutions to the one-dimensional QHD system. Here we focus on Schrödinger-generated hydrodynamic states.

For this purpose, we consider a finite energy solution $\psi \in C(\mathbf{R}; H^1(\mathbf{R}^d))$ to the following Cauchy problem,

$$\begin{cases} i\partial_t\psi = -\frac{1}{2}\Delta\psi + f'(|\psi|^2)\psi \\ \psi(0) = \psi_0, \end{cases} \tag{1.59}$$

where $f = f(\rho)$ is given by (1.32) with $\rho_\star = 0$. From Theorem 1.3, we also know that if $\psi_0 \in H^2$, then actually $\psi \in C(\mathbf{R}; H^2(\mathbf{R}^d)) \cap C^1(\mathbf{R}; L^2(\mathbf{R}^d))$, and it satisfies the bounds (1.30). As we will see in the next section, for instance, Theorem 1.5 below, the time-dependent hydrodynamic state $(\sqrt{\rho}, \Lambda)$ associated to ψ defines a finite energy weak solution to QHD (1.31). By recalling the Definition of the generalized chemical potential given in (1.49), we have

$$\lambda = \frac{\xi}{\sqrt{\rho}}\mathbf{1}_{\{\sqrt{\rho}>0\}}, \tag{1.60}$$

where $\xi = -\frac{1}{4}\Delta\rho + \frac{1}{2}|\nabla\sqrt{\rho}|^2 + \frac{1}{2}|\Lambda|^2 + \rho f'(\rho)$. We now claim that, for such solution (ρ, J) generated by ψ solution to (1.59), we have

$$\lambda = -\operatorname{Im}(\bar{\phi}\partial_t\psi), \tag{1.61}$$

where ϕ is a (time-dependent) polar factor for ψ. First, we have that $\lambda \in L^\infty(\mathbf{R}; L^2(\mathbf{R}^d))$. Moreover, a direct computation gives

$$\sqrt{\rho}\lambda = -\operatorname{Im}(\bar{\psi}\partial_t\psi) = -\frac{1}{4}\Delta\rho + \frac{1}{2}|\nabla\sqrt{\rho}|^2 + \frac{1}{2}|\Lambda|^2 + \rho f'(\rho).$$

Finally, by using again the NLS equation (1.59), we also have that

$$\lambda = -\frac{1}{2}\operatorname{Im}(\bar{\phi}\Delta\psi) + \sqrt{\rho} f'(\rho).$$

By recalling that $\psi \in C_t H_x^2$ and by invoking again Theorem 1.4, we then infer $\lambda = 0$ a.e. on $\{\rho = 0\}$. Thus, λ defined in (1.61) coincides with the generalized chemical potential defined in (1.60).

Let us now focus on the entropy functional $I(t)$ defined in (1.47). We claim that, for Schrödinger-generated solutions, it holds

$$I(t) = \int_{\mathbf{R}^d} \frac{1}{2}|\partial_t\psi|^2\, dx. \tag{1.62}$$

Indeed, by (1.47) and the continuity equation, we write

$$I(t) = \int \frac{1}{2}\lambda^2 + \frac{1}{2}(\partial_t\sqrt{\rho})^2\, dx.$$

On the other hand, by using the polar factorization, we also have that

$$|\partial_t\psi|^2 = |\bar{\phi}\partial_t\psi|^2 = |\operatorname{Im}(\bar{\phi}\partial_t\psi)|^2 + |\operatorname{Re}(\bar{\phi}\partial_t\psi)|^2 = \lambda^2 + (\partial_t\sqrt{\rho})^2,$$

a.e. $x \in \mathbf{R}^d$.

Resuming, we see that for Schrödinger-generated solutions to QHD system, there is a quite immediate characterization of the generalized chemical potential and the entropy functional provided using the polar factorization. Moreover, given the identity (1.62), we observe that the class of GCP hydrodynamic states corresponds to the H^2-regularity class of wave functions. However, Definition 1.6 in the previous section is valid for general hydrodynamic states. In the one-dimensional case, a more general characterization of λ for arbitrary hydrodynamic states is provided; see [14, Section 7].

1.5.3 Wave Function Lifting and Instabilities

In the previous subsection, we discussed the polar factorization method that defines the hydrodynamic state $(\sqrt{\rho}, \Lambda)$ associated with a given wave function ψ. The definition also entails that $\nabla\sqrt{\rho} = \Lambda = 0$ a.e. in the vacuum region $\{\rho = 0\}$, see also the Remark 1.8.

In this subsection, we try to address the opposite question, namely, when it is possible to define a wave function associated with a given hydrodynamic state.

The problem in its full generality remains widely open. Besides the non-uniqueness issues related to the Pauli problem, even the existence can be proved only in special cases. Clearly, not all hydrodynamic states can generate a wave function. Formally, if we want to reconstruct its phase, then we have to require the velocity field to be irrotational. Thus, a necessary condition for a hydrodynamic state to be associated with a wave function is that it satisfies the generalized irrotationality condition (1.38). Moreover, not all irrotational hydrodynamic states may be lifted to some wave function, and a further quantization condition on the vorticity will be needed. A partial result in this direction will be presented in Proposition 1.6 below, see also the recent paper by Bianchini [27] for a generalization of these results.

The main part of this subsection deals with the one-dimensional case, where we can give a more precise characterization of the wave function lifting at different levels. The main advantages of working in one space dimension are two: first of all, the generalized irrotationality condition becomes trivial, and moreover, the finite energy assumption on the hydrodynamic state $(\sqrt{\rho}, \Lambda)$ readily entails the continuity of the mass density. This clearly implies that the non-vacuum set of ρ, namely $\{\rho > 0\}$ is the countable union of open disjoint intervals,

$$\{\rho > 0\} = \cup_j (a_j, b_j). \tag{1.63}$$

The content of this subsection summarizes some of the recent results in [14] and [15].

Hereafter, the discussion is focused on the one-dimensional case. We will state a few results on the multi-dimensional case in Sect. 1.5.3.1. Having in mind the Definition 1.9, we provide a somewhat converse definition.

Definition 1.10 Let $\Omega \subset \mathbf{R}$ be an arbitrary set. Given the finite energy state $(\sqrt{\rho}, \Lambda) \in H^1(\Omega) \times L^2(\Omega)$, we say that the wave function $\psi \in H^1(\Omega)$ is associated to $(\sqrt{\rho}, \Lambda)$ if

$$\sqrt{\rho} = |\psi|, \qquad \Lambda = \mathrm{Im}(\bar{\phi}\partial_x \psi),$$

where $\phi \in P(\psi)$ is a polar factor for ψ.

Remark 1.9 The polar factorization in Proposition 1.2 above shows that, given a wave function $\psi \in H^1(\mathbf{R}^d)$, then the state $(\sqrt{\rho}, \Lambda)$ is uniquely defined. As we already said at the beginning of this section, the converse is not true; it is easy to consider a hydrodynamic state for which more than one associated wave function exists. Therefore, while in Definition 1.9, we refer to $(\sqrt{\rho}, \Lambda)$ as the state associated with ψ, on the contrary, in Definition 1.10, the wave function $\psi \in H^1$ may not be unique.

The following proposition gives sufficient and necessary conditions for the hydrodynamic state $(\sqrt{\rho}, \Lambda)$ to have an associated wave function $\psi \in H^1(\mathbf{R})$.

Proposition 1.3 *Let $(\sqrt{\rho}, \Lambda)$ be a hydrodynamic state. There exists an associated wave function $\psi \in H^1(\mathbf{R})$ in the sense of the Definition 1.10, if and only if $(\sqrt{\rho}, \Lambda) \in H^1(\mathbf{R}) \times L^2(\mathbf{R})$ and $\Lambda = 0$, a.e. $x \in \{\rho = 0\}$.*

Proof The "only if" implication is straightforward because of the polar factorization of Proposition 1.2.

To prove the converse statement, let us consider a sequence of approximating mass densities given by

$$\sqrt{\rho}_n(x) = \sqrt{\rho}(x) + \frac{1}{n} e^{-|x|^2/2}.$$

In this way, $\{(\sqrt{\rho}_n, \Lambda)\}$ is a sequence of uniformly bounded finite energy states such that $\sqrt{\rho}_n > 0$ and $\sqrt{\rho}_n$ is uniformly bounded away from zero on compact sets. Therefore, it is possible to define the approximating velocity field by $v_n = \frac{\Lambda}{\sqrt{\rho}_n}$ which satisfies $v_n \in L^1_{loc}(\mathbf{R})$, and consequently the phase

$$S_n(x) = \int_0^x v_n(x)\, dx.$$

This defines a sequence of approximating wave functions $\{\psi_n\}$ given by

$$\psi_n(x) = \sqrt{\rho_n}(x)e^{iS_n(x)}.$$

It is now straightforward to see that the sequence $\{\psi_n\}$ is uniformly bounded in $H^1(\mathbf{R})$, since by construction we have $\partial_x\psi_n = e^{iS_n}(\partial_x\sqrt{\rho_n}+i\Lambda)$ and consequently

$$\|\psi_n\|_{H^1}^2 = \|\sqrt{\rho_n}\|_{H^1}^2 + \|\Lambda\|_{L^2}^2 \le C.$$

Therefore, there exists a subsequence, not relabeled, such that $\psi_n \rightharpoonup \psi$ in $H^1(\mathbf{R})$ and $e^{iS_n} \overset{*}{\rightharpoonup} \phi$ in $L^\infty(\mathbf{R})$, for some $\phi \in L^\infty$. By the strong convergence of $\sqrt{\rho}_n \to \sqrt{\rho}$ we infer that $\phi \in P(\psi)$ and consequently,

$$\partial_x\psi_n = e^{iS_n}\left(\partial_x\sqrt{\rho}_n + i\Lambda\right) \rightharpoonup \phi\left(\partial_x\sqrt{\rho}+i\Lambda\right), \quad \text{in } L^2(\mathbf{R})$$

so that $\partial_x\psi = \phi\left(\partial_x\sqrt{\rho}+i\Lambda\right)$. Hence, ψ is a wave function associated with the state $(\sqrt{\rho}, \Lambda)$. □

The non-uniqueness of the wave function associated with a given hydrodynamic state leads to some instability properties of the method, as opposed to the polar factorization, which on the contrary, enjoys suitable stability properties in $H^1(\mathbf{R}^d)$.

More precisely, in what follows, we will see that the associated wave function can not be uniquely defined up to arbitrary phase shifts in the connected components of the non-vacuum set $\{\rho > 0\}$.

We first state an elementary lemma for hydrodynamic states with positive mass densities.

Lemma 1.3 *Let $I \subset \mathbf{R}$ be an arbitrary interval and $\psi_1, \psi_2 \in H^1(I)$ be two wave functions associated to the same hydrodynamic data $(\sqrt{\rho}, \Lambda) \in H^1(I) \times L^2(I)$. Furthermore, we assume that $\rho > 0$ on the interval I. Then, there exists a constant $\theta \in [0, 2\pi)$ such that*

$$\psi_2 = e^{i\theta}\,\psi_1. \tag{1.64}$$

Proof By our assumptions and the polar factorization, see Proposition 1.2, we have

$$\psi_j = \sqrt{\rho}\phi_j, \qquad \partial_x\psi_j = (\partial_x\sqrt{\rho}+i\Lambda)\phi_j, \quad j = 1, 2,$$

where $\phi_j \in P(\psi_j)$, $j = 1, 2$. Since $\rho > 0$ on the interval I, the polar factor ϕ_j is uniquely defined by

$$\phi_j = \frac{\psi_j}{\sqrt{\rho}},$$

and moreover $\phi_j \in H^1_{loc}(I)$. By defining $\Psi = \bar{\psi}_1\psi_2$, then we see that $\partial_x\Psi = \bar{\phi}_1\phi_2\partial_x\rho$ and consequently $\rho\partial_x(\bar{\phi}_1\phi_2) = 0$. By the positivity of ρ on I, we obtain that $\bar{\phi}_1\phi_2$ is a constant. □

The arbitrary choice of phase shifts in the connected components of the non-vacuum set can also be exploited to show a form of instability for the wave function lifting by the following straightforward example. Let us consider the interval $[-1, 1]$ and the wave function

$$\psi_0(x) = e^{iH(x)\frac{\pi}{2}}|x|,$$

where $H(x)$ is the Heaviside function. Then $\psi_0 \in H^1(-1, 1)$ and the associated hydrodynamic state is $(\sqrt{\rho_0}, \Lambda_0) = (|x|, 0)$. If we now consider the approximating hydrodynamic states $(\sqrt{\rho_n}, \Lambda_n)$ given by $\sqrt{\rho_n} = (|x| + \frac{1}{n})$ and $\Lambda_n = 0$, then by Lemma 1.3, for any $n \in \mathbf{N}$ the only associated wave functions are $\psi_n = e^{i\theta_n}(|x| + 1/n)$, where θ_n can be an arbitrary constant $\theta_n \in [0, 2\pi)$. In this way, we see that no sequence of associated wave functions $\{\psi_n\}$ can converge to ψ_0 in $H^1(-1, 1)$.

We notice that this instability stems from the combination of two facts: the former one, being the indetermination of the phase in the connected components of $\{\rho > 0\}$, and the latter one, which is the change of the number of connected components of the non-vacuum regions in the limit. In the dynamical case, a similar kind of instability has also been used to show the non-uniqueness of solutions for a QHD system, see [82].

On the other hand, the previous example also suggests that it is possible to extract a subsequence from the family of approximating wave functions with a strong limit in $H^1(\mathbf{R})$. This indeed shall be proven in Proposition 1.4 below.

Before studying this kind of stability property for the wave function lifting, let us recall some further basic facts that will be useful later. First of all, if g is a function satisfying $g \in H^1(a, b)$ and $g \in H^1(b, c)$, then $g \in H^1(a, c)$ if and only if g is continuous at $x = b$, in which case we have

$$\|g\|_{H^1(a,c)} = \|g\|_{H^1(a,b)} + \|g\|_{H^1(b,c)}.$$

The second elementary fact already mentioned at the beginning of this section is that if $g \in H^1(\Omega)$ with $\Omega \subset \mathbf{R}$ open, then the set $\{g \neq 0\}$ is a countable union of disjoint open intervals,

$$\{x; g \neq 0\} = \cup_j(a_j, b_j), \quad g(a_j) = g(b_j) = 0. \tag{1.65}$$

Those elementary facts also yield a further fundamental property, which is that arbitrary phase shifts in the connected components of the non-vacuum set do not affect the $H^1(\mathbf{R})$ regularity of a function.

Lemma 1.4 *Let $g \in H^1(\Omega)$ with $\{g \neq 0\} = \cup_j(a_j, b_j)$ and let $\Theta \in L^\infty(\Omega)$ be a picecewise constant function of the form*

$$\Theta(x) = \sum_j \theta_j \mathbf{1}_{(a_j,b_j)}(x), \quad \theta_j \in [0, 2\pi). \tag{1.66}$$

Then we have $e^{i\Theta}g \in H^1(\Omega)$ *and*

$$\partial_x(e^{i\Theta}g) = e^{i\Theta}\partial_x g. \tag{1.67}$$

Proof The proof is elementary and can be found in [14, Lemma 21]. □

Proposition 1.4 *Let* $\{(\sqrt{\rho_n}, \Lambda_n)\} \subset H^1(\mathbf{R}) \times L^2(\mathbf{R})$ *be a sequence of finite energy hydrodynamic states, strongly converging to* $(\sqrt{\rho_0}, \Lambda_0) \in H^1(\mathbf{R}) \times L^2(\mathbf{R})$*. Assume* ψ_n *and* ψ_0 *are wave functions associated to* $(\sqrt{\rho_n}, \Lambda_n)$ *and* $(\sqrt{\rho_0}, \Lambda_0)$*, respectively. Then there exists a subsequence* $\{\psi_{n_k}\}$ *and a piecewise constant function* $\Theta_0 \in L^\infty(\mathbf{R})$ *defined as in* (1.66)*, with* $\{\psi_0 \neq 0\} = \cup_j (a_j, b_j)$*, such that*

$$\lim_{k\to\infty} \|\psi_{n_k} - e^{i\Theta_0}\psi_0\|_{H^1(\mathbf{R})} = 0.$$

Proof Given ψ_n associated to $(\sqrt{\rho_n}, \Lambda_n)$, we know that

$$\partial_x\psi_n = \phi_n\left(\partial_x\sqrt{\rho_n} + i\Lambda_n\right),$$

with $\phi_n \in P(\psi_n)$ and clearly $\{\psi_n\} \subset H^1(\mathbf{R})$ is uniformly bounded. Hence there exists a subsequence $\{n_k\}$ such that $\phi_{n_k} \overset{*}{\rightharpoonup} \widetilde{\phi}_0 \in L^\infty(\mathbf{R})$ and consequently $\psi_{n_k} \rightharpoonup \widetilde{\psi}_0 = \sqrt{\rho_0}\widetilde{\phi}_0 \in H^1(\mathbf{R})$, with $\widetilde{\phi}_0 \in P(\widetilde{\psi}_0)$. Moreover, by using identity (1.51) we have that

$$\lim_{k\to\infty} \|\psi_{n_k}\|_{H^1}^2 = \lim_{k\to\infty}\left(\|\sqrt{\rho_{n_k}}\|_{H^1}^2 + \|\Lambda_{n_k}\|_{L^2}^2\right) = \|\widetilde{\psi}_0\|_{H^1}^2.$$

Consequently, the subsequence converges strongly. It only remains to notice that, since both ψ_0 and $\widetilde{\psi}_0$ are associated to the same hydrodynamic state $(\sqrt{\rho_0}, \Lambda_0)$, by Lemma 1.3 we have that on each connected component (a_j, b_j),

$$\widetilde{\psi}_0(x) = e^{i\theta_j}\psi_0(x), \quad x \in (a_j, b_j),$$

so that $\Theta_0(x) = \sum_j \theta_j \mathbf{1}_{(a_j,b_j)}(x)$. □

Remark 1.10 Let us notice that in the proof of the previous lemma, the phase shift determined by Θ_0 depends on the subsequence; thus, different subsequences yield different phase shifts. Consequently, one could think of "tuning" the whole sequence by applying to each element a different phase shift to allow the strong convergence of the entire sequence towards some wave function associated with $(\sqrt{\rho_0}, \Lambda_0)$. However, to do this, we would also need to control how the non-vacuum region changes along the sequence. For this reason, we can only find a sequence of wave functions whose associated hydrodynamic states are arbitrarily close to $\{(\sqrt{\rho}_n, \Lambda_n)\}$, but may vary due to the variations of the non-vacuum region.

In what follows, we are going to prove a wave function lifting result related to H^2 wave functions. As remarked at the end of Sect. 1.5.2.1, the H^2–regularity class for wave functions corresponds to the class of GCP hydrodynamic states, see Definition 1.7. In particular, the Sobolev regularity for the wave function does not correspond to further regularity for the hydrodynamical quantities. For example if we consider $\psi(x) = x$ on $[-1, 1]$, then $\psi \in C^\infty$, but on the other hand $\sqrt{\rho}$ is only Lipschitz, with $\partial_x^2\sqrt{\rho} = \delta_0$. On the other hand, the H^2 regularity for a wave function yields information on the total energy density e defined in (1.46) and on the generalized chemical potential, see the Definition 1.6. In what follows, we deal with the general case and consider time-independent wave functions and hydrodynamic states so that the definition does not depend on the prescribed dynamics. Recall that, by (1.61), we defined $\lambda = -\operatorname{Im}(\bar{\phi}\partial_t\psi)$.

Another important fact that needs to be highlighted is that in one space dimension, since for $\psi \in H^2$ we have that $\partial_x\psi$ is continuous, we cannot allow for arbitrary phase shifts on the connected components of $\{\rho > 0\}$. This means that given a GCP state $(\sqrt{\rho}, \Lambda)$, the wave function $\psi \in H^2$ will be constructed in such a way that $\partial_x\psi$ cannot have jump discontinuities between two adjacent components of the non-vacuum set. This is somehow opposed to the H^1 case, where arbitrary phase shifts on the connected components of the non-vacuum region do not affect the H^1 regularity, see Lemma 1.3.

Proposition 1.5 (**H^2 Wave Function Lifting**) *Let $(\sqrt{\rho}, \Lambda)$ be a hydrodynamic state. There exists a wave function $\psi \in H^2(\mathbf{R})$, associated to $(\sqrt{\rho}, \Lambda)$ in the sense of Definition 1.10, if and only if*

- *$(\sqrt{\rho}, \Lambda)$ is a GCP state;*
- *the total energy density e defined in (1.46) is continuous.*

Moreover, from the bounds

$$
\begin{aligned}
\|\sqrt{\rho}\|_{H^1} + \|\Lambda\|_{L^2} &\leq M_1 \\
\|\mathbf{1}_{\{\rho>0\}}\partial_x J/\sqrt{\rho}\|_{L^2(\mathbf{R})} + \|\lambda\|_{L^2(\mathbf{R})} &\leq M_2,
\end{aligned}
\tag{1.68}
$$

where λ is defined in (1.49), it follows

$$
\|\psi\|_{H^2(\mathbf{R})} \leq C(M_1, M_2). \tag{1.69}
$$

In order to prove Proposition 1.5, we first state a lemma which shows the connection between the hydrodynamic quantities λ, $\frac{\partial_x J}{\sqrt{\rho}}\mathbf{1}_{\{\rho>0\}}$ and the H^2 regularity of the associated wave function ψ.

Lemma 1.5 *Let $(\sqrt{\rho}, \Lambda)$ be a finite energy hydrodynamic state.*

(1) If the hydrodynamic state $(\sqrt{\rho}, \Lambda)$ *is associated to a wave function* $\psi \in H^2(\mathbf{R})$*, then* $(\sqrt{\rho}, \Lambda)$ *is a GCP state and we have*

$$\frac{\partial_x J}{\sqrt{\rho}} \mathbf{1}_{\{\rho>0\}} = \operatorname{Im}(\bar{\phi}\partial_x^2\psi), \quad \lambda - f'(\rho)\sqrt{\rho} = -\frac{1}{2}\operatorname{Re}(\bar{\phi}\partial_x^2\psi), \tag{1.70}$$

for a.e. $x \in \mathbf{R}$*, where* $\phi \in P(\psi)$ *is a polar factor of* ψ*.*

(2) Let us further assume that $(\sqrt{\rho}, \Lambda)$ *is a GCP state and let* $\psi \in H^1(\mathbf{R})$ *be a wave function associated to* $(\sqrt{\rho}, \Lambda)$*, then* $\psi \in H^2(a_j, b_j)$*, for all connected components* (a_j, b_j) *of the set* $\{\rho > 0\}$ *as defined in* (1.63)*, and we have*

$$\partial_x^2\psi = \left[2(f'(\rho)\sqrt{\rho} - \lambda) + i\frac{\partial_x J}{\sqrt{\rho}}\right]\phi, \quad \text{a.e. } x \in \{\rho > 0\}, \tag{1.71}$$

where $\phi \in P(\psi)$*.*

Proof Let $\psi \in H^2(\mathbf{R})$, then on the non-vacuum set $\{\rho > 0\}$ we can compute

$$\operatorname{Im}(\bar{\phi}\partial_x^2\psi) = \frac{\partial_x \operatorname{Im}(\bar{\psi}\partial_x\psi)}{|\psi|} = \frac{\partial_x J}{\sqrt{\rho}}.$$

Analogously, by using the definition (1.49) for λ, we also have that on $\{\rho > 0\}$

$$\begin{aligned} -\frac{1}{2}\operatorname{Re}(\bar{\phi}\partial_x^2\psi) &= -\frac{1}{2|\psi|}\left[\partial_x \operatorname{Re}(\bar{\psi}\partial_x\psi) - |\partial_x\psi|^2\right] \\ &= -\frac{1}{2\sqrt{\rho}}\left[\frac{1}{2}\partial_x^2\rho - (\partial_x\sqrt{\rho})^2 - \Lambda^2\right] \\ &= \frac{1}{\sqrt{\rho}}(\xi - \rho f'(\rho)) = \lambda - f'(\rho)\sqrt{\rho}. \end{aligned}$$

Let us emphasize that the previous computations are rigorous since $\psi \in H^2(\mathbf{R})$ and $\{\rho > 0\}$ is an open set. Moreover, from Theorem 1.4 we have $\partial_x^2\psi = 0$ a.e. x in $\{\rho = 0\}$, thus identities (1.70) hold almost everywhere on $\mathbf{R}$.

To prove the second statement of this lemma, let $(\sqrt{\rho}, \Lambda)$ now be a GCP state, and let $\psi \in H^1(\mathbf{R})$ be a wave function associated to $(\sqrt{\rho}, \Lambda)$. By using the polar factorization, it follows

$$\psi = \sqrt{\rho}\phi, \ \partial_x\psi = \phi(\partial_x\sqrt{\rho} + i\Lambda),$$

where $\phi \in P(\psi)$. On each connected component (a_j, b_j) of $\{\rho > 0\}$, see (1.63), the polar factor $\phi \in P(\psi)$ is uniquely defined by $\phi = \psi/|\psi| \in H^1_{loc}(a_j, b_j)$ and a direct computation gives that

$$\partial_x\phi = \frac{1}{|\psi|}\left(\partial_x\psi - \partial_x|\psi|\frac{\psi}{|\psi|}\right) = \frac{i\Lambda}{\sqrt{\rho}}\phi, \quad \text{a.e. } x \in (a_j, b_j).$$

Using the identity above and the definition of the hydrodynamic quantities, we obtain

$$\left[2(f'(\rho)\sqrt{\rho}-\lambda)+i\frac{\partial_x J}{\sqrt{\rho}}\right]\phi=\left(\partial_x^2\sqrt{\rho}+i\partial_x\Lambda+i\partial_x\sqrt{\rho}\frac{\Lambda}{\sqrt{\rho}}-\frac{\Lambda^2}{\sqrt{\rho}}\right)\phi$$
$$=\partial_x\left[(\partial_x\sqrt{\rho}+i\Lambda)\phi\right]=\partial_x(\partial_x\psi),$$

which proves (1.71). The bounds given in the Definition 1.7 also yield $\psi \in H^2(a_j,b_j)$. □

The previous lemma shows that it is possible to improve the regularity of the wave function ψ on the connected components of $\{\rho > 0\}$ for a GCP state. However, in general, ψ is not in $H^2(\mathbf{R})$ since $\partial_x\psi$ may possibly experience jump discontinuities at vacuum boundaries due to the fact that the H^1 regularity allows for arbitrary phase shifts on connected components, see Lemma 1.4. On the contrary, the H^2 regularity requires compatibility conditions for the phase shift at vacuum boundaries. We will show that the continuity assumption for the total energy density given in Proposition 1.5 allows for a consistent choice of the phase shifts on each connected component. Let us recall the definition of the energy density,

$$e=\frac{1}{2}(\partial_x\sqrt{\rho})^2+\frac{1}{2}\Lambda^2+f(\rho).$$

By Proposition 1.3, we know that a hydrodynamic state satisfying the assumptions of Proposition 1.5 in particular provides a finite energy wave function $\psi \in H^1(\mathbf{R})$. Thus, to prove Proposition 1.5, we will construct another wave function $\widetilde{\psi} \in H^2(\mathbf{R})$, associated with the same hydrodynamic state.

More precisely, we will construct a piecewise constant function Θ satisfying $\operatorname{supp}\Theta \subset \{\rho > 0\} = \cup_j(a_j,b_j)$ and such that

$$\Theta=\sum_j \theta_j \mathbf{1}_{(a_j,b_j)},\quad \theta_j\in[0,2\pi). \tag{1.72}$$

In this way, the new wave function will be given by $\widetilde{\psi}=e^{i\Theta}\psi \in H^2(\mathbf{R})$.

Before proving Proposition 1.5, we also need the next lemma regarding the total energy density. In what follows, we denote $\partial V = \partial\{\rho = 0\}$ the boundary of the vacuum region. Let us notice that, since from (1.63) we defined $\{\rho > 0\} = \cup_j(a_j,b_j)$, then we have $\partial V = \cup_j\{a_j,b_j\}$.

Lemma 1.6 *Let $(\sqrt{\rho},\Lambda)$ be a GCP hydrodynamic state, and let us assume that the total energy density e is continuous. If $x^\star$ is an accumulation point for the set ∂V, then $e(x^\star)=0$.*

Proof By using Proposition 1.3, there exists a $\psi \in H^1(\mathbf{R})$ associated to $(\sqrt{\rho},\Lambda)$, consequently the total energy density can be expressed as

$$e = \frac{1}{2}|\partial_x \psi|^2 + f(|\psi|^2).$$

In particular, the continuity of e implies that $|\partial_x \psi|$ is also a continuous function. Let $x^\star$ be an accumulation point of the set ∂V, hence there exists a sequence of intervals (a_{j_k}, b_{j_k}) such that $\lim_{k\to\infty} a_{j_k} = \lim_{k\to\infty} b_{j_k} = x^\star$. Let us define $x_{j_k} = \frac{1}{2}(a_{j_k} + b_{j_k})$, then we claim that $\partial_x \psi(x_{j_k}) \to 0$, as $k \to \infty$. If the claim holds, the continuity of $|\partial_x \psi|$ implies that $|\partial_x \psi(x^\star)| = 0$. Since a_{j_k}, b_{j_k} satisfy $\psi(a_{j_k}) = \psi(b_{j_k}) = 0$, then

$$0 = \int_{a_{j_k}}^{b_{j_k}} \partial_x \psi(y)\, dy.$$

Moreover, by Lemma 1.5, we know that $\psi \in H^2(a_{j_k}, b_{j_k})$ so that we can write

$$\begin{aligned} 0 =& \int_{a_{j_k}}^{b_{j_k}} \partial_x \psi(x_{j_k}) + \int_{x_{j_k}}^{y} \partial_{xx} \psi(z)\, dz dy \\ =&(b_{j_k} - a_{j_k})\partial_x \psi(x_{j_k}) + \int_{a_{j_k}}^{b_{j_k}} \int_{x_{j_k}}^{y} \partial_{xx} \psi(z)\, dz dy. \end{aligned}$$

Consequently

$$(b_{j_k} - a_{j_k})|\partial_x \psi(x_{j_k})| \leq (b_{j_k} - a_{j_k})^{3/2} \|\partial_{xx} \psi\|_{L^2(a_{j_k}, b_{j_k})},$$

which proves the claim. □

It is now possible to prove Proposition 1.5.

Proof of Proposition 1.5 Let us consider the hydrodynamic state $(\sqrt{\rho}, \Lambda)$, determined by an associated wave function $\psi \in H^2(\mathbf{R})$, then by the property (1) of Lemma 1.5 we know that $(\sqrt{\rho}, \Lambda)$ is a GCP state. Moreover, by recalling the definition of the energy density given in (1.46) and by the polar factorization (1.51), we have

$$e = \frac{1}{2}(\partial_x \sqrt{\rho})^2 + \frac{1}{2}\Lambda^2 + f(\rho) = \frac{1}{2}|\partial_x \psi|^2 + f(|\psi|^2),$$

and therefore, e is a continuous function.

Now let us assume that $(\sqrt{\rho}, \Lambda)$ is a GCP state satisfying the bounds (1.68) and such that the total energy density e is continuous. As in (1.63), we denote

$$\{\rho > 0\} = \cup_{j\in\mathbf{Z}}(a_j, b_j), \quad \rho(a_j) = \rho(b_j) = 0. \tag{1.73}$$

The wave function lifting stated in Proposition 1.3 shows that there exists a wave function $\psi \in H^1(\mathbf{R})$ associated to $(\sqrt{\rho}, \Lambda)$. Furthermore by Lemma 1.5, $\psi \in$

$H^2(a_j, b_j)$ for all connected components (a_j, b_j) in (1.63) and the identity (1.71) implies

$$\|\partial_x^2 \psi\|_{L^2(a_j,b_j)}^2 = 4\|\lambda - f'(\rho)\sqrt{\rho}\|_{L^2(a_j,b_j)}^2 + \|\frac{\partial_x J}{\sqrt{\rho}}\|_{L^2(a_j,b_j)}^2. \tag{1.74}$$

In the remaining part of the proof, starting from ψ, we will construct a new wave function $\widetilde{\psi}$ such that $\widetilde{\psi} \in H^2(\mathbf{R})$ and $\widetilde{\psi}$ is associated with the same hydrodynamic state $(\sqrt{\rho}, \Lambda)$. As anticipated before, to do that we are going to adjust the phase of the wave function on each connected component of (1.73), in such a way that $\widetilde{\psi}$ does not jump at the vacuum boundary. More precisely we will define $\Theta = \sum_j \theta_j \mathbf{1}_{(a_j,b_j)}$, such that $\widetilde{\psi} = e^{i\Theta}\psi \in H^2$ and also $\widetilde{\psi}$ is associated to $(\sqrt{\rho}, \Lambda)$.

Before constructing Θ, we discuss some properties of the vacuum boundary ∂V. First of all, we decompose ∂V into the union of three subsets,

$$\partial V = \partial(V^\circ) \cup (\partial V)_p \cup (\partial V)_s.$$

Here V° is the interior of V, hence $x_0 \in \partial(V^\circ)$ if either $\rho|_{(x_0-\varepsilon,x_0]} \equiv 0$ or $\rho|_{[x_0,x_0+\varepsilon)} \equiv 0$, for some $\varepsilon > 0$ sufficiently small. The subset $(\partial V)_p$ consists in the set of isolated vacuum points, namely $x_0 \in (\partial V)_p$ if $\rho(x_0) = 0$ and $\rho(x) > 0$ for every $x \in (x_0 - \varepsilon, x_0) \cup (x_0, x_0 + \varepsilon)$, for some $\varepsilon > 0$ sufficiently small. Finally, $(\partial V)_s$ denotes the set of accumulation points of ∂V which do not belong to $\partial(V^\circ)$. By construction, the set

$$W = \{\rho > 0\} \cup (\partial V)_p$$

is open, and hence, it may be represented as the union of disjoint open intervals,

$$W = \cup_j (\bar{a}_j, \bar{b}_j), \tag{1.75}$$

where $\bar{a}_j, \bar{b}_j \in \partial(V^\circ) \cup (\partial V)_s$. By using the above definition and Lemma 1.6 we then infer that $\partial_x \psi(\bar{a}_j)| = |\partial_x \psi(\bar{b}_j)| = 0$.

Let us now consider an arbitrary interval $(\bar{a}_j, \bar{b}_j)$ defined in (1.75). Let $\{x_k\} \subset (\bar{a}_j, \bar{b}_j)$ be the vacuum points in the interval $(\bar{a}_j, \bar{b}_j)$, hence we straightforwardly have $\{x_k\} \subset (\partial V)_p$. By construction, the only possible accumulation points of the set $\{x_k\}$ are $\bar{a}_j$ or $\bar{b}_j$, hence it is possible to assume $x_k < x_{k+1}$ for any k. We will construct Θ on $(\bar{a}_j, \bar{b}_j)$ by using an induction argument on k. Let us start with (x_0, x_1) by fixing $\Theta|_{(x_0,x_1)} \equiv 0$. Now by assuming we already constructed the phase shift on (x_{k-1}, x_k), say $\Theta|_{(x_{k_1},x_k)} = \theta_{k-1}$, $k > 0$ (the argument for $k < 0$ is completely analogous), we want to construct θ_k such that $\Theta|_{(x_k,x_{k+1})} = \theta_k$. First of all we notice that, since both (x_{k-1}, x_k) and (x_k, x_{k+1}) are two connected components of the set $\{\rho > 0\}$, by Lemma 1.5 and the assumptions on $(\sqrt{\rho}, \Lambda)$, we have that $\psi \in H^2$ on both (x_{k-1}, x_k) and (x_k, x_{k+1}). Consequently, the right and left limits $\partial_x \psi(x_k-)$ and $\partial_x \psi(x_k+)$ exist, and moreover, by the continuity assumption on the total energy density e, we also have

$$|\partial_x \psi(x_k-)| = |\partial_x \psi(x_k+)|.$$

Consequently there exists $\theta_k \in [0, 2\pi)$ such that

$$e^{i\theta_{k-1}} \partial_x \psi(x_k-) = e^{i\theta_k} \partial_x \psi(x_k+).$$

Therefore we may set $\Theta|_{(x_k, x_{k+1})} = \theta_k$ so that the inductive step is proved. In this way we constructed a piecewise constant function Θ on $(\bar{a}_j, \bar{b}_j)$ such that

$$\Theta|_{(\bar{a}_j, \bar{b}_j)} = \sum_k \theta_k \mathbf{1}_{(x_k, x_{k+1})}.$$

Now, let us also recall that, by construction, $\partial_x \psi(\bar{a}_j) = \partial_x \psi(\bar{b}_j) = 0$, for any j. Hence, $\partial_x \psi$ cannot experience jumps on $\bar{a}_j$ or $\bar{b}_j$ because of the continuity assumption on the total energy density. This means that, on each connected component $(\bar{a}_j, \bar{b}_j)$ of W we can apply the previous construction, whereas on $W^c = \partial(V^\circ) \cup (\partial V)_s$ we just set $\Theta|_{W^c} = 0$. In this way the piecewise constant function Θ is constructed globally and it satisfies $\operatorname{supp} \Theta \subset \{\rho > 0\} = \cup_j (a_j, b_j)$, with $\Theta = \sum_j \theta_j \mathbf{1}_{(a_j, b_j)}$ (upon relabeling the phase shifts θ_j's). Moreover, by construction we have that $\widetilde{\psi} = e^{i\Theta}\psi \in H^2(\mathbf{R})$ and by Lemma 1.4, also $\widetilde{\psi}$ is a wave function associated to the GCP state $(\sqrt{\rho}, \Lambda)$. □

1.5.3.1 The Multidimensional Case

The wave function lifting developed in the previous section turns out to be useful for studying solutions to one-dimensional QHD systems. In the multi-dimensional case, a general result is not available. One cannot expect to prove a wave function lifting result for general hydrodynamic states $(\sqrt{\rho}, \Lambda)$. A necessary condition for $(\sqrt{\rho}, \Lambda)$ to be associated with a wave function ψ is that $(\sqrt{\rho}, \Lambda)$ must satisfy the generalized irrotationality condition

$$\nabla \wedge J = 2\nabla\sqrt{\rho} \wedge \Lambda. \tag{1.76}$$

Moreover, this is a local condition, hence not sufficient to define a lifting. Indeed, we can consider the following example in $\mathbf{R}^2$. We define $\rho \in C_c^1(\mathbf{R}^2)$, $\rho(0) = 0$, and $J(x) = \rho(x)\frac{\alpha x^\perp}{|x|^2}$, with $\alpha \notin \mathbf{Z}$, so that we have $v(x) = \frac{\alpha x^\perp}{|x|^2}$. The generalized irrotationality condition is satisfied, but it is not possible to solve $\nabla\theta = v$ globally.

This implies that we will require a quantization condition on the vorticity. In what follows, we provide a wave function lifting result in the two-dimensional case; see Proposition 1.6 below. Its proof can be found in [15, Proposition 23]. Let us remark that a more general result was given in [27].

Proposition 1.6 *Let $(\sqrt{\rho}, \Lambda) \in H^1(\mathbf{R}^2) \times L^2(\mathbf{R}^2)$ be a finite energy hydrodynamic state satisfying the following assumptions:*

- *the mass density is continuous, $\rho \in C(\mathbf{R}^2)$, and the vacuum set $V = \{\rho = 0\}$ consists is isolated points, namely $V = \{x_1, x_2, \dots, x_j, \dots\} \subset \mathbf{R}^2$, where, in the case that V is an infinite set, we have $\inf_{j \neq k} |x_j - x_k| =: \alpha > 0$;*
- *the velocity field $v = J/\rho$ (which can be defined a.e. because of the previous assumption), is such that $v \in \mathcal{D}'(\mathbf{R}^2)$ and moreover the vorticity satisfies*

$$\nabla \wedge v = 2\pi \sum_j k_j \delta_{x_j}, \quad k_j \in \mathbf{Z}. \tag{1.77}$$

Then there exists a finite energy wave function $\psi \in H^1(\mathbf{R}^2)$ associated to $(\sqrt{\rho}, \Lambda)$.

1.6 Existence Results for Solutions to Some QHD Systems

In this section, we will provide some existence results for weak solutions to some QHD systems by exploiting the tools developed in the previous section.

Our strategy exploits the analogy with the wave function dynamics given by an associated nonlinear Schrödinger equations. Before stating the rigorous result, we first present the main ideas by providing a sketch of the proof for the prototype model (1.31) in the case where the pressure equation of state is explicitly given by

$$p(\rho) = \frac{\gamma - 1}{\gamma} \rho^\gamma, \tag{1.78}$$

with $1 < \gamma < \frac{d}{(d-2)_+}$. We claim that the associated NLS dynamics is given by

$$i\partial_t \psi = -\frac{1}{2}\Delta\psi + |\psi|^{2(\gamma-1)}\psi. \tag{1.79}$$

By recalling Theorem 1.3 above, we see the Cauchy problem associated to (1.79) is globally well-posed in $H^1(\mathbf{R}^d)$, see the Definition 1.3. Moreover, Eq. (1.79) also enjoys a propagation of regularity property, namely if $\psi_0 \in H^2(\mathbf{R}^d)$, then we also have $\psi \in C(\mathbf{R}; H^2(\mathbf{R}^d))$. With such existence and stability properties at hand, we can now show the existence of weak solutions to (1.31) using the polar factorization method.

Thus, for $\psi_0 \in H^1(\mathbf{R}^d)$, we define its associated momenta (hydrodynamic quantities) $\rho_0 = |\psi_0|^2$ and $J_0 = \text{Im}(\bar{\psi}_0 \nabla \psi_0)$. Let ψ be the solution to the Cauchy problem associated to (1.79) with initial datum $\psi(0) = \psi_0 \in H^1(\mathbf{R}^d)$. Analogously, we define its associated mass density $\rho(t, x) = |\psi(t, x)|^2$, and we differentiate it with respect to time. By using the fact that ψ solves (1.79), we have

$$\partial_t \rho = 2\,\mathrm{Re}\left\{\bar{\psi}\left(\frac{i}{2}\Delta\psi - i|\psi|^{2(\gamma-1)}\psi\right)\right\} = -\,\mathrm{div}\,J, \tag{1.80}$$

where the current density is given by $J(t,x) = \mathrm{Im}(\bar{\psi}\nabla\psi)(t,x)$. We can now differentiate J with respect to time. Analogously, after some simple algebraic calculations, we get

$$\begin{aligned}\partial_t J &= \mathrm{Im}\left(\partial_t\bar{\psi}\nabla\psi + \bar{\psi}\nabla\partial_t\psi\right) \\ &= \mathrm{Im}\left((-\frac{i}{2}\Delta\bar{\psi} + i|\psi|^{2(\gamma-1)}\bar{\psi})\nabla\psi + \bar{\psi}\nabla(\frac{i}{2}\Delta\psi - i|\psi|^{2(\gamma-1)}\psi\right) \\ &= \frac{1}{4}\nabla\Delta\rho - \mathrm{div}\left(\mathrm{Re}(\nabla\bar{\psi}\otimes\nabla\psi)\right) - \rho\nabla\rho^{\gamma-1},\end{aligned} \tag{1.81}$$

so that the following conservation law is obtained,

$$\partial_t J + \mathrm{div}\left(\mathrm{Re}(\nabla\bar{\psi}\otimes\nabla\psi)\right) + \nabla p(\rho) = \frac{1}{4}\nabla\Delta\rho, \tag{1.82}$$

with $p = p(\rho)$ given by (1.78). Let us recall the bilinear identity (1.51) proved in Proposition 1.2, we have

$$\mathrm{Re}(\nabla\bar{\psi}\otimes\nabla\psi) = \nabla\sqrt{\rho}\otimes\nabla\sqrt{\rho} + \Lambda\otimes\Lambda, \quad \text{a.e. } x\in\mathbf{R}^d,$$

and consequently (1.82) yields

$$\partial_t J + \mathrm{div}(\Lambda\otimes\Lambda) + \nabla p(\rho) = \frac{1}{4}\nabla\Delta\rho - \mathrm{div}(\nabla\sqrt{\rho}\otimes\nabla\sqrt{\rho}),$$

which proves that (ρ, J) is a solution to (1.31), in the sense of the Definition 1.4.

Let us emphasize that the proof above is formal because we don't have sufficient regularity to justify all the passages in the derivation of the equation for the momentum density. If we consider finite energy solutions $\psi \in C(\mathbf{R}; H^1(\mathbf{R}^d)) \cap C^1(\mathbf{R}; H^{-1}(\mathbf{R}^d))$, then the right hand side in (1.81) is not a well-defined distribution. Therefore, to make this calculation rigorous, we first need to regularize our initial datum ψ_0 by considering a sequence of more regular initial data, say $\{\psi_{0,n}\} \subset H^2$. By the persistence of regularity property, the solutions emanated from the regular data are also smoother, namely $\psi_n \in C(\mathbf{R}; H^2(\mathbf{R}^d))$. In this way, the calculations above can be justified in the distributional sense for H^2 solutions. Finally, to recover the limit solution, we exploit the continuous dependence of the NLS equation with respect to the initial data and H^1-stability of the polar factorization stated in Proposition 1.2.

By following the above strategy, it is easy to be convinced that we may prove the existence of global in time finite energy weak solutions to QHD systems, as long as the corresponding NLS type dynamics satisfies suitable existence and stability properties, namely global well-posedness and persistence of H^2 regularity.

In particular, we can consider NLS-type evolutions whose nonlinearities satisfy the Assumptions 1.1. Thus, we consider the following Cauchy problem,

$$\begin{cases} \partial_t \rho + \operatorname{div} J = 0 \\ \partial_t J + \operatorname{div}\left(\frac{J \otimes J}{\rho}\right) + \nabla p(\rho) + \rho\nabla(V + \widetilde{V}) = \frac{1}{4}\nabla\Delta\rho - \operatorname{div}(\nabla\sqrt{\rho} \otimes \nabla\sqrt{\rho}) \\ \rho(0) = \rho_0, \quad J(0) = J_0. \end{cases} \tag{1.83}$$

The total energy of the system (1.83) is determined by

$$E(t) = \int_{\mathbf{R}^d} \frac{1}{2}|\nabla\sqrt{\rho}|^2 + \frac{1}{2}|\Lambda|^2 + f(\rho) + \rho(V + \widetilde{V})\, dx, \tag{1.84}$$

where $f = f(\rho)$ is the internal energy defined by the pressure from formula (1.32) with $\rho^* = 0$.

For the sake of clarity, we collect the assumptions on the various terms appearing in (1.83). This is the analog of Assumption 1.1 given for the nonlinear Schrödinger equation (1.24).

Assumption 1.2 The external potential V can be split into a time-independent and a time-dependent part, $V = V_1 + V_2$ such that

- $V_1 = V_1(x)$ satisfies $V_1, \nabla V_1 \in L^\delta(\mathbf{R}^d) + L^\infty(\mathbf{R}^d)$, for some $\delta \geq 1$, $\delta > d/2$;
- $V_2 = V_2(t, x)$ satisfies $V_2 \in L^p(\mathbf{R}; W^{1,s}(\mathbf{R}^d))$ with $\frac{2}{p} + \frac{d}{s} = 2$, $(p, s) \neq (\infty, \frac{d}{2})$, $d \geq 3$ and moreover $\partial_t V \in L^1_{loc}(\mathbf{R}; L^\eta(\mathbf{R}^d) + L^\infty(\mathbf{R}^d))$, with $\max\{1, \frac{d}{2}\} \geq \eta < \infty$, $\eta > 1$ for $d = 2$.

The non-local interaction term $\widetilde{V}$ is given by $\widetilde{V} = W * |\psi|^2$, where the even, interaction kernel $W = W(x)$ satisfies $W \in L^\sigma(\mathbf{R}^d) + L^\infty(\mathbf{R}^d)$ for some $\sigma \geq 1, \sigma > d/4$ and its negative part $W^- := \min\{0, W\}$ satisfies $W^- \in L^\theta(\mathbf{R}^d) + L^\infty(\mathbf{R}^d)$ for some $\theta \geq \max\{1, d/2\}$ ($\theta > 1$ for $d = 2$).

The internal energy $f : \mathbf{R}_+ \to \mathbf{R}$ satisfies:

- $f \in C^1([0, \infty)) \cap C^2((0, \infty))$, $f(0) = 0$;
- $|f'(\rho)|$, $|\rho f''(\rho)| \lesssim 1 + \rho^\alpha$, with $0 < \alpha < \infty$ for $d = 1, 2$ and $0 < \alpha < \frac{2}{d-2}$ for $d \geq 3$;
- $f(\rho) \gtrsim -\rho - \rho^\beta$, with $1 < \beta < 1 + \frac{2}{d}$.

Remark 1.11 Let us emphasize that we do not claim these are the optimal assumptions that ensure an existence result for global-in-time finite energy weak solutions to (1.83). However, this collection provides some interesting examples that have physical relevance.

Theorem 1.5 *Let the terms in system* (1.83) *satisfy the Assumption 1.2. Let us assume the initial data* ρ_0, J_0 *are given by* $\rho_0 = |\psi_0|^2$, $J_0 = \operatorname{Im}(\bar{\psi}_0 \nabla \psi_0)$, *for some* $\psi_0 \in H^1(\mathbf{R}^d)$. *Then, there exists a global in time finite energy weak solution* (ρ, J) *in the sense of Definition 1.5. Moreover, the solution satisfies the following properties:*

- *the time-dependent hydrodynamic state satisfies* $(\sqrt{\rho}, \Lambda) \in C(\mathbf{R}; H^1(\mathbf{R}^d) \times L^2(\mathbf{R}^d))$;
- *the total energy, defined by*

$$E(t) = \int_{\mathbf{R}^d} \frac{1}{2}|\nabla\sqrt{\rho}|^2 + \frac{1}{2}|\Lambda|^2 + f(\rho) + \rho(V + \widetilde{V})\,dx$$

  satisfies the following balance equation

$$E(t) = E(0) + \int_0^t \int_{\mathbf{R}^d} \rho \partial_t V\,dxdt';$$

- *for any* $0 < T < \infty$ *and* (q, r) *admissible pair of exponents, we have*

$$\|\sqrt{\rho}\|_{L^q(0,T;W^{1,r}(\mathbf{R}^d))} + \|\Lambda\|_{L^q(0,T;L^r(\mathbf{R}^d))} \leq C(T, \|\sqrt{\rho_0}\|_{H^1}, \|\Lambda_0\|_{L^2}). \tag{1.85}$$

Proof Let us consider the Cauchy problem (1.24) with $\psi_0 \in H^1(\mathbf{R}^d)$. From Theorem 1.3, we know there exists a unique global solution $\psi \in C(\mathbf{R}; H^1(\mathbf{R}^d))$, Let us consider a sequence $\{\psi_{0,n}\} \subset H^2(\mathbf{R}^d)$ such that $\psi_{0,n} \to \psi_0 \in H^1(\mathbf{R}^d)$. By the persistence of regularity property, we know that the solutions ψ_n emanated from such initial data satisfy $\psi_n \in C(\mathbf{R}; H^2(\mathbf{R}^d))$. Moreover, from the continuous dependence with respect to the initial data, we also have that, for any $0 < T < \infty$,

$$\|\psi_n - \psi\|_{L^\infty(0,T;H^1(\mathbf{R}^d))} \to 0, \quad n \to \infty. \tag{1.86}$$

We define the (time-dependent) hydrodynamic state $(\sqrt{\rho_n}, \Lambda_n)$ associated to ψ_n, namely for $t \in \mathbf{R}$ we set $\sqrt{\rho_n}(t) = |\psi_n(t)|$ and $\Lambda_n(t) = \operatorname{Im}(\bar{\phi}_n \nabla \psi_n)(t)$, where $\phi_n \in L^\infty(\mathbf{R}_t \times \mathbf{R}^d_x)$ is a polar factor associated to ψ_n, i.e. for a.e. $t \in \mathbf{R}$, $\phi_n(t) \in P(\psi_n(t))$. We now claim that $(\rho_n, J_n) = ((\sqrt{\rho_n})^2, \sqrt{\rho_n}\Lambda_n)$ is a global in time finite energy weak solution to (1.83). Indeed, for any $0 < T < \infty$ the following identities hold in $\mathcal{D}'([0, T] \times \mathbf{R}^d)$,

$$\begin{aligned} &\partial_t \rho_n + \operatorname{div} J_n = 0 \\ &\partial_t J_n + \operatorname{div}(\Lambda_n \otimes \Lambda_n) + \nabla p(\rho_n) + \rho_n \nabla(V + \widetilde{V}_n) = \frac{1}{4}\nabla\Delta\rho_n \\ &\quad - \operatorname{div}(\nabla\sqrt{\rho_n} \otimes \nabla\sqrt{\rho_n}), \end{aligned} \tag{1.87}$$

where $\widetilde{V}_n = W * \rho_n$. In particular, it is straightforward to see that for H^2 solutions, the passages in (1.81) can be justified in the sense of distributions. We now pass to the limit as $n \to \infty$. By using the uniform limit in (1.86) and the H^1 stability for the polar factorization as stated in Proposition 1.2, we have that

$$\|\sqrt{\rho}_n - \sqrt{\rho}\|_{L^\infty(0,T;H^1(\mathbf{R}^d))} + \|\Lambda_n - \Lambda\|_{L^\infty(0,T;L^2(\mathbf{R}^d))} \to 0, \quad n \to \infty, \tag{1.88}$$

where $(\sqrt{\rho}, \Lambda)$ is the (time-dependent) hydrodynamic state associated to ψ. It is now straightforward to see that (1.88) yields the strong convergence in the weak formulation of (1.87), which implies that $(\rho, J) = ((\sqrt{\rho})^2, \sqrt{\rho}\Lambda)$ is a finite energy weak solution to the Cauchy problem (1.83).

The continuity in time of $(\sqrt{\rho}, \Lambda)$ comes from Corollary 1.1.

By using the identity (1.51) and the Strichartz estimates (1.29) for ψ, we also obtain (1.85). Finally, the energy balance equation comes from (1.28). □

Let us remark again that the above existence result does not require any smallness and/or regularity assumptions in the initial data other than being of finite energy and mass. On the other hand, the theorem above cannot provide uniqueness for such solutions, which is generally not expected; see, for instance, [40, 82]. Moreover, it is not even clear how to determine a possible uniqueness class for solutions to (1.83).

More importantly, the main drawback of Theorem 1.5 is that the initial data must be generated by a wave function. Given the fact that such models describe quantum fluids, this assumption is reasonable. However, it is relevant to see whether such a constraint can be neglected. This kind of generalization of the theory is motivated by the fact that for the more general class of Euler-Korteweg systems, see (1.22), we are not able to exploit the corresponding wave function dynamics, which in this case would be given by a quasilinear Schrödinger equation, see (1.23) and also the discussion in Sect. 1.8. We shall see in Theorems 1.7 and 1.8 below that for the one-dimensional QHD system, we can remove this assumption and consider a general class of initial data.

Again, by exploiting the NLS dynamics, we can also show an existence result for more regular weak solutions. In particular, for such solutions, we are able to define the generalized chemical potential, see the Definition 1.6, and derive the conservation law for the total energy density. To simplify the exposition, in the following theorem, we only consider the prototype system (1.31), for which the corresponding wave function dynamics is given by (1.79). We also recall the preliminary discussion in Sect. 1.5.2.1.

Theorem 1.6 *Let (ρ_0, J_0) be given by $\rho_0 = |\psi_0|^2$, $J_0 = \mathrm{Im}(\bar{\psi}_0\nabla\psi_0)$, where $\psi_0 \in H^2(\mathbf{R}^d)$. Then the solution (ρ, J) to* (1.31), *constructed in Theorem 1.5, satisfies the further following properties;*

- *there exists $\lambda \in C(\mathbf{R}; L^2(\mathbf{R}^d))$ satisfying identity* (1.49) *for any $t \in \mathbf{R}$;*
- *$\rho \in C^1(\mathbf{R}; L^p(\mathbf{R}^d))$ and $\xi = \sqrt{\rho}\lambda \in C(\mathbf{R}; L^p(\mathbf{R}^d))$, where $1 \le p \le 2$ for $1 \le d \le 3$, $1 \le p < 2$ for $d = 4$ and $1 \le p \le \frac{d-2}{d}$ for $d \ge 5$;*

- *the functional* $I(t)$ *defined in* (1.47) *is bounded on compact time intervals and we have*

$$I(t) \leq C(T, I(0)); \tag{1.89}$$

- *the total energy density* (1.46) *satisfies the following equation in the distributional sense, i.e.*

$$\partial_t e + \operatorname{div}\left(\Lambda\lambda - \nabla\sqrt{\rho}\partial_t\sqrt{\rho}\right) = 0, \quad \text{in } \mathcal{D}'(\mathbf{R} \times \mathbf{R}^d). \tag{1.90}$$

Proof If $\psi_0 \in H^2(\mathbf{R}^d)$, then by Theorem 1.3 the solutions to (1.24) further satisfies $\psi \in C(\mathbf{R}; H^2(\mathbf{R}^d)) \cap C^1(\mathbf{R}; L^2(\mathbf{R}^d))$. We proceed as in Sect. 1.5.2.1. By defining

$$\lambda = -\operatorname{Im}(\bar{\phi}\partial_t\psi),$$

we have that the bounds (1.30) yield $\lambda \in L^\infty(0, T; L^2(\mathbf{R}^d))$, for any $0 < T < \infty$. Moreover, the stability property of the polar factorization also gives $\lambda \in C(\mathbf{R}; L^2(\mathbf{R}^d))$. By Theorem 1.4 we also have $\lambda = 0$ a.e. on $\{\rho = 0\}$. Moreover, by the polar factorization,

$$I(t) = \int \frac{1}{2}\lambda^2 + \frac{1}{2}(\partial_t\sqrt{\rho})^2\,dx = \int \frac{1}{2}|\partial_t\psi|^2\,dx.$$

Consequently, the bounds (1.30) also yield the estimate (1.89). It remains to show the identity (1.90). Here, we only sketch the proof. By the polar factorization, the total energy density reads

$$e = \frac{1}{2}|\nabla\psi|^2 + f(|\psi|^2).$$

By differentiating it with respect to time and by using the Eq. (1.79), we obtain

$$\partial_t e - \operatorname{div}\left(\operatorname{Re}(\nabla\bar{\psi}\partial_t\psi)\right) = 0, \quad \text{in } \mathcal{D}'(\mathbf{R} \times \mathbf{R}^d).$$

Again, we use the polar factorization to express the energy flux in terms of the hydrodynamic variables, namely

$$\operatorname{Re}(\nabla\bar{\psi}\partial_t\psi) = \nabla\sqrt{\rho}\partial_t\sqrt{\rho} - \Lambda\lambda, \quad \text{a.e. } x \in \mathbf{R}^d,$$

so that (1.90) is satisfied. To rigorously justify the above calculation, we need to proceed similarly to what was done for the proof of Theorem 1.5. We omit the details here. □

We conclude this section by resuming the strategy adopted to construct finite energy weak solutions to QHD systems. The main tool is to exploit the analogy, established through the Madelung transformation and the polar factorization, between (1.83) and an NLS-type equation, see (1.24) for instance. The QHD equations may be derived by following the passages in (1.80), (1.81), that may be justified in the sense of distributions only for regular (say H^2) solutions to (1.24). For finite energy weak solutions, the above procedure may be justified by exploiting a standard density argument. In particular, we need to exploit two main properties of the underlying NLS evolution. Namely, we use the persistence of regularity (H^2 initial data emanate continuous H^2 solutions) and the continuity of the solution map with respect to the initial data. Let us mention that there is an alternative approach that does not exploit any approximation argument but rather the integral formulation (Duhamel's formula) associated with (1.24); see [3] for more details. This approach is particularly useful when the underlying wave function dynamics does not satisfy suitable well-posedness (or stability) properties in the energy space; see, for instance, [16] for an application to the quantum Euler-Maxwell system.

1.6.1 One Dimensional Case

In the one-dimensional case, we are able to provide a more general existence result. In particular, by exploiting the wave function lifting of Sect. 1.5.3, we remove the assumption for the initial data to be generated by a wave function. The results presented in this section first appeared in [14].

For the sake of clarity in the exposition, here we focus on the following one-dimensional system, analogous to (1.31), already written consistently with the weak formulation given in the Definition 1.4,

$$\begin{cases} \partial_t \rho + \partial_x J = 0 \\ \partial_t J + \partial_x \left(\Lambda^2 + (\partial_x \sqrt{\rho})^2 + p(\rho) \right) = \frac{1}{4} \partial_{xxx} \rho, \end{cases} \tag{1.91}$$

and again the pressure term is given by $p(\rho) = \frac{\gamma - 1}{\gamma} \rho^\gamma$, $1 < \gamma < \infty$. More general QHD systems can be treated similarly as presented in Theorem 1.5.

Theorem 1.7 *Let $(\sqrt{\rho_0}, \Lambda_0)$ be a finite energy hydrodynamic state with*

$$\|\sqrt{\rho_0}\|_{H^1} + \|\Lambda_0\|_{L^2} \le M_1 \tag{1.92}$$

and let us further assume that $\Lambda_0 = 0$ a.e. on the set $\{\rho_0 = 0\}$. Then there exists a global in time finite energy weak solution to the Cauchy problem (1.91) *which conserves the total energy for all times, $E(t) = E(0)$. In particular, we have*

$$\|\sqrt{\rho}\|_{L^\infty(0,T;H^1(\mathbf{R}))} + \|\Lambda\|_{L^\infty(0,T;L^2(\mathbf{R}))} \le C(M_1). \tag{1.93}$$

Proof The proof is a consequence of the wave function lifting in Proposition 1.3 and of the previous existence Theorem 1.5, so we omit it. We refer to [14, Theorem 1] for more details. □

Analogously, we can also exploit the wave function lifting of Proposition 1.5 for H^2 wave functions to provide an existence result for more regular weak solutions. More precisely, we show that for a GCP hydrodynamic state, see the Definition 1.7, such that its total energy density is continuous. An analogous result as in Theorem 1.6 holds. The main advantage here is that as in Theorem 1.7, the initial data are not required to be determined by a given wave function.

Theorem 1.8 *Let us consider a pair of finite energy initial data* $(\sqrt{\rho_0}, \Lambda_0)$ *satisfying the same hypotheses of Theorem 1.7 and let us further assume:*

- $\partial_x^2 \rho_0 \in L^1_{loc}(\mathbf{R})$, $\partial_x J_0 \in L^1_{loc}(\mathbf{R})$;
- *the initial total energy density* $e_0 := \frac{1}{2}(\partial_x \sqrt{\rho_0})^2 + \frac{1}{2}\Lambda_0^2 + f(\rho_0)$ *is continuous;*
- *the initial data satisfies the bounds*

$$\|(\frac{\Lambda_0^2}{\sqrt{\rho_0}} - \partial_x^2 \sqrt{\rho_0})\mathbf{1}_{\{\rho_0>0\}}\|_{L^2} + \|\frac{\partial_x J_0}{\sqrt{\rho_0}}\mathbf{1}_{\{\rho_0>0\}}\|_{L^2} \leq M_2. \tag{1.94}$$

Then the solution to (1.91) *constructed in Theorem 1.7 further enjoys the following further properties:*

- *there exists* $\lambda \in C(\mathbf{R}; L^2(\mathbf{R}))$ *satisfying the identity* (1.49) *for a.e.* $t \in \mathbf{R}$ *and moreover for any* $0 < T < \infty$ *the following bound is satisfied,*

$$\|\partial_t \sqrt{\rho}\|_{L^\infty(0,T;L^2(\mathbf{R}))} + \|\lambda\|_{L^\infty(0,T;L^2(\mathbf{R}))} \leq C(T, M_1, M_2); \tag{1.95}$$

- *for any* $0 < T < \infty$ *we have*

$$\|\rho\|_{L^\infty(0,T;H^2(\mathbf{R}))} + \|J\|_{L^\infty(0,T;H^1(\mathbf{R}))} + \|\sqrt{e}\|_{L^\infty(0,T;H^1(\mathbf{R}))} \leq C(T, M_1, M_2), \tag{1.96}$$

  where e is the total energy density defined in (1.46)*;*
- *the conservation law for the total energy density* (1.90) *holds in the distributional sense.*

Proof As for Theorem 1.7, also this proof is a consequence of the wave function lifting in Proposition 1.5 and the existence Theorem 1.6. Here, we only prove the estimates in (1.96); for more details, we refer to [14, Theorem 2]. The first two bounds are straightforward, since $\partial_{xx}\rho = 2\,\mathrm{Re}(\bar{\psi}\partial_{xx}\psi) + 2|\partial_x\psi|^2$ and $\partial_x J = \mathrm{Im}(\bar{\psi}\partial_{xx}\psi)$, remember that $\psi \in C(\mathbf{R}; H^2(\mathbf{R}))$. The third one is consequence of the fact that $e = \frac{1}{2}|\partial_x\psi|^2 + f(|\psi|^2)$ and moreover $\partial_x|\partial_x\psi| \leq |\partial_{xx}\psi|$ a.e. $x \in \mathbf{R}$. □

1.7 Dissipative QHD Systems

This section is devoted to the analysis of global-in-time finite energy solutions to a dissipative QHD system, given by

$$\begin{cases} \partial_t \rho + \operatorname{div} J = 0 \\ \partial_t J + \operatorname{div}\left(\frac{J \otimes J}{\rho}\right) + \nabla p(\rho) + \rho \nabla V = \frac{1}{2}\rho \nabla\left(\frac{\Delta\sqrt{\rho}}{\sqrt{\rho}}\right) - \alpha J \\ -\Delta V = \rho - C. \end{cases} \tag{1.97}$$

The above system is particularly relevant in modeling semiconductor devices [1, 2, 44], where ρ is the electron charge density and J is the momentum density. V is the electrostatic potential, satisfying the Poisson equation where the given function $C = C(x)$ describes the doping profile of the semiconductor. The term $-\alpha J$, with $\alpha \geq 0$, introduces a dissipation in the system. From the modeling point of view, it describes the collisions between electrons, in which case α is inversely proportional to the momentum relaxation time [21].

The relevance of semiconductor device modeling in engineering applications contributed to the wide interest received by the system (1.97) in the mathematical literature [45, 59, 60, 65, 66, 78], see also the monograph [63] for a more complete description of the various models.

The energy associated to the system (1.97) is given by

$$E(t) = \int_{\mathbf{R}^d} \frac{1}{2}|\nabla\sqrt{\rho}|^2 + \frac{1}{2}|\Lambda|^2 + f(\rho) + \frac{1}{2}|\nabla V|^2 \, dx$$

and it is formally dissipated along the flow of solutions through the following balance equation

$$E(t) + \alpha \int_0^t \int |\Lambda(t', x)|^2 \, dx dt' = E(0). \tag{1.98}$$

Besides the typical applications related to semiconductor device modeling, the dissipative term in the equation for the momentum density in (1.97) may also describe an interaction of the quantum fluid with an external environment. For instance, in the Landau two-fluid model [6, 71], the superfluid part interacts with the normal fluid part. Hence, in this context, the term $-\alpha J$ roughly imitates the term determining the momentum exchange between the superfluid and normal fluid part. In this respect, the study of system (1.97) can be viewed as a preliminary step towards a more comprehensive study of more complex dynamics. See [6] where this approach is undertaken, and also [62] where a different two-fluid model is studied.

From the mathematical point of view, the collisional term in (1.97) somehow destroys the analogy with the NLS-type dynamics. Indeed, in this case, the associated NLS-type dynamics is formally given by

$$\begin{cases} i\partial_t\psi = -\frac{1}{2}\Delta\psi + f'(|\psi|^2)\psi + V\psi + \widetilde{V}\psi \\ -\Delta V = |\psi|^2 - C, \end{cases} \tag{1.99}$$

where the self-consistent potential $\widetilde{V}$ is given by

$$\widetilde{V} = \frac{\alpha}{2i}\log\left(\psi/\bar{\psi}\right).$$

Similar dissipative Schrödinger equations arise also in other contexts, see for example, the Schrödinger-Langevin equation [75]; however, to our knowledge, there is no satisfactory well-posedness result in the energy space for the Cauchy problem associated to (1.99), due to the singular behavior of the potential $\widetilde{V}$ close to vacuum. A partial result in this direction (existence of global-in-time solutions bounded away from vacuum in the one-dimensional torus) is given in the recent preprint [19]. For this reason, we cannot directly exploit the analysis developed in Sect. 1.6 to study (1.97). Here, we use a different strategy based on a fractional-step argument.

The next theorem presents the main result of the present section, see also [4]. In order to simplify the presentation, we set the doping profile $C(x) \equiv 0$ and consider the pressure to be

$$p(\rho) = c\rho^\gamma, \qquad 1 < \gamma < \frac{d}{(d-2)_+}. \tag{1.100}$$

For an analog result in the two-dimensional case, we also refer to [5].

Theorem 1.9 *Let $\psi_0 \in H^1(\mathbf{R}^3)$ and let us define $\rho_0 := |\psi_0|^2$, $J_0 := \mathrm{Im}(\bar{\psi}_0\nabla\psi_0)$. Then there exists a global in time finite energy weak solution to* (1.97) *such that $\sqrt{\rho} \in L^\infty(\mathbf{R}; H^1(\mathbf{R}^3))$, $\Lambda \in L^\infty(\mathbf{R}; L^2(\mathbf{R}^3))$, and we have the following energy inequality*

$$E(t) + \alpha\int_0^t\int_{\mathbf{R}^3}|\Lambda|^2\,dxdt' \le CE(0), \quad \textit{for a.e. } t > 0. \tag{1.101}$$

Remark 1.12 Let us notice that in (1.101), we have $C > 1$, namely for the solutions constructed in the previous theorem, we are not able to obtain (1.98) in the limit. This is due to the errors produced at each step of the approximating procedure. See Proposition 1.7 below.

As we are going to discuss below, the proof of Theorem 1.9 differs substantially from the one of Theorem 1.5. This is due to the lack of a satisfactory well-posedness theory for Eq. (1.99). For this reason, here we proceed by constructing a sequence of approximating solutions based on an operator-splitting argument.

More precisely, let us fix a small parameter $\tau > 0$ that will play the role of the time discretization. After dividing the time half-line $\mathbf{R}_+ = \cup_{k=0}^{\infty}[k\tau, (k+1)\tau)$, we split our evolutionary problem into two parts, which will be solved alternately on each subinterval. In the former step, we solve the non-collisional dynamics, given by system (1.97) with $\alpha = 0$. This step can be studied by relying on the analysis developed in the previous section, that is, by solving the associated wave function dynamics given by a Schrödinger-Poisson system. In the latter step, we solve the collisional part, namely, we consider the following linear system of ODE's,

$$\begin{cases} \partial_t \rho = 0 \\ \partial_t J + \alpha J = 0. \end{cases} \tag{1.102}$$

However, we can not implement the collisional step in this direct way. The subsequent step in the argument would require the definition of a wave function in order to initiate again with the Schrödinger-Poisson dynamics. As we discussed in Sect. 1.5.3.1, in general, it is not clear how to determine a wave function, given a hydrodynamic state. This means that we need to implement the collisional step at the wave function level.

Heuristically, by recalling the Madelung transformation $\psi = \sqrt{\rho}e^{iS}$ and formula (1.7) to define the momentum density as $J = \rho\nabla S$, we see that (1.102) would require to set the phase density as $S(t) = e^{-\alpha t}S_o$, $t \in (0, \tau)$, where S_o would come from the previous step, while the mass density would remain constant. Approximating at the first order, we formally consider

$$S_n = (1 - \alpha\tau)S_o \tag{1.103}$$

in this step.

In order to make rigorous the previous heuristics, we exploit the polar factorization presented in Sect. 1.5.1. In particular, we would like to define the updated wave function as follows,

$$\psi_{new} = \sqrt{\rho_{old}}\phi_{old}^{1-\alpha\tau},$$

where $\psi_{old} = \sqrt{\rho_{old}}\phi_{old}$, with $\phi_{old} \in P(\psi_{old})$. Notice that the updating procedure through the polar decomposition method is consistent with (1.103). More precisely, the collisional step will be implemented by means of Lemma 1.2.

With this setup for the fractional step, we are able to define $\psi^\tau \in L^\infty([0, \infty); H^1)$. The polar factorization method allows us to determine the time-dependent hydrodynamic state $(\sqrt{\rho}^\tau, \Lambda^\tau) \in L^\infty([0, \infty); H^1 \times L^2)$, which in turn defines the approximate solution (ρ^τ, J^τ). The convergence will then be shown by exploiting the dispersive properties associated with the system.

In what follows, we explain in detail the construction of the approximating solutions. Thus, let $\psi_0 \in H^1(\mathbf{R}^d)$ be the initial datum, and let $\tau > 0$ be a fixed small parameter. We construct $\psi^\tau \in L^\infty([0, k\tau); H^1(\mathbf{R}^d))$ by induction on $k \in \mathbf{N}$. In the first step, $k = 1$, we solve the following Cauchy problem

$$\begin{cases} i\partial_t \psi = -\frac{1}{2}\Delta\psi + f'(|\psi|^2)\psi + V\psi, & (t,x) \in [0,\tau) \times \mathbf{R}^d \\ -\Delta V = |\psi|^2, & (t,x) \in [0,\tau) \times \mathbf{R}^d \\ \psi(0) = \psi_{in}, & x \in \mathbf{R}^d, \end{cases} \tag{1.104}$$

with $\psi_{in} = \psi_0$. On the space-time slab $[0, \tau) \times \mathbf{R}^d$, we define ψ^τ to be the solution to (1.104), see Theorem 1.3, so that $\psi^\tau \in C([0, \tau); H^1(\mathbf{R}^d))$. Let us now assume we already constructed ψ^τ on $[(k-1)\tau, k\tau) \times \mathbf{R}^d$, with $k \geq 1$, we are going to define ψ^τ in the next space-time slab $[k\tau, (k+1)\tau) \times \mathbf{R}^d$. In order to do this, we use Lemma 1.2 with $\Phi := \psi^\tau(k\tau-)$, $\varepsilon = \tau 2^{-k}\|\psi_0\|_{H^1}$, and we define $\psi^\tau(k\tau) = \Psi_\star$, where $\Psi_\star$ is given by Lemma 1.2. More precisely, we know that $\psi^\tau(k\tau)$ and its associated hydrodynamic state $(\sqrt{\rho}^\tau(k\tau), \Lambda^\tau(k\tau))$ satisfy

$$\sqrt{\rho}^\tau(k\tau) = \sqrt{\rho}^\tau(k\tau-) + r_{k,\tau} \tag{1.105}$$

$$\Lambda^\tau(k\tau) = (1 - \alpha\tau)\Lambda^\tau(k\tau-) + \Gamma_{k,\tau}, \tag{1.106}$$

with

$$\|r_{k,\tau}\|_{H^1} + \|\Gamma_{k,\tau}\|_{L^2} \leq \tau 2^{-k}\|\psi_0\|_{H^1}, \tag{1.107}$$

and moreover

$$\begin{aligned} \psi^\tau(k\tau) &= \psi^\tau(k\tau-) + R^1_{k,\tau} \\ \nabla\psi^\tau(k\tau) &= \nabla\psi^\tau(k\tau-) - i\alpha\tau\phi^\tau_k \Lambda^\tau(k\tau-) + R^2_{k,\tau}, \end{aligned} \tag{1.108}$$

for some ϕ^τ_k with $\|\phi^\tau_k\|_{L^\infty} \leq 1$ and

$$\|R^1_{k,\tau}\|_{L^2} + \|R^2_{k,\tau}\|_{L^2} \leq C\left(\tau\|\psi^\tau(k\tau-)\|_{H^1} + \tau 2^{-k}\|\psi_0\|_{H^1}\right) \leq \tau C E_0^{1/2}.$$

Notice that (1.105), (1.106) provide a first-order approximation of the collisional step (1.102). By using $\psi^\tau(k\tau)$ as the initial datum for the next iteration in the fractional step, we can then start again with the Schrödinger-Poisson equation. We solve system (1.104) with $\psi_{in} = \psi^\tau(k\tau)$. The definition of ψ^τ on the space-time slab $[k\tau, (k+1)\tau) \times \mathbf{R}^d$ is consequently given by

$$\psi^\tau(t) = \psi(t - k\tau), \quad t \in [k\tau, (k+1)\tau),$$

where $\psi \in C([0, \tau); H^1(\mathbf{R}^3))$ is the solution to (1.104) with $\psi_{in} = \psi^\tau(k\tau)$.

The induction step is thus constructed. In this way, for any $\tau > 0$ we defined $\psi^\tau \in L^\infty(\mathbf{R}_+; H^1(\mathbf{R}^3))$. By using the polar factorization we construct also $(\sqrt{\rho}^\tau, \Lambda^\tau) \in L^\infty_t H^1_x \times L^\infty_t L^2_x$. Consequently, we define $\{(\rho^\tau, J^\tau)\}_{\tau>0}$ to be our sequence of approximating solutions for system (1.97).

We now need to show that the sequence $\{(\rho^\tau, J^\tau)\}$ has a (strongly) converging subsequence and that the limit indeed solves the QHD system (1.97).

We first derive some a priori bounds for the sequence $\{(\rho^\tau, J^\tau)\}$.

Proposition 1.7 *Let $\{(\rho^\tau, J^\tau)\}$ be the sequence of approximating solutions constructed via the fractional step method, then we have*

$$M^\tau(t) \lesssim M(0) + \tau, \tag{1.109}$$

$$E^\tau(t) + \alpha\tau \sum_{k=1}^{[t/\tau]} \|\Lambda^\tau(k\tau-)\|_{L^2(\mathbf{R}^3)} \lesssim E(0) + \tau.. \tag{1.110}$$

Let us notice that (1.110) provides an approximated version of the energy balance (1.98).

Proof It is straightforward to see that by construction, both the total mass and the total energy are conserved on each subinterval $[k\tau, (k+1)\tau)$. Thus we only need to estimate the variation of $M^\tau(t)$ and $E^\tau(t)$ at the end of each interval. Let us first consider the mass, from (1.105) we have

$$\begin{aligned} M^\tau(k\tau) - M^\tau(k\tau-) =& 2\int \sqrt{\rho}^\tau(k\tau-) r_{k,\tau}\, dx + \|r_{k,\tau}\|_{L^2}^2 \\ \le& \delta_k M^\tau(k\tau-) + \left(1 + \frac{1}{\delta_k}\right)\|r_{k,\tau}\|_{L^2}^2, \end{aligned}$$

for some $0 < \delta_k \le 1$ to be chosen later, and consequently

$$M^\tau(k\tau) \le (1+\delta_k) M^\tau(k\tau-) + \left(1 + \frac{1}{\delta_k}\right)\|r_{k,\tau}\|_{L^2}^2, \tag{1.111}$$

Now, we only need to apply (1.111) iteratively. For any $0 < t < \infty$, let $k_* = \left[\frac{t}{\tau}\right]$ be the integer part of t/τ, then by choosing $\delta_k = \frac{1}{k_*}$ for any $k = 1, \ldots, k_*$ and by exploiting the estimate (1.107), we obtain

$$M^\tau(t) \lesssim \left(1 + \frac{1}{k_*}\right)^{k_*} M(0) + \tau \sum_{k=1}^{k_*} k 2^{-k},$$

which yields (1.109).

Let us now consider the energy. By exploiting (1.106), we compute

$$\begin{aligned} E^\tau(k\tau) - E^\tau(k\tau-) =& E_{int}^\tau(k\tau) - E_{int}^\tau(k\tau-) - \alpha\tau\left(1 - \frac{\alpha\tau}{2}\right)\|\Lambda^\tau(k\tau-)\|_{L^2}^2 \\ &+ (1-\alpha\tau)\int_{\mathbf{R}^3} \Lambda^\tau(k\tau-)\cdot\Gamma_{k,\tau}\, dx + \frac{1}{2}\|\Gamma_{k,\tau}\|_{L^2}^2, \end{aligned}$$

where in the expression on the right-hand side we used the identity (1.106), and we defined

$$E^{\tau}_{int} = \int_{\mathbf{R}^3} \frac{1}{2}|\nabla\sqrt{\rho^{\tau}}|^2 + \frac{1}{2}|\nabla V^{\tau}|^2 + f(\rho^{\tau})\,dx.$$

By the assumptions on the pressure (1.100) and the fact that the electrostatic potential is given by $V^{\tau} = (-\Delta)^{-1}\rho^{\tau}$, it is straightforward to check that

$$|E^{\tau}_{int}(k\tau) - E^{\tau}_{int}(k\tau-)| \lesssim \delta E^{\tau}_{int}(k\tau-) + (1 + C_{\delta})\|r_{k,\tau}\|^2_{H^1}\left(1 + \|r_{k,\tau}\|^{2(\gamma-1)}_{H^1}\right),$$

where again we used (1.105). For the variation in the kinetic energy part, we can proceed analogously as for the total mass to finally obtain

$$\begin{aligned} &E^{\tau}(k\tau) - E^{\tau}(k\tau-) \\ &\leq \delta_k E^{\tau}_{int}(k\tau-) + \frac{\delta_k + \alpha^2\tau^2}{2}\|\Lambda^{\tau}(k\tau-)\|^2_{L^2} - \alpha\tau\|\Lambda^{\tau}(k\tau-)\|^2_{L^2} \\ &\quad + C(1 + C_{\delta_k})\left(\|r_{k,\tau}\|^2_{H^1} + \|r_{k,\tau}\|^{2\gamma}_{H^1} + \|\Gamma_{k,\tau}\|^2_{L^2}\right). \end{aligned}$$

We finally obtain (1.110) by iterating the previous estimate. □

The estimates (1.109) and (1.110) for the total mass and energy, respectively, imply the following uniform bounds

$$\|\sqrt{\rho^{\tau}}\|_{L^{\infty}(\mathbf{R}_+;H^1(\mathbf{R}^3))} + \|\Lambda^{\tau}\|_{L^{\infty}(\mathbf{R}_+;L^2(\mathbf{R}^3))} \leq C(E(0), M(0)), \tag{1.112}$$

where C does not depend on $\tau \in [0, 1]$. However, those bounds are not sufficient to pass to the limit as $\tau \to 0$ because they can not rule out possible concentrations for the quadratic terms

$$\nabla\sqrt{\rho^{\tau}} \otimes \nabla\sqrt{\rho^{\tau}} + \Lambda^{\tau} \otimes \Lambda^{\tau},$$

appearing in the equation for the momentum in (1.97). To gain compactness we are going to exploit the dispersive effects encoded in the system. More precisely, by exploiting the fact that on each space-time slab, the dynamics is given by the Schrödinger-Poisson system (1.104), we can infer improved space-time bounds on $(\sqrt{\rho^{\tau}}, \Lambda^{\tau})$ by deriving dispersive-type estimates for ψ^{τ}. To do this we derive a Duhamel type formula for the approximating wave function ψ^{τ} constructed in the fractional step argument.

Lemma 1.7 *Let ψ^{τ} be the approximate solution constructed above, then we have*

$$\psi^{\tau}(t) = e^{\frac{i}{2}t\Delta}\psi_0 - i\int_0^t e^{\frac{i}{2}(t-s)\Delta}\mathcal{N}(\psi^{\tau})(s)\,ds + \sum_{k=1}^{[t/\tau]} e^{\frac{i}{2}(t-k\tau)\Delta}R^1_{k,\tau}, \tag{1.113}$$

and

$$\begin{aligned}\nabla\psi^\tau(t) =& e^{\frac{i}{2}t\Delta}\nabla\psi_0 - i\int_0^t e^{\frac{i}{2}(t-s)\Delta}\nabla\mathcal{N}(\psi^\tau)(s)\,ds \\ & - i\alpha\tau\sum_{k=1}^{[t/\tau]} e^{\frac{i}{2}(t-k\tau)\Delta}\left[\phi_k^\tau\Lambda^\tau(k\tau-)\right] + \sum_{k=1}^{[t/\tau]} e^{\frac{i}{2}(t-k\tau)\Delta}R^2_{k,\tau},\end{aligned} \tag{1.114}$$

where

$$\mathcal{N}(\psi^\tau) = f'(|\psi^\tau|^2)\psi^\tau + V^\tau\psi^\tau,$$

and ϕ_k^τ, $R^1_{k,\tau}$, $R^2_{k,\tau}$ *are defined in* (1.108).

Proof The lemma is a straightforward consequence of the fact that ψ^τ solves (1.104) on each $[k\tau,(k+1)\tau)\times\mathbf{R}^3$ and of the implementation of the collisional step which yields (1.108). □

By having the approximate Duhamel's formulae (1.113) and (1.114) at hand, it is now possible to exploit the Strichartz estimates. More precisely, by applying the Strichartz estimates of Theorem 1.1 and the local smoothing estimates of Theorem 1.2, we can show the compactness for the sequence $\{\psi^\tau\}_{\tau>0}$. The following proposition collects all the needed a priori estimates, its proof can be found in [4].

Proposition 1.8 *Let* $0 < T < \infty$ *be a finite time. Then for any admissible pair* (q,r) *we have*

$$\|\nabla\psi^\tau\|_{L^q(0,T;L^r(\mathbf{R}^d))} \le C(E_0, \|\rho_0\|_{L^1}, T).$$

Furthermore,

$$\|\nabla\psi^\tau\|_{L^2([0,T];H^{1/2}_{loc}(\mathbf{R}^d))} \le C(E_0, \|\rho_0\|_{L^1}, T).$$

With those estimates at hand we can now use a Aubin-Lions type lemma to extract a subsequence (which we will also call $\{\psi^\tau\}$) which has a strong limit. More precisely, we make use of a result by Rakotoson and Temam [91] (see [4] for more details).

Proposition 1.9 *For any finite time* $0 < T < \infty$, *the sequence* $\nabla\psi^\tau$ *is relatively compact in* $L^2(0,T;L^2_{loc}(\mathbf{R}^d))$. *More precisely, there exists* $\psi \in L^2(0,T;H^1_{loc}(\mathbf{R}^d))$ *such that*

$$\psi = s - \lim_{\tau\to 0}\psi^\tau, \quad \text{in } L^2(0,T;H^1_{loc}(\mathbf{R}^d)).$$

As a consequence,

$$\begin{aligned} \sqrt{\rho^\tau} &\to \sqrt{\rho} \quad \text{in } L^2(0,T;H^1_{loc}(\mathbf{R}^d)) \\ \Lambda^\tau &\to \Lambda \quad \text{in } L^2(0,T;L^2_{loc}(\mathbf{R}^d)). \end{aligned}$$

We now need to prove that the limit obtained in the previous proposition provides a finite energy weak solution to (1.97). We do this by showing that (ρ^τ, J^τ) are approximating solutions for $\tau \to 0$, indeed.

Proposition 1.10 *Let $\{(\rho^\tau, J^\tau)\}$ be the sequence of approximating solutions constructed via the fractional step method and let $0 < T < \infty$. Then for any $\eta \in C_c^\infty([0,T) \times \mathbf{R}^3)$ we have*

$$\int_0^T \int_{\mathbf{R}^3} \rho^\tau \partial_t \eta + J^\tau \cdot \nabla \eta \, dxdt + \int_{\mathbf{R}^3} \rho_0(x)\eta(0,x)\, dx = o(1), \quad \text{as } \tau \to 0.$$

For any $\zeta \in C_c^\infty([0,T) \times \mathbf{R}^3; \mathbf{R}^3)$ we have

$$\begin{aligned} \int_0^T \int_{\mathbf{R}^3} & J^\tau \cdot \partial_t \zeta + \left(\Lambda^\tau \otimes \Lambda^\tau + \nabla\sqrt{\rho^\tau} \otimes \nabla\sqrt{\rho^\tau}\right) : \nabla\zeta + p(\rho^\tau)\operatorname{div}\zeta \\ & - \frac{1}{4}\rho^\tau \Delta \operatorname{div}\zeta - \alpha J^\tau \cdot \zeta \, dxdt \\ & + \int_{\mathbf{R}^3} J_0(x) \cdot \zeta(0,x)\, dx = o(1), \quad \text{as } \tau \to 0. \end{aligned} \tag{1.115}$$

Proof In what follows, we only prove (1.115); the analog result for the continuity equation is similar and will be omitted here. Let us consider the sequence of approximating solutions $\{(\rho^\tau, J^\tau)\}$, we know that on each space-time slab $[k\tau, (k+1)\tau) \times \mathbf{R}^3$, (ρ^τ, J^τ) solves the non-collisional QHD, namely system (1.97) with $\alpha = 0$, in the sense of distributions. Thus, for $\zeta \in C_c^\infty([0,T) \times \mathbf{R}^3; \mathbf{R}^3)$ and on $[k\tau, (k+1)\tau] \subset [0,T)$, we have

$$\begin{aligned} \int_{k\tau}^{(k+1)\tau} \int_{\mathbf{R}^3} & J^\tau \cdot \partial_t \zeta + \left(\Lambda^\tau \otimes \Lambda^\tau + \nabla\sqrt{\rho^\tau} \otimes \nabla\sqrt{\rho^\tau}\right) : \nabla\zeta - \rho^\tau \nabla V^\tau \cdot \zeta \\ & + p(\rho^\tau)\operatorname{div}\zeta - \frac{1}{4}\rho^\tau \Delta \operatorname{div}\zeta \, dxdt \\ & + \int_{\mathbf{R}^3} \zeta(k\tau, x) \cdot J^\tau(k\tau, x)\, dx \\ & - \int_{\mathbf{R}^3} \zeta((k+1)\tau, x) \cdot J^\tau((k+1)\tau, x)\, dx = 0. \end{aligned}$$

Let $k_\star = \lfloor T/\tau \rfloor$ be the integer part of T/τ, by summing up over all space-time slabs in $[0,T) \times \mathbf{R}^3$, we then have

$$\int_0^T \int_{\mathbf{R}^3} J^\tau \cdot \partial_t \zeta + \left(\Lambda^\tau \otimes \Lambda^\tau + \nabla\sqrt{\rho}^\tau \otimes \nabla\sqrt{\rho}^\tau\right) : \nabla\zeta - \rho^\tau \nabla V^\tau \cdot \zeta$$
$$+ p(\rho^\tau)\operatorname{div}\zeta - \frac{1}{4}\rho^\tau \Delta \operatorname{div}\zeta \, dxdt + \int_{\mathbf{R}^3} \zeta(0,x) \cdot J_0(x)\, dx$$
$$+ \sum_{k=1}^{k_\star} \int_{\mathbf{R}^3} \zeta((k+1)\tau) \cdot (J^\tau(k\tau) - J^\tau(k\tau-))\, dx = 0\, dx = 0.$$

Let us notice that, by using the identities (1.105) and (1.106), we have

$$\int \zeta((k+1)\tau) \cdot (J^\tau(k\tau) - J^\tau(k\tau-))\, dx$$
$$= -\alpha\tau \int_{\mathbf{R}^3} J^\tau(k\tau-) \cdot \zeta(k\tau)\, dx + \int R^{tot}_{k,\tau} \cdot \zeta(k\tau)\, dx,$$

with

$$R^{tot}_{k,\tau} = (1 - \alpha\tau) r_{k,\tau} \Lambda^\tau(k\tau-) + (\sqrt{\rho}^\tau(k\tau-) + r_{k,\tau}\Gamma_{k,\tau}.$$

By using (1.107) and (1.112) we have that $\|R^{tot}_{k,\tau}\|_{L^1(\mathbf{R}^3)} \lesssim \tau 2^{-k}$. To prove (1.115), it only remains to show that

$$\int_0^T \int_{\mathbf{R}^3} J^\tau \cdot \zeta\, dxdt - \tau \sum_{k=1}^{k_\star} J^\tau(k\tau-) \cdot \zeta(k\tau)\, dx = o(1), \quad \text{as } \tau \to 0,$$

but this is a straightforward consequence of the integrability properties of J^τ. □

Combining the strong convergence in Proposition 1.9 with the consistency result of Proposition 1.10 proves the existence of global in time finite energy weak solutions. Moreover, by using the strong convergence and the approximate energy estimate (1.110), we obtain (1.101) as $\tau \to 0$. Thus, Theorem 1.9 is proved.

Remark 1.13 We should observe here that, despite the fact ψ is the strong limit of the sequence $\{\psi^\tau\}$ and the hydrodynamic quantities $(\sqrt{\rho}, \Lambda)$ associated to ψ solve the QHD system (1.97), it is not clear whether the wave function ψ solve any nonlinear Schrödinger equation. Indeed, if we consider the approximate Duhamel's formulae (1.113) and (1.114), it is not possible to determine the contributions of $R^1_{k,tau}$ and $R^2_{k,tau}$ in the limit as $\tau \to 0$.

We conclude this section by mentioning that, by following a similar approach, it is possible to consider a more general class of collisional terms; see [7].

1.8 Dispersive Estimates

In the previous sections, we discussed the existence of solutions to QHD systems, whose construction relied on the analysis of a corresponding wave function dynamics given by a NLS type equations. As a byproduct of this strategy, the constructed solution inherits all the dispersive estimates available for NLS-type equations, see for instance (1.85), (1.89), or even Proposition 1.8 used to gain compactness for the sequence of approximating solutions in the previous section.

Some of these estimates are not known to be satisfied by general solutions to the QHD system. The dispersive estimates for NLS-type evolutions usually come from suitable smoothing properties of the free propagator (see Theorems 1.1 and 1.2, for instance), which are then transferred to semilinear equations by a perturbative argument. The QHD system is quasilinear, and in general, it is not possible to derive dispersive estimates similarly.

In this section, we derive some dispersive estimates for QHD systems, which may be inferred from the monotonicity properties of some suitable functionals. The motivation for our analysis stems from the fact that the tools presented in the previous sections do not provide any stability results for general solutions. Moreover, let us also remark that the previous global existence results cannot be extended to the more general class of Euler-Korteweg (EK) systems unless we consider small, regular perturbations of constant states [20, 26]. We recall that the EK system is given by

$$\begin{cases} \partial_t \rho + \operatorname{div}(\rho v) = 0 \\ \partial_t(\rho v) + \operatorname{div}(\rho v \otimes v) + \nabla p(\rho) = \rho \nabla \left(\operatorname{div}(k(\rho)\nabla\rho) - \frac{1}{2} K'(\rho)|\nabla\rho|^2 \right). \end{cases} \tag{1.116}$$

The total energy associated with (1.116) is given by

$$E_{EK}(t) = \int \frac{1}{2}\rho|v|^2 + \frac{1}{2}k(\rho)|\nabla\rho|^2 + f(\rho)\,dx, \tag{1.117}$$

where $k = k(\rho) \geq 0$ is the capillarity coefficient. Let us notice that the QHD system can be recast as a particular case of (1.116) when we set $k(\rho) = \frac{1}{4\rho}$.

As for the QHD system, also the capillary term on the right-hand-side of the momentum equation may be written as the divergence of a (capillary) stress tensor, namely

$$\rho\nabla\left(\operatorname{div}(k(\rho)\nabla\rho) - \frac{1}{2}k'(\rho)|\nabla\rho|^2 \right) = \operatorname{div}\mathbb{K},$$

where the capillarity tensor is given by

$$\mathbb{K} = \left(\operatorname{div}(\rho k(\rho)\nabla\rho) - \frac{1}{2}(\rho k'(\rho) + k(\rho))|\nabla\rho|^2\right)\mathbb{I} - k(\rho)\nabla\rho\otimes\nabla\rho.$$

Moreover, for the EK system, it is also possible to formally write related wave function dynamics. However, in this case, it is given by a quasilinear Schrödinger equation, which can also be conveniently written as

$$i\partial_t\psi = -\frac{1}{2}\Delta\psi - \operatorname{div}\left(\widetilde{k}(\rho)\nabla\rho\right)\psi + \frac{1}{2}\widetilde{k}'(\rho)|\nabla\rho|^2\psi + f'(\rho)\psi, \tag{1.118}$$

where clearly $\rho = |\psi|^2$ and we denoted $\widetilde{k}(\rho) = k(\rho) - \frac{1}{4\rho}$. At present, no satisfactory theory for (1.118) in the space of energy is available hence in order to study (1.116) we can not exploit the strategy developed in the previous sections.

For these reasons, as a preliminary step towards the analysis of the aforementioned problems, it is relevant to initiate an intrinsically hydrodynamic approach for the study of QHD systems. In this section, we are going to collect some a priori dispersive estimates for general solutions to the QHD system, not necessarily Schrödinger-generated. As an application, in the next section, we are going to show a (sequential) stability result for weak solutions to the one-dimensional QHD system. The results in this section are presented for the following QHD system

$$\begin{cases} \partial_t\rho + \operatorname{div}(\rho v) = 0 \\ \partial_t(\rho v) + \operatorname{div}(\rho v\otimes v) + \nabla p(\rho) = \frac{1}{2}\rho\nabla\left(\frac{\Delta\sqrt{\rho}}{\sqrt{\rho}}\right), \end{cases} \tag{1.119}$$

whose total energy reads

$$E = \int \frac{1}{2}\rho|v|^2 + \frac{1}{2}|\nabla\sqrt{\rho}|^2 + f(\rho)\,dx, \tag{1.120}$$

where $f = f(\rho)$ is determined by (1.32), with $\rho^* = 0$. We recall that, as usual, when dealing with finite energy weak solutions to (1.119), we set up our analysis in terms of the hydrodynamic state $(\sqrt{\rho}, \Lambda)$, rather than the physical quantities (ρ, v). Moreover, we remark that some of the results presented here also apply with minor modifications to different QHD models. In fact, they are also valid for weak solutions to the EK system (1.116). In what follows, we are going to specify the estimates associated with (1.116).

The first dispersive estimate we are going to present exploits the convexity (in time) of the momentum of inertia functional, given by

$$\int_{\mathbf{R}^d} \frac{|x|^2}{2}\rho_0(x)\,dx < \infty.$$

In particular, we consider the following functional

$$V(t)=\int_{\mathbf{R}^d}\frac{|x|^2}{2}\rho(t,x)\,dx-t\int_{\mathbf{R}^d}x\cdot J(t,x)\,dx+t^2E(t).$$

By recalling the definition of the total energy (1.120), we also see that $V(t)$ is non-negative and indeed may be written as

$$V(t)=\int_{\mathbf{R}^d}\frac{t^2}{2}|\Lambda-\frac{x}{t}\sqrt{\rho}|^2+\frac{t^2}{2}|\nabla\sqrt{\rho}|^2+t^2f(\rho)\,dx. \tag{1.121}$$

First, we show a monotonicity property for $V(t)$.

Proposition 1.11 *Let (ρ, J) be a finite energy weak solution to (1.119) on $[0,T)\times\mathbf{R}^d$, such that the total momentum of inertia at initial time is finite,*

$$\int_{\mathbf{R}^d}|x|^2\rho_0(x)\,dx<\infty \tag{1.122}$$

and the energy is non-increasing in time, i.e. $E(t)\le E(s)$ for $0\le s\le t$. Then we have

$$V(t)+\int_0^t\int_{\mathbf{R}^d}s\,(dp(\rho)-2f(\rho))\,dxds\le\int_{\mathbf{R}^d}\frac{|x|^2}{2}\rho_0(x)\,dx. \tag{1.123}$$

Remark 1.14 Let us notice that, to prove the identity (1.123), we need to assume that the total energy is non-increasing in time. This, in general, does not hold true for arbitrary finite energy weak solutions, for which we only assumed $E(t)\le E(0)$ in Definition 1.5. On the other hand, if the solutions to the QHD system are such that the total energy density e satisfies the weak entropy inequality

$$\partial_t e+\operatorname{div}\left(\lambda\Lambda-\partial_t\sqrt{\rho}\nabla\sqrt{\rho}\right)\le 0,$$

then the assumption is straightforwardly satisfied. The weak entropy inequality will be further discussed in the next section, see Lemma 1.8.

Estimates similar to (1.123) related to the total momentum of inertia are ubiquitous in the study of applied PDE's. For instance, in [51], Glassey uses a similar calculation to show the existence of blowing-up solutions for NLS equations. Estimate (1.123) has an analog in the theory of NLS equations, and it is related to the pseudo-conformal identity. See, for instance, [48], where the pseudo-conformal energy is used to study the asymptotic behavior of solutions. See also [22] where a similar calculation is used to prove the non-existence of non-trivial scattering states for the NLS equation with a mass-subcritical nonlinearity.

In classical compressible fluid dynamics, Sideris [95] exploits the analysis of some virial functionals to show the formation of singularities for strong solutions to the three-dimensional Euler equations, see also [104] for a similar result for the compressible Navier-Stokes equations. A calculation similar to (1.123) and especially to the analysis performed in Proposition 1.12 below was also used to show dispersive properties for the compressible Euler equations, see [35] and later also [52, 94]. Indeed such arguments apply also to Euler-Korteweg systems, this was already noticed in [32]. Moreover, similar functionals are also related to the moments associated to solutions for kinetic equations, see [23, 61, 87]. We also mention that those arguments are also related to the method of commuting vector fields, see [73] for applications to nonlinear wave equations, [56, 57] for nonlinear Schrödinger equations, [42] for kinetic type equations.

Proof Formally, the proposition can be proved by differentiating the functional with respect to time and by exploiting Eq. (1.119) to obtain

$$\frac{d}{dt}V(t) = -t\int_{\mathbf{R}^d} dp(\rho) - 2f(\rho)\,dx.$$

The statement would then follow by just integrating the above identity in time. To prove the proposition for arbitrary finite energy weak solutions, we exploit the weak formulation of (1.119), with a suitable choice of the test functions in (1.36), (1.37).

We first define some suitable cut-off functions. Given $R > 0$, we denote

$$\Phi_R(x) = \frac{|x|^2}{2}\chi_R(|x|)e^{-\frac{|x|}{R^2}},$$

where $\chi_R \in C_c^\infty([0,\infty))$ is a bump function satisfying $\chi_R(r) \equiv 1$ for $r \leq R$ and $\operatorname{supp}\chi_R \subset \{r \leq 2R\}$. In this way, we have that $|\chi'| \lesssim 1/R$. By using the above definition, we also have that

$$\begin{aligned}
\nabla\Phi_R(x) =& x\chi_R(|x|)e^{-\frac{|x|}{R^2}} + O\Big(\frac{1}{R}\Big)\\
\nabla^2\Phi_R(x) =& \mathbb{I}\chi_R(|x|)e^{-\frac{|x|}{R^2}} + O\Big(\frac{1}{R^2}\Big)\\
\Delta^2\Phi_R(|x|) =& O(1/R^4).
\end{aligned}$$

Let us fix $t \in [0,T)$, and let $\varepsilon > 0$ be sufficiently small. We define $\widetilde{\chi}_{t,\varepsilon} \in C_c^\infty([0,T))$ such that $\widetilde{\chi}_{t,\varepsilon}(s) = t - s$ for $s \in [0, t-\varepsilon]$ and $\operatorname{supp}\widetilde{\chi}_{t,\varepsilon} \subset [0,t]$. We choose $\eta(s,x) = \widetilde{\chi}_{t,\varepsilon}(s)\Phi_R(x)$ as a test function in the weak formulation for the continuity Eq. (1.36). By observing that $\widetilde{\chi}'_{t,\varepsilon}(0) = -1$ we obtain

$$\begin{aligned}
\int_0^T\int_{\mathbf{R}^d} &\widetilde{\chi}''_{t,\varepsilon}(s)\Phi_R(x)\rho(s,x) + \widetilde{\chi}'_{t,\varepsilon}(s)\nabla\Phi_R(x)\\
&\cdot J(s,x)\,dxds - \int_{\mathbf{R}^d}\Phi_R(x)\rho_0(x)\,dx = 0.
\end{aligned} \tag{1.124}$$

Now, in the weak formulation of the equation for the momentum density (1.37), we choose $\zeta(s,x) = \left(\widetilde{\chi}_{t,\varepsilon}(s) + t\widetilde{\chi}'_{t,\varepsilon}(s)\right)\nabla\Phi_R(x)$ as a test function, by observing that $\widetilde{\chi}_{t,\varepsilon}(0) + t\widetilde{\chi}'_{t,\varepsilon}(0) = 0$ we have

$$\begin{aligned}
&\int_0^T \int_{\mathbf{R}^d} \left(\widetilde{\chi}'_{t,\varepsilon}(s) + t\widetilde{\chi}''_{t,\varepsilon}(s)\right)\nabla\Phi_R(x)\cdot J(s,x) \\
&\quad + \left(\widetilde{\chi}_{t,\varepsilon}(s) + t\widetilde{\chi}'_{t,\varepsilon}(s)\right)\nabla^2\Phi_R(x) \\
&: \left(\Lambda\otimes\Lambda + \nabla\sqrt{\rho}\otimes\nabla\sqrt{\rho} + p(\rho)\mathbb{I} - \frac{1}{4}\nabla^2\rho\right)(s,x)\,dxds \\
&= 0.
\end{aligned} \tag{1.125}$$

By subtracting formula (1.125) to (1.124) we obtain

$$\begin{aligned}
&\int_0^T \int_{\mathbf{R}^d} \widetilde{\chi}''_{t,\varepsilon}(s)\Phi_R(x)\rho(s,x) - t\widetilde{\chi}''_{t,\varepsilon}(s)\nabla\Phi_R(x)\cdot J(s,x) \\
&\quad - \left(\widetilde{\chi}_{t,\varepsilon}(s) + t\widetilde{\chi}'_{t,\varepsilon}(s)\right)\nabla^2\Phi_R(x) \\
&\quad : \left(\Lambda\otimes\Lambda + \nabla\sqrt{\rho}\otimes\nabla\sqrt{\rho} + p(\rho)\mathbb{I} - \frac{1}{4}\nabla^2\rho\right)(s,x)\,dxds \\
&\quad - \int_{\mathbf{R}^d} \Phi_R(x)\rho_0(x)\,dx = 0.
\end{aligned}$$

By letting $\varepsilon \to 0$ and $R \to \infty$ we obtain

$$\begin{aligned}
&\int_{\mathbf{R}^d} \frac{|x|^2}{2}\rho(t,x)\,dx - t\int_{\mathbf{R}^d} x\cdot J(t,x)\,dx + \int_0^t s\int_{\mathbf{R}^d} |\Lambda|^2 + |\nabla\sqrt{\rho}|^2 \\
&\quad + dp(\rho)\,dxds \\
&= \int_{\mathbf{R}^d} \frac{|x|^2}{2}\rho_0(x)\,dx.
\end{aligned}$$

By using the definition of the energy in (1.40) and the assumption that the energy is non-increasing, $E(t) \leq E(s)$ for $s \leq t$, we then obtain inequality (1.123). □

A similar identity can also be obtained from the Euler-Korteweg system. Indeed, by recalling (1.117), we define

$$\begin{aligned}
V_{EK}(t) &= \int \frac{|x|^2}{2}\rho\,dx - t\int x\cdot J\,dx + t^2 E_{EK}(t) \\
&= \int \frac{t^2}{2}\left|\Lambda - \frac{x}{t}\sqrt{\rho}\right|^2 + \frac{t^2}{2}k(\rho)|\nabla\rho|^2 + t^2 f(\rho)\,dx.
\end{aligned} \tag{1.126}$$

Then, the EK analogue of estimate (1.123) reads

$$V_{EK}(t) + \int_0^t s \int d\rho f'(\rho) - (d+2) f(\rho) + \frac{d}{2}(\rho k'(\rho) + k(\rho))|\nabla\rho|^2 \, dx ds$$
$$\leq \int \frac{|x|^2}{2} \rho_0(x) \, dx.$$

Inequality (1.123) was derived for an arbitrary choice of the pressure law. However, if we assume $p(\rho) = \frac{\gamma-1}{\gamma}\rho^\gamma$, $\gamma > 1$, then by (1.32) we have that the internal energy is given by $f(\rho) = \frac{1}{\gamma}\rho^\gamma$, so that (1.123) becomes

$$V(t) + \frac{d(\gamma-1)-2}{\gamma} \int_0^t \int_{\mathbf{R}^d} s\rho^\gamma \, dx ds \leq \int_{\mathbf{R}^d} \frac{|x|^2}{2} \rho_0(x) \, dx. \tag{1.127}$$

The analysis of inequality (1.127), together with formula (1.121), provides an explicit dispersive estimate for finite energy weak solutions to (1.119) when the pressure term is given by a $\gamma-$law.

Proposition 1.12 *Let* (ρ, J) *be a finite energy weak solution to* (1.119) *with* $p(\rho) = \frac{\gamma-1}{\gamma}\rho^\gamma$, *with* $1 < \gamma < \frac{d}{(d-2)_+}$. *Then, under the same assumptions of Proposition 1.11, we have*

$$V(t) \leq C t^{2(1-\sigma)}, \tag{1.128}$$

where $\sigma = \min\{1, \frac{d}{2}(\gamma-1)\}$ *and* $C = C(V(0), E(0))$. *In particular, we have*

$$\int_{\mathbf{R}^d} |\nabla\sqrt{\rho}(t,x)|^2 \, dx \lesssim t^{-2\sigma} \tag{1.129}$$

$$\int_{\mathbf{R}^d} |\Lambda(t,x) - \frac{x}{t}\sqrt{\rho}(t,x)|^2 \, dx \lesssim t^{-2\sigma} \tag{1.130}$$

$$\int_{\mathbf{R}^d} \rho^\gamma(t,x) \, dx \lesssim t^{-2\sigma}. \tag{1.131}$$

Proof In the case $1 + \frac{2}{d} \leq \gamma \leq \frac{d}{(d-2)_+}$ the statement straightforwardly comes from (1.127) and the definition of σ. We thus consider the case $1 < \gamma < 1 + \frac{2}{d}$ by defining

$$F(t) = \frac{t^2}{\gamma} \int_{\mathbf{R}^d} \rho^\gamma(t,x) \, dx,$$

from (1.127) we infer

$$F(t) \leq (2 - d(\gamma - 1)) \int_0^t \frac{1}{s} F(s)\, ds + \int_{\mathbf{R}^d} \frac{|x|^2}{2} \rho_0(x)\, dx. \tag{1.132}$$

From the finiteness of the energy we have

$$F(1) \lesssim E(0) + \int_{\mathbf{R}^d} \frac{|x|^2}{2} \rho_0(x)\, dx.$$

Again by using (1.132) we infer

$$F(t) \leq (2 - d(\gamma - 1)) \int_1^t \frac{1}{s} F(s)\, ds + F(1),$$

so that Gronwall's inequality yields

$$F(t) \leq t^{2-d(\gamma-1)} F(1).$$

By plugging the previous estimate into (1.127) we finally obtain

$$V(t) \leq \left(1 + (2 - d(\gamma - 1)) \int_1^t s^{1-d(\gamma-1)}\, ds\right) F(1),$$

which proves the statement of the proposition. □

Remark 1.15 Let us remark that the assumption on the pressure can be relaxed to obtain a result similar to the one in Proposition 1.12 even for more general pressure laws. In this case, the decay rate depends on the term $dp(\rho) - 2f(\rho)$ appearing in (1.123). For example, if $\int dp(\rho) - 2f(\rho) \geq 0$, then we still have $\sigma = 1$ in (1.128). Otherwise, let $c^\star \in [0, 2]$ be such that $\int dp(\rho) - 2f(\rho) \geq -c^\star \int f(\rho)$, then we have $\sigma = 1 - c^\star/2$. This approach can be further generalized; however, we will not address this issue here.

In the case of Euler–Korteweg type systems, a discussion concerning non-monotone pressure laws can be found in [46].

The dispersive estimates in Proposition 1.12 have several consequences. First, we see that, even if the total mass is conserved, the mass density locally converges to zero. Moreover, we can also see that, formally, the velocity field v asymptotically approaches a rarefaction wave, namely $v(t, x) \sim x/t$ as $t \to \infty$.

The estimates (1.129), (1.131) also imply that, for large time, the total energy is concentrating on the kinetic part. Finally, we see how the dispersive estimates derived in Proposition 1.12 are consistent with the dispersive estimate associated to the Schrödinger propagator (1.27).

Proposition 1.13 *Let the assumptions of Proposition 1.12 be satisfied. If $\gamma \geq 1+\frac{2}{d}$, then we have*

$$\|\sqrt{\rho}(t)\|_{L^p(\mathbf{R})} \lesssim t^{-\left(\frac{1}{2}-\frac{1}{p}\right)}, \quad p \in [2,\infty],\ d=1,$$
$$\|\sqrt{\rho}(t)\|_{L^p(\mathbf{R}^2)} \lesssim t^{-2\left(\frac{1}{2}-\frac{1}{p}\right)}, \quad p \in [2,\infty),\ d=2$$
$$\|\sqrt{\rho}(t)\|_{L^p(\mathbf{R}^d)} \lesssim t^{-d\left(\frac{1}{2}-\frac{1}{p}\right)}, \quad p \in [2, \frac{2d}{d-2}],\ d \geq 3.$$

If $1 < \gamma < 1+\frac{2}{d}$, then we have

$$\|\sqrt{\rho}(t)\|_{L^p(\mathbf{R}^d)} \lesssim t^{-d\left(\frac{1}{2}-\frac{1}{p}\right)}, \quad p \in [2, 2\gamma].$$

Notice how the decay estimates stated in the previous proposition match the dispersive estimate for the linear Schrödinger estimate (1.27) on a restricted range of Lebesgue exponents.

Let us notice that here the constant depends on the total inertia at initial time, while in (1.27) it depends on the $L^{p'}(\mathbf{R}^d)$ norm of the initial datum.

The dispersive estimate derived in Proposition 1.12 holds for weak solutions having a suitable decay at space infinity, as the momentum of inertia needs to be finite. In what follows, we derive some further estimates for arbitrary finite energy weak solutions, without assumptions on the decay at space infinity. As a drawback, these estimates are weaker. While (1.128) implies a pointwise decay in time, we derive some global space-time bounds in the next proposition.

Such estimates may be regarded as the hydrodynamical analog of the so-called interaction Morawetz estimates [36, 83, 96]. Indeed, the fact that the interaction Morawetz estimates can be viewed as a purely hydrodynamical estimate is also inferred from the approach in [49], see also [90] where a similar argument is performed by using the Radon transform.

Proposition 1.14 (Interaction Morawetz Estimates) *Let (ρ, J) be a finite energy weak solution to* (1.119). *Then we have*

$$\||\nabla|^{\frac{3-d}{2}}\rho\|^2_{L^2(\mathbf{R}\times\mathbf{R}^d)} + \||\nabla|^{\frac{1-d}{2}}\sqrt{\rho p(\rho)}\|_{L^2(\mathbf{R}\times\mathbf{R}^d)} \leq C(E(0), M(0)).$$

As for the dispersive estimates presented above, also the interaction Morawetz can be proved for finite energy weak solutions to the Euler-Korteweg system, indeed by denoting $K(\rho) = \int_0^\rho s k(s)\,ds$, where $k = k(\rho)$ is the capillarity coefficient, then the following estimate holds,

$$\||\nabla|^{\frac{3-d}{2}}\sqrt{\rho K(\rho)}\|^2_{L^2(\mathbf{R}\times\mathbf{R}^d)} + \||\nabla|^{\frac{1-d}{2}}\sqrt{\rho p(\rho)}\|_{L^2(\mathbf{R}\times\mathbf{R}^d)} \leq C(E(0), M(0)).$$

A proof of the interaction Morawetz estimate for the Euler-Korteweg system may be found in [13].

Proof To highlight the analogies with the approach for NLS equations, in this proof, we closely follow the calculations presented in [49, Proposition 2.1].

Let $h : \mathbf{R}^d \to \mathbf{R}_+$ be an even function, and let us define the following functional

$$V_{Mor}(t) = \frac{1}{2}\int_{\mathbf{R}^d\times\mathbf{R}^d} \rho(t,x)h(x-y)\rho(t,y)\,dxdy.$$

By differentiating it with respect to time, we have

$$\frac{d}{dt}V_{Mor}(t) = \int \rho(t,y)\nabla_x h(x-y)\cdot J(t,x)\,dxdy.$$

In what follows, we omit the time dependence of the hydrodynamic quantities to simplify the exposition. By differentiating V_{Mor} twice, we obtain

$$\begin{aligned}\frac{d^2}{dt^2}V_{Mor}(t) = &-\int J(y)\cdot\nabla_x^2 h(x-y)\cdot J(x)\,dxdy\\ &+\int \rho(y)\nabla_x^2 h(x-y) : \left(\Lambda\otimes\Lambda + \nabla\sqrt{\rho}\otimes\nabla\sqrt{\rho}\right)(x)\,dxdy\\ &+\int \rho(y)\Delta_x h(x-y)\left(p(\rho)-\frac{1}{4}\Delta\rho\right)(x)\,dxdy.\end{aligned} \tag{1.133}$$

We now split the right-hand side into several contributions. First of all, we consider the terms

$$-\int J(y)\cdot\nabla_x^2 h(x-y)\cdot J(x)\,dxdy + \int \rho(y)\nabla_x^2 h(x-y) : (\Lambda\otimes\Lambda)(x)\,dxdy.$$

By using the symmetry of h, we write the two contributions above as follows

$$\begin{aligned}&\frac{1}{2}\int \left(\sqrt{\rho}(x)\Lambda(y)-\sqrt{\rho}(y)\Lambda(x)\right)\cdot\nabla_x^2 h(x-y)\\ &\quad\cdot\left(\sqrt{\rho}(x)\Lambda(y)-\sqrt{\rho}(y)\Lambda(x)\right)\,dxdy\\ &\quad=: Q(\sqrt{\rho},\Lambda)\end{aligned}$$

A similar calculation gives the following identity

$$\begin{aligned}&\int \rho(y)\nabla_x^2 h(x-y) : \left(\nabla\sqrt{\rho}\otimes\nabla\sqrt{\rho}\right)(x) =\\ &\quad\frac{1}{4}\int \nabla\rho(y)\cdot\nabla_x^2 h(x-y)\cdot\nabla\rho(x)\,dxdy + Q(\sqrt{\rho},\nabla\sqrt{\rho})\end{aligned}$$

and we further notice that

$$\frac{1}{4}\int \nabla\rho(y)\cdot\nabla_x^2 h(x-y)\cdot\nabla\rho(x)\,dxdy = \tfrac{1}{4}\int \Delta_x h(x-y)\nabla\rho(y)\cdot\nabla\rho(x)\,dxdy$$
$$= -\tfrac{1}{4}\int \rho\Delta_x h(x-y)\Delta\rho(x)\,dxdy.$$

Consequently, by summing up all the contributions appearing on the right-hand-side of (1.133), we obtain

$$\begin{aligned}\frac{d^2}{dt^2}V_{Mor}(t) &= Q(\sqrt{\rho},\Lambda)+Q(\sqrt{\rho},\nabla\sqrt{\rho})\\ &+\frac{1}{2}\int \Delta_x h(x-y)\nabla\rho(y)\cdot\nabla\rho(x)\,dxdy\\ &+\int \rho\Delta_x h(x-y)p(\rho)(x)\,dxdy.\end{aligned}$$

If we set $h(x)=|x|$, then $\nabla h(x)=\frac{x}{|x|}$ and

$$\begin{cases}\nabla^2 h(x)=\frac{1}{|x|}\left[\mathbb{I}-\frac{x\otimes x}{|x|^2}\right], & d\geq 2\\ h''(x)=\delta_x, & d=1.\end{cases}$$

Notice that the choice of the weight h yields $\|\nabla h\|_{L^\infty}\leq 1$, and consequently

$$\left|\frac{d}{dt}V_{Mor}(t)\right|\leq \|\rho(t)\|_{L^1}\|J(t)\|_{L^1}\leq C(M(0),E(0)).$$

Moreover, $\nabla^2 h\geq 0$, which implies $Q(\sqrt{\rho},\Lambda)\geq 0$, $Q(\sqrt{\rho},\nabla\sqrt{\rho})\geq 0$, and we have that $\Delta h = \frac{d-1}{|x|}$ is the integral kernel associated to the operator $c(-\Delta)^{-\frac{d-1}{2}}$, for a suitable constant $c>0$. Consequently,

$$\frac{d^2}{dt^2}V_{Mor}(t)\geq \int \rho(y)\frac{d-1}{|x-y|}p(\rho)(x)\,dxdy + c\int \left|(-\Delta)^{-\frac{d-1}{4}}\nabla\rho\right|^2 dx.$$

Also, the contribution coming from the pressure term can be symmetrized in such a way that the following inequality holds

$$\int \rho(y)\frac{d-1}{|x-y|}p(\rho)(x)\,dxdy\geq c\int \left|(-\Delta)^{-\frac{d-1}{4}}\sqrt{\rho p(\rho)}\right|^2 dx.$$

Consequently

$$\frac{d^2}{dt^2}V_{Mor}(t) \geq c\int \left|(-\Delta)^{-\frac{d-1}{4}}\sqrt{\rho p(\rho)}\right|^2 dx + c\int \left|(-\Delta)^{-\frac{d-1}{4}}\nabla\rho\right|^2 dx.$$

By integrating in time, we then prove the proposition. To have a rigorous proof, we need to use the weak formulation of the continuity Eq. (1.36) and the equation for the momentum density (1.37) by choosing suitable test functions. The details are left to the reader. □

Remark 1.16 Let us notice that in the one-dimensional case it is possible to prove Proposition 1.14 by performing an alternative calculation, see Theorem 31 in [14, Theorem 31]. A similar approach will also be used later in Proposition 1.16 to prove further integrability properties for the total energy density in the one-dimensional case.

1.8.1 Entropic Solutions for the One-dimensional QHD System

The dispersive estimates presented in the previous section mostly provide information about the qualitative behavior of solutions for large times. However, they are still not sufficient to show further stability properties, such as sequential stability.

In this section, we derive some novel, higher-order a priori estimates. Those bounds will be related to the a priori estimates given by the functional $I(t)$ introduced in (1.47); moreover we will see that the weak entropy inequality (1.135) below will play a prominent role in the proof of Proposition 1.16, that improves on the integrability properties for the total energy density.

In what follows, we first discuss the role of entropic solutions and prove the boundedness of the functional $I(t)$, defined in (1.47), along the flow of smooth, positive solutions. Moreover, in the one-dimensional case we will introduce a novel Morawetz-type estimate valid for weak solutions with bounded generalized chemical potential, see Proposition 1.16. This new a priori estimate will enable us to prove a stability property for the one-dimensional QHD system in Theorem 1.10.

We now focus on the weak entropy inequality. In Theorem 1.6 we proved the existence of weak solutions satisfying the Eq. (1.90) in the sense of distributions. More precisely, any solution to the QHD system that is constructed from a H^2 solution to the corresponding NLS dynamics satisfies the Eq. (1.90). In what follows, we provide an alternative proof of (1.90), which holds for smooth, positive solutions, not necessarily Schrödinger-generated. Let us recall that the generalized chemical potential λ was introduced in Sect. 1.4.1.

Lemma 1.8 *Let (ρ, J) be a smooth solution to* (1.119) *such that $\rho > 0$. Then the total energy density e, defined in* (1.46), *satisfies the following conservation law*

$$\partial_t e + \operatorname{div}(\Lambda\lambda - \partial_t\sqrt{\rho}\nabla\sqrt{\rho}) = 0. \tag{1.134}$$

Proof By assuming the positivity and smoothness of solutions, it is possible to rewrite the system (1.119) as

$$\begin{cases} \partial_t \rho + \operatorname{div}(\rho v) = 0 \\ \rho \partial_t v + \rho v \cdot \nabla v + \rho \nabla f'(\rho) = \dfrac{1}{2} \rho \nabla \left(\dfrac{\Delta \sqrt{\rho}}{\sqrt{\rho}} \right). \end{cases}$$

Moreover, from the generalized irrotationality condition (1.38) and the strict positivity of the mass density, we have the velocity field is irrotational $\nabla \wedge v = 0$, see also Remark 1.5. Therefore, the equation for the momentum density can be written equivalently as

$$\rho \partial_t v + \rho \nabla \mu = 0,$$

where μ denotes the chemical potential defined in (1.43). Similarly, the total energy density satisfies

$$e = \frac{1}{2} (\nabla \sqrt{\rho})^2 + \frac{1}{2} \rho |v|^2 + f(\rho),$$

and hence

$$\begin{aligned} \partial_t e =& \nabla \sqrt{\rho} \cdot \nabla \partial_t \sqrt{\rho} + \left(\frac{1}{2} |v|^2 + f'(\rho) \right) \partial_t \rho + \rho v \cdot \partial_t v \\ =& \operatorname{div} \left(\partial_t \sqrt{\rho} \nabla \sqrt{\rho} \right) + \left(-\frac{1}{2} \frac{\Delta \sqrt{\rho}}{\sqrt{\rho}} + \frac{1}{2} |v|^2 + f'(\rho) \right) \partial_t \rho + \rho v \cdot \partial_t v. \end{aligned}$$

By using the previous identities and the definition (1.43), it follows

$$\begin{aligned} \partial_t e =& \operatorname{div} \left(\partial_t \sqrt{\rho} \nabla \sqrt{\rho} \right) - \mu \operatorname{div}(\rho v) - \rho v \cdot \nabla \mu \\ =& \operatorname{div} \left(\partial_t \sqrt{\rho} \nabla \sqrt{\rho} - \rho v \mu \right). \end{aligned}$$

□

By resuming, we can prove the Eq. (1.90) for two particular classes of solutions, namely Schrödinger-generated solutions and smooth, positive solutions. However, it is not clear whether arbitrary weak solutions satisfy the entropy equation, as the answer to this question could be negative. A reasonably expected result is that, for a (rather) general class of weak solutions, its weaker version given by the entropy inequality

$$\partial_t e + \operatorname{div}(\Lambda \lambda - \partial_t \sqrt{\rho} \nabla \sqrt{\rho}) \leq 0, \tag{1.135}$$

is satisfied.

In the one-dimensional case, we prove that the weak entropy inequality (1.135) provides the additional estimate (1.141) below, which in turn yields the compactness result in Theorem 1.10. Hence, it could be guessed that the weak entropy inequality could play the role of an admissibility criterion for stability properties of weak solutions. At the moment, this is not clear. However, as we will see in the next proposition, the Eq. (1.134) for smooth, positive solutions also yields the boundedness of the functional $I(t)$ on compact time intervals.

Let us remark that also, in this case, the uniform bound for $I(t)$ can be easily proved for Schrödinger-generated solutions, see (1.89). Indeed, in this case, the bound is a direct consequence of the Strichartz estimates for H^2 solutions to the NLS equation. In what follows, we provide an alternative proof valid for smooth, positive solutions to the one-dimensional QHD system (1.91).

Proposition 1.15 *Let (ρ, J) be a smooth solution to* (1.91) *with finite mass and finite energy, such that $\rho > 0$. Then we have*

$$\sup_{t\in[0,T]} I(t) \le C(T)I(0). \tag{1.136}$$

Proof Again, by exploiting the positivity and smoothness of solutions to (1.119), we can rewrite the functional I(t) in the following way,

$$I(t) = \frac{1}{2}\int_{\mathbf{R}} \rho(\mu^2 + \sigma^2)\, dx,$$

where μ is the chemical potential introduced in (1.43) and we define $\sigma := \partial_t \log\sqrt{\rho}$. Now, by using the identity

$$\rho\mu = -\frac{1}{4}\partial_x^2\rho + e + p(\rho)$$

we have

$$\begin{aligned}\partial_t(\rho\mu) =& \partial_t\left(e - \frac{1}{4}\partial_x^2\rho + p(\rho)\right)\\ =& \partial_x(\partial_x\sqrt{\rho}\partial_t\sqrt{\rho} - \rho v\mu) - \frac{1}{4}\partial_t\partial_x^2\rho + \partial_t p(\rho).\end{aligned}$$

Again, by using the continuity equation, we have

$$\rho\partial_t\mu + \rho v\partial_x\mu = \partial_x(\partial_x\sqrt{\rho}\partial_t\sqrt{\rho}) - \frac{1}{4}\partial_t\partial_x^2\rho + \partial_t p(\rho). \tag{1.137}$$

Now, we are going to derive the equation for σ. By using the continuity equation and by multiplying it by $1/2\rho$, we obtain

$$\partial_t \log \sqrt{\rho} + v\partial_x \log \sqrt{\rho} + \frac{1}{2}\partial_x v = 0,$$

then by using the equation $\partial_t v = -\partial_x \mu$, we obtain

$$\partial_t \sigma + v\partial_x \sigma - \nabla\mu\partial_x \log \sqrt{\rho} - \frac{1}{2}\partial_x^2 \mu = 0,$$

hence, by multiplying the previous equation by ρ, we get

$$\rho\partial_t \sigma + \rho v\partial_x \sigma = \frac{1}{2}\partial_x(\rho\partial_x \mu). \tag{1.138}$$

Now we can exploit the Eqs. (1.137) and (1.138) to compute the time derivative of the functional $I(t)$. We have

$$\frac{d}{dt}I(t) = \int \mu\partial_t p(\rho)\, dx = 2\int p'(\rho)\partial_t \sqrt{\rho}\lambda\, dx$$

By using the fact that, in the one-dimensional case, we have $\|\sqrt{\rho}(t)\|_{L^\infty} \leq C(M(0), E(0))$, then by Gronwall's inequality we obtain 1.136. □

We now go back to the study of finite energy weak solutions to the one-dimensional QHD system (1.91), without strict positivity assumptions for the mass density. The following proposition shows that GCP solutions to QHD (see Definition 1.8) satisfy an additional integrability property for the energy density.

Proposition 1.16 *Let (ρ, J) be a GCP weak solution to* (1.91)*, namely it satisfies the bounds*

$$\begin{aligned} &\|\sqrt{\rho}\|_{L^\infty(0,T;H^1(\mathbf{R}))} + \|\Lambda\|_{L^\infty(0,T;L^2(\mathbf{R}))} \leq M_1 \\ &\|\partial_t\sqrt{\rho}\|_{L^\infty(0,T;L^2(\mathbf{R}))} + \|\lambda\|_{L^\infty(0,T;L^2(\mathbf{R}))} \leq M_2. \end{aligned} \tag{1.139}$$

Furthermore, let us assume that the weak entropy inequality

$$\partial_t e + \partial_x(\Lambda\lambda - \partial_t\sqrt{\rho}\partial_x\sqrt{\rho}) \leq 0 \tag{1.140}$$

is satisfied in the sense of distribution. Then we have

$$\|e\|_{L^2_{t,x}} + \|\partial_x^2\rho\|_{L^2_{t,x}} + \|\partial_x J\|_{L^\infty_t L^2_x} \leq C(M_1, M_2)(1+T)^{\frac{1}{2}}. \tag{1.141}$$

Proof By Sobolev embedding, (1.139) implies ρ is continuous in space and $\|\sqrt{\rho}\|_{L^\infty_{t,x}} \leq C(M_1)$. Then the continuity equation and the bounds (1.139), we have

$$\|\partial_x J\|_{L^\infty_t L^2_x} = \|\partial_t\rho\|_{L^\infty_t L^2_x} \leq 2\|\sqrt{\rho}\|_{L^\infty_{t,x}}\|\partial_t\sqrt{\rho}\|_{L^\infty_t L^2_x} \leq C(M_1, M_2).$$

Let us define

$$G(t,x) = \int_{-\infty}^{x} J(t,s)ds + C_1,$$

where C_1 is a suitable constant such that $G \geq 0$, for instance it is sufficient to choose $C_1 > \|J\|_{L_t^\infty L_x^1}$. By integrating in space the momentum equation in (1.91), it follows

$$\partial_t G + \Lambda^2 + (\partial_x\sqrt{\rho})^2 + p(\rho) = \frac{1}{4}\partial_x^2\rho,$$

namely

$$\partial_t G + (\sqrt{\rho}\lambda + e) - 2f(\rho) = 0. \tag{1.142}$$

The bounds in (1.139) imply that $\partial_t G \in L_t^\infty L_x^1$.

Let us multiply (1.142) by $(\sqrt{\rho}\lambda + e)$, then

$$\begin{aligned} \int_0^T\int_{\mathbf{R}}(\sqrt{\rho}\lambda + e)^2 dxdt = &-\int_0^T\int_{\mathbf{R}}(\sqrt{\rho}\lambda + e)\partial_t G\, dxdt \\ &+2\int_0^T\int_{\mathbf{R}}(\sqrt{\rho}\lambda + e)f(\rho)dxdt. \end{aligned} \tag{1.143}$$

We estimate the right-hand side of (1.143) in the following way.

The first term on the right-hand side satisfies

$$\begin{aligned} -\int_0^T\int_{\mathbf{R}}(\sqrt{\rho}\lambda + e)\partial_t G\, dxdt = &-\int_{\mathbf{R}}(\sqrt{\rho}\lambda + e)(t)G(t)dx\Big|_{t=0}^{T} \\ &+\int_0^T\int_{\mathbf{R}} G\,\partial_t(\sqrt{\rho}\lambda + e)\, dxdt \\ = &\int_{\mathbf{R}}(\sqrt{\rho}\lambda + e)(t)G(t)dx\Big|_{t=0}^{T} \\ &+\int_0^T\int_{\mathbf{R}} G\,\partial_t(-\frac{1}{4}\partial_x^2\rho + 2e + p(\rho))\, dxdt, \end{aligned}$$

where the boundary terms can be estimated as follows

$$\begin{aligned} \int_{\mathbf{R}}(\sqrt{\rho}\lambda + e)(t)G(t)dx \leq &\|G\|_{L_{t,x}^\infty}(\|e\|_{L_t^\infty L_x^1} \\ &+ \|\sqrt{\rho}\|_{L_t^\infty L_x^2}\|\lambda\|_{L_t^\infty L_x^2}) \leq C(M_1, M_2). \end{aligned}$$

Now, we estimate the integral

$$\int_0^T \int_{\mathbf{R}} G\, \partial_t(-\frac{1}{4}\partial_x^2\rho + 2e + p(\rho))\, dxdt. \tag{1.144}$$

By using the continuity equation and integration by parts in space, we have

$$\begin{aligned}\int_0^T \int_{\mathbf{R}} G\, \partial_t(-\frac{1}{4}\partial_x^2\rho)dxdt &= -\frac{1}{4}\int_0^T \int_{\mathbf{R}} \partial_x^2 G\, \partial_t\rho\, dxdt \\ &= \frac{1}{4}\int_0^T \int_{\mathbf{R}} (\partial_x J)^2 dxdt \leq C(M_1, M_2)T.\end{aligned}$$

By using the entropy inequality (1.140) and $G \geq 0$, we obtain

$$\begin{aligned}\int_0^T \int_{\mathbf{R}} G\, \partial_t e\, dxdt \leq &-\int_0^T \int_{\mathbf{R}} G\, \partial_x(\lambda\Lambda - \partial_x\sqrt{\rho}\partial_t\sqrt{\rho})\, dxdt \\ = &\int_0^T \int_{\mathbf{R}} J\, (\lambda\Lambda - \partial_x\sqrt{\rho}\partial_t\sqrt{\rho})\, dxdt \\ \leq &T\, \|J\|_{L^\infty_{t,x}} \Big(\|\lambda\|_{L^\infty_t L^2_x} \|\Lambda\|_{L^\infty_t L^2_x} \\ &+ \|\partial_x\sqrt{\rho}\|_{L^\infty_t L^2_x} \|\partial_t\sqrt{\rho}\|_{L^\infty_t L^2_x} \Big) \leq C(M_1, M_2)T.\end{aligned}$$

Next, we have

$$\begin{aligned}\int_0^T \int_{\mathbf{R}} G\partial_t p(\rho)dxdt = &2\int_0^T \int_{\mathbf{R}} G\, p'(\rho)\sqrt{\rho}\partial_t\sqrt{\rho}\, dxdt \\ \leq &2T\|G\|_{L^\infty_{t,x}} \|p'(\rho)\sqrt{\rho}\|_{L^\infty_t L^2_x} \|\partial_t\sqrt{\rho}\|_{L^\infty_t L^2_x} \leq C(M_1, M_2)T.\end{aligned}$$

Thus we control (1.144) by $C(M_1, M_2)T$.

For the last term on the right hand side of (1.143), by using the $L^\infty_{t,x}$ bound of $\sqrt{\rho}$ and (1.139), it is straightforward to obtain that

$$\int_0^T \int_{\mathbf{R}} (\sqrt{\rho}\lambda + e) f(\rho)dxdt \leq C(M_1, M_2)T.$$

Summarising the previous estimates, we get

$$\|\sqrt{\rho}\lambda + e\|^2_{L^2_{t,x}} \leq C(M_1, M_2)(1 + T),$$

therefore

$$\begin{aligned}\|e\|_{L^2_{t,x}}^2 &\leq \|\sqrt{\rho}\lambda + e\|_{L^2_{t,x}}^2 + \|\sqrt{\rho}\lambda\|_{L^2_{t,x}}^2 \\ &\leq \|\sqrt{\rho}\lambda + e\|_{L^2_{t,x}}^2 + T\|\sqrt{\rho}\|_{L^\infty_{t,x}}^2 \|\lambda\|_{L^\infty_t L^2_x}^2 \leq C(M_1, M_2)(1+T).\end{aligned}$$

We conclude our argument by estimating $\partial_x^2 \rho$. By using the definition of λ and the bounds above, it follows

$$\|\partial_x^2 \rho\|_{L^2_{t,x}} \leq 4\|\sqrt{\rho}\lambda + e\|_{L^2_{t,x}} + 4\|p(\rho)\|_{L^2_{t,x}} \leq C(M_1, M_2)(1+T)^{\frac{1}{2}}.$$

□

1.9 Stability of Weak Solutions for the One-dimensional Case

In this section, we exploit some a priori estimates previously derived to prove a stability result for weak solutions to the one-dimensional QHD system (1.91). This result was first presented in [14]. More precisely, we are going to prove that, for a family of GCP solutions with uniform bounds, there exists a subsequence that strongly converges to a finite energy weak solution to (1.91).

To our knowledge, this is the first stability result for this type of hydrodynamic equation in this framework. Indeed, for QHD systems (and more generally Euler-Korteweg systems), the only available results consider small, regular perturbations of constant states [20, 26], or non-constant stationary solutions in bounded domains [59, 60, 65, 78]. On the other hand, if the QHD system is augmented by a viscous stress tensor, then it is possible to show a compactness result in the framework of finite energy weak solutions, see for instance [10] where the case of degenerate viscosity coefficient is considered (the so-called quantum Navier-Stokes equations), see also [11] for a similar model. The presence of dissipative effects entails further a priori estimates, which, in the case of viscous flows with degenerate viscosities, correspond to the so-called Bresch-Desjardins entropy estimates [30].

We need to exploit the additional estimates developed in the previous Sect. 1.8.1 for quantum fluid systems. A key point will be to exploit the bounds inferred in Proposition 1.16.

In what follows, we consider a sequence of GCP solutions $\{(\rho_n, J_n)\}$, see Definition 1.8. More precisely, we assume that $\{(\rho_n, J_n)\}$ satisfies

$$\begin{aligned}&\|\sqrt{\rho_n}\|_{L^\infty(0,T;H^1(\mathbf{R}))} + \|\Lambda_n\|_{L^\infty(0,T;L^2(\mathbf{R}))} \leq M_1 \\ &\|\partial_t\sqrt{\rho_n}\|_{L^\infty(0,T;L^2(\mathbf{R}))} + \|\lambda_n\|_{L^\infty(0,T;L^2(\mathbf{R}))} \leq M_2,\end{aligned} \tag{1.145}$$

where the generalized chemical potential λ_n satisfies the identity

$$\sqrt{\rho_n}\lambda_n = -\frac{1}{4}\partial_x^2 \rho_n + e_n + p(\rho_n), \tag{1.146}$$

and e_n is the total energy density given by

$$e_n = \frac{1}{2}(\partial_x \sqrt{\rho_n})^2 + \frac{1}{2}\Lambda_n^2 + f(\rho_n). \tag{1.147}$$

Moreover, we will also assume that for any $n \in \mathbf{N}$, the solution (ρ_n, J_n) satisfies the weak entropy inequality (1.135), namely

$$\partial_t e_n + \partial_x(\Lambda_n \lambda_n - \partial_x \sqrt{\rho}_n \partial_t \sqrt{\rho}_n) \leq 0, \quad \text{in } \mathcal{D}'(\mathbf{R}_t \times \mathbf{R}_x). \tag{1.148}$$

Theorem 1.10 ([14]) *Let* $\{(\rho_n, J_n)\}$ *be a sequence of finite energy weak solutions to* (1.91) *satisfying the uniform bounds* (1.145) *and the weak entropy inequality* (1.148). *Then, there exists a subsequence (not relabeled) such that*

$$\sqrt{\rho}_n \to \sqrt{\rho}, \quad L^2(0, T; H^1_{loc}(\mathbf{R}))$$
$$\Lambda_n \to \Lambda, \quad L^2(0, T; L^2_{loc}(\mathbf{R})),$$

where $(\rho, J) = ((\sqrt{\rho})^2, \sqrt{\rho}\Lambda)$ *is a finite energy weak solution to* (1.91).

The remaining part of this section is devoted to proving Theorem 1.10 above.

First of all, from the bounds (1.145), we can infer that, up to subsequences, we have

$$\sqrt{\rho_n} \rightharpoonup \sqrt{\rho}, \quad L^\infty(0, T; H^1(\mathbf{R})) \cap W^{1,\infty}(0, T; L^2(\mathbf{R})), \tag{1.149}$$

$$\Lambda_n \rightharpoonup \Lambda, \quad L^\infty(0, T; L^2(\mathbf{R})), \tag{1.150}$$

$$\lambda_n \rightharpoonup \lambda, \quad L^\infty(0, T; L^2(\mathbf{R})). \tag{1.151}$$

Moreover, from Proposition 1.16 we also obtain the additional bounds,

$$\|e_n\|_{L^2_{t,x}} + \|\partial_x^2 \rho_n\|_{L^2_{t,x}} + \|\partial_x J_n\|_{L^\infty_t L^2_x} \leq C(1+T)^{\frac{1}{2}}. \tag{1.152}$$

In particular, up to subsequences, we also have

$$e_n \rightharpoonup \nu, \quad L^2(0, T; L^2(\mathbf{R})). \tag{1.153}$$

One of the main points in proving Theorem 1.10 consists in identifying the weak limit ν; namely, we claim that

$$\nu = \frac{1}{2}(\partial_x \sqrt{\rho})^2 + \frac{1}{2}\Lambda^2 + f(\rho),$$

where $(\sqrt{\rho}, \Lambda)$ is the weak limit in (1.149), (1.150). This is shown in the next proposition. As a consequence, we are also able to prove the strong convergence of $\{(\sqrt{\rho}_n, \Lambda_n)\}$. Moreover, we also prove that ν and Λ vanish almost everywhere in the vacuum set $\{\rho = 0\}$, which matches the physical interpretation of the total energy density.

Proposition 1.17 *Let $\{(\rho_n, J_n)\}$ satisfy the assumptions of Theorem 1.10. Then the weak limits* (1.149), (1.150) *and* (1.153) *satisfy the following nonlinear identity*

$$\nu = \frac{1}{2}(\partial_x \sqrt{\rho})^2 + \frac{1}{2}\Lambda^2 + f(\rho)$$

for a.e. $(t, x) \in [0, T] \times \mathbf{R}$, and $\nu = 0$ a.e. $(t, x) \in \{\rho = 0\}$. Furthermore, we have the following strong local convergence results

$$\begin{aligned} \partial_x \sqrt{\rho_n} &\to \partial_x \sqrt{\rho}, \quad L^2(0, T; L^2_{loc}(\mathbf{R})), \\ \Lambda_n &\to \Lambda, \qquad\quad L^2(0, T; L^2_{loc}(\mathbf{R})). \end{aligned}$$

Proof By using identity (1.147), we have

$$\rho_n e_n = \frac{1}{8}(\partial_x \rho_n)^2 + \frac{1}{2}J_n^2 + f(\rho_n)\rho_n. \tag{1.154}$$

Moreover, the uniform bounds (1.152) imply that, up to subsequences,

$$\begin{aligned} \rho_n &\rightharpoonup \rho, \quad L^2(0, T; H^1_{loc}(\mathbf{R})) \\ J_n &\rightharpoonup J, \quad L^2(0, T; L^2_{loc}(\mathbf{R})). \end{aligned} \tag{1.155}$$

Thus, by passing to the limit in (1.154), we obtain

$$\rho\nu = \frac{1}{8}(\partial_x \rho)^2 + \frac{1}{2}J^2 + f(\rho), \quad \text{a.e.} \tag{1.156}$$

Since $J_n = \sqrt{\rho_n}\Lambda_n$, then again by (1.150) and (1.155) we have $J = \sqrt{\rho}\Lambda$ a.e. in $\mathbf{R}$. Therefore that we can write (1.156) as

$$\rho\left(\nu - \frac{1}{2}(\partial_x \sqrt{\rho})^2 - \frac{1}{2}\Lambda^2 - f(\rho)\right) = 0, \quad \text{a.e.} \tag{1.157}$$

We now claim that $\nu = 0$ a.e. on $\{\rho = 0\}$. By passing to the limit in the identity (1.146) we obtain

$$\nu = \sqrt{\rho}\lambda + \frac{1}{4}\partial_x^2 \rho - p(\rho).$$

Since $\rho \in L^2(0, T; H^2(\mathbf{R}))$, we apply Theorem 1.4 to infer that $\partial_x^2 \rho = 0$ a.e. on $\{\rho = 0\}$. Consequently, the previous identity implies $\nu = 0$ a.e. on $\{\rho = 0\}$. We now have to show that also the weak limit Λ, see (1.150), satisfies $\Lambda = 0$ a.e. on $\{\rho = 0\}$. For this purpose, for any $R > 0$ we define $V_R = \{\rho = 0\} \cap ([0, T] \times (-R, R))$, then by (1.145), (1.147) and (1.153) it follows

$$\begin{aligned}\int\int_{V_R} \frac{1}{2}(\partial_x\sqrt{\rho})^2 + \frac{1}{2}\Lambda^2\,dxdt &\leq \liminf_{n\to\infty} \int\int_{V_R} \frac{1}{2}(\partial_x\sqrt{\rho_n})^2 + \frac{1}{2}\Lambda_n^2\,dxdt \\ &= \lim_{n\to\infty} \int\int_{V_R} e_n - f(\rho_n)\,dxdt \\ &= \int\int_{V_R} \nu - f(\rho)\,dxdt = 0.\end{aligned}$$

Hence, we obtained that $\Lambda = 0$ a.e. on $\{\rho = 0\}$. Consequently, from (1.157) we identify the weak limit of the energy density,

$$\nu = \frac{1}{2}(\partial_x\sqrt{\rho})^2 + \frac{1}{2}\Lambda^2 + f(\rho), \qquad \text{a.e. } x \in \mathbf{R}.$$

We conclude by proving the local strong convergence. Because of the previous results, for any $R > 0$ it follows,

$$\lim_{n\to\infty} \int_0^T \int_{-R}^R e_n - f(\rho_n)dxdt = \int_0^T \int_{-R}^R \nu - f(\rho)\,dxdt,$$

therefore

$$\begin{aligned}\int_0^T \int_{-R}^R \frac{1}{2}(\partial_x\sqrt{\rho})^2 + \frac{1}{2}\Lambda^2\,dxdt &\leq \liminf_{n\to\infty} \int_0^T \int_{-R}^R \frac{1}{2}(\partial_x\sqrt{\rho_n})^2 + \frac{1}{2}\Lambda_n^2\,dxdt \\ &= \lim_{n\to\infty} \int_0^T \int_{-R}^R e_n - f(\rho_n)\,dxdt \\ &= \int_0^T \int_{-R}^R \nu - f(\rho)\,dxdt \\ &= \int_0^T \int_{-R}^R \frac{1}{2}(\partial_x\sqrt{\rho})^2 + \frac{1}{2}\Lambda^2\,dxdt,\end{aligned}$$

and we conclude

$$\|\partial_x\sqrt{\rho}\|_{L_t^2 L_{x,loc}^2} = \lim_{n\to\infty} \|\partial_x\sqrt{\rho_n}\|_{L_t^2 L_{x,loc}^2}, \qquad \|\Lambda\|_{L_t^2 L_{x,loc}^2} = \lim_{n\to\infty} \|\Lambda_n\|_{L_t^2 L_{x,loc}^2}.$$

Hence, by using the convergence of norms, we can improve the weak convergence to a strong one. □

Proof of Theorem 1.10 To finish the proof of Theorem 1.10, we only need to show that the strong limit $(\sqrt{\rho}, \Lambda)$ obtained in Proposition 1.17 is indeed a weak solution of the QHD system. This is a direct consequence of the compactness of $\{(\sqrt{\rho_n}, \Lambda_n)\}$. □

Acknowledgments This manuscript has been written as lecture notes for the EMS Summer School "Mathematical Aspects of Fluid Flows" (Kácov, May 12–17, 2024). The first author would like to thank all the organizers for the opportunity of giving a course there, and in particular for the unique atmosphere during the entire week of the school.
This work was partially supported by the Italian Ministry of University and Research (MUR) through the PRIN project with grant number 2022YXWSLR z "Boundary Analysis for Dispersive and Viscous Fluids" and the Excellence Department Project awarded to GSSI, CUP D13C22003740001 (P.A.). The two authors acknowledge partial support from INdAM-GNAMPA. P.A. is also partially supported by PRIN project 20204NT8W4 "Nonlinear evolution PDEs, fluid dynamics and transport equations: theoretical foundations and applications".

References

1. M. Ancona, G. Iafrate, Quantum correction to the equation of state of an electron gas in a seminconductor. Phys. Rev. B **39**, 9536 (1989)
2. M. Ancona, H. Tiersten, Macroscopic physics of the silicon inversion layer. Phys. Rev. B **35**, 7959 (1987)
3. P. Antonelli, Remarks on the derivation of finite energy weak solutions to the QHD system. Proc. Am. Math. Soc. **149**, 1985–1997 (2021)
4. P. Antonelli, P. Marcati, On the finite energy weak solutions to a system in quantum fluid dynamics. Commun. Math. Phys. **287**, 657–686 (2008)
5. P. Antonelli, P. Marcati, The quantum hydrodynamics system in two space dimensions. Arch. Rat. Mech. Anal. **203**, 499–527 (2012)
6. P. Antonelli, P. Marcati, Finite energy global solutions to a two-fluid model arising in superfluidity. Bull. Inst. Math. Acad. Sin. (N.S.) **10**, 349–373 (2015)
7. P. Antonelli, P. Marcati, Quantum hydrodynamics with nonlinear interactions. Disc. Cont. Dyn. Sys. Ser. S **9**, 1–13 (2016)
8. P. Antonelli, P. Marcati, Some results for quantum fluids, in *Recent Advances in Partial Differential Equations and Applications*. Contemporary Mathematics, vol. 666 (American Mathematical Society, Providence, 2016), pp. 41–54
9. P. Antonelli, S. Spirito, Global existence of finite energy weak solutions of quantum Navier-Stokes equations. Arch. Ration. Mech. Anal. **225**, 1161–1199 (2017)
10. P. Antonelli, S. Spirito, On the compactness of finite energy weak solutions to the quantum Navier-Stokes equations. J. Hyper. Differ. Equ. **15**, 133–147 (2018)
11. P. Antonelli, S. Spirito, On the compactness of weak solutions to the Navier-Stokes-Korteweg equations for capillary fluids. Nonlinear Anal. **187**, 110–124 (2019)
12. P. Antonelli, S. Spirito, Global existence of weak solutions to the Navier-Stokes-Korteweg equations. Ann. I.H.P. C Anal. Non Linéaire **39**, 171–200 (2022)
13. P. Antonelli, P. Marcati, H. Zheng, Dispersive estimates for the QHD system and Euler-Korteweg equations. Methods Appl. Anal. **28**, 2113–2161 (2021)
14. P. Antonelli, P. Marcati, H. Zheng, Genuine hydrodynamic analysis to the 1–d QHD system: existence, dispersion and stability. Commun. Math. Phys. **383**, 85–103 (2021)
15. P. Antonelli, P. Marcati, H. Zheng, An intrinsically hydrodynamic approach to multidimensional QHD systems. Arch. Ration. Mech. Anal. **247**, Paper no. 24 (2023)

16. P. Antonelli, P. Marcati, R. Scandone, Exixtence and stability of almost finite energy weak solutions to the quantum Euler-Maxwell system. J. Math. Pures Appl. (9) **191**, Paper no. 103629 (2024)
17. P. Antonelli, D. Bresch, S. Spirito, Global weak solutions of the Navier-Stokes-Korteweg equations in one dimension. Preprint, https://arxiv.org/abs/2502.17147
18. P. Antonelli, L.E. Hientzsch, P. Marcati, On the Cauchy problem for the QHD system with infinite mass and energy: applications to quantum vortex dynamics. In preparation
19. P. Antonelli, P. Marcati, H. Zheng, The time-relaxation limit for weak solutions to the quantum hydrodynamics system. Preprint, https://arxiv.org/abs/2412.16800
20. C. Audiard, B. Haspot, Global well-posedness of the Euler-Korteweg system for small irrotational data. Commun. Math. Phys. **351**, 201–247 (2017)
21. G. Baccarani, M. Wordeman, An investigation of steady-state velocity overshoot in silicon. Sold-State Electron. **28**, 407–416 (1985)
22. J.E. Barab, Nonexistence of asymptotically free solutions for a nonlinear Schrödinger equation. J. Math. Phys. **25**, 3270–3273 (1984)
23. C. Bardos, P. Degond, Global existence for the Vlasov-Poisson equation in 3 space variables with small initial data. Ann. I.H.P. Anal. Non Linéaire **2**, 101–118 (1985)
24. C.F. Barenghi, L. Skrbek, K.R. Sreenivasan, *Quantum Turbulence* (Cambridge University Press, Cambridge, 2023)
25. S. Benzoni-Gavage, *Propagating Phase Boundaries and Capillary Fluids*. Lecture Notes CIRM - Levico (2010). Available at https://math.univ-lyon1.fr/~benzoni/Levico.pdf
26. S. Benzoni-Gavage, R. Danchin, S. Descombes, On the well-posedness for the Euler-Korteweg model in several space dimensions. Indiana Univ. Math. J. **56**, 1499–1579 (2007)
27. S. Bianchini, Exact integrability conditions for cotangent vector fields. Manuscr. Math. **173**, 293–340 (2024)
28. J. Bourgain, *Global Solutions of Nonlinear Schrödinger Equations*. American Mathematical Society Colloquium Publications, vol. 46 (American Mathematical Society, Providence, 1999)
29. Y. Brenier, Polar factorization and monotone rearrangement of vector-valued functions. Commun. Pure Appl. Math. **44**, 375–417 (1991)
30. D. Bresch, B. Desjardins, Existence of global weak solutions for a 2D viscous shallow water equations and convergence to the quasi-geostrophic model. Commun. Math. Phys. **238**, 211–223 (2003)
31. D. Bresch, A.F. Vasseur, C. Yu, Global existence of entropy-weak solutions to the compressible Navier-Stokes equations with non-linear density dependent viscosities. J. Eur. Math. Soc. **24**, 1791–1837 (2022)
32. R. Carles, R. Danchin, J.-C. Saut, Madelung, Gross-Pitaevskii and Korteweg. Nonlinearity **25**, 2843–2873 (2012)
33. I. Carusotto, C. Ciuti, Quantum fluids of light. Rev. Modern Phys. **85**, 299 (2013)
34. T. Cazenave, *Semilinear Schrödinger Equations*. Courant Lecture Notes in Mathematics, vol. 10 (New York University, Courant Institute of Mathematical Sciences/American Mathematical Society, Providence/New York, 2003)
35. J.-Y. Chemin, Dynamique des gaz à masse totale finie. Asympt. Anal. **3**, 215–220 (1990)
36. J. Colliander, M. Keel, G. Staffilani, H. Takaoka, T. Tao, Global existence and scattering for rough solutions of a nonlinear Schrödinger equation on $\mathbf{R}^3$. Commun. Pure Appl. Math. **57**, 987–1014 (2004)
37. P. Constantin, J.-C. Saut, Local smoothing properties of dispersive equations. J. Am. Math. Soc. **1**, 413–439 (1988)
38. C. Dafermos, *Hyperbolic Conservation Laws in Continuum Physics*. Grundlehren der Mathematischen Wissenschaften [Fundamental Principles of Mathematical Sciences], vol. 325, 4th edn. (Springer-Verlag, Berlin, 2016)
39. P. Degond, C. Ringhofer, Quantum moment hydrodynamics and the entropy principle. J. Stat. Phys. **112**, 587–628 (2003)

40. D. Donatelli, E. Feireisl, P. Marcati, Well/ill posedness for the Euler-Korteweg-Poisson system and related problems. Commun. Partial Differ. Equ. **40**, 1314–1335 (2015)
41. J.E. Dunn, J. Serrin, On the thermomechanics of interstitial working. Arch. Ration. Mech. Anal. **88**, 95–133 (1985)
42. D. Fajman, J. Joudioux, J. Smulevici, A vector field method for relativistic transport equations with applications. Anal. PDE **10**, 1539–1612 (2017)
43. R.P. Feynman, Application of quantum mechanics to liquid helium, in *Progress in Low Temperature Physics*, vol. 1 (Elsevier, Amsterdam, 1955), pp. 17–53
44. C.L. Gardner, The quantum hydrodynamic model for semiconductor devices. SIAM J. Appl. Math. **54**, 409–427 (1994)
45. I. Gasser, P. Markowich, Quantum hydrodynamics, Wigner transforms and the classical limit. Asymptotic Anal. **14**, 97–116 (1997)
46. J. Giesselmann, C. Lattanzio, A.E. Tzavaras, Relative energy for the Korteweg theory and related Hamiltonian flows in gas dynamics. Arch. Ration. Mech. Anal. **223**, 1427–1484 (2017)
47. J. Ginibre, G. Velo, The global Cauchy problem for the nonlinear Schrödinger equation revisited. Ann. I.H.P. Anal. Non Linéaire **2**, 309–327 (1985)
48. J. Ginibre, G. Velo, Scattering theory in the energy space for a class of nonlinear Schrödinger equations. J. Math. Pures Appl. (9) **64**, 363–401 (1985)
49. J. Ginibre, G. Velo, Quadratic Morawetz inequalities and asymptotic completeness in the energy space for nonlinear Schrödinger and Hartree equations. Q. Appl. Math. **68**, 113–134 (2010)
50. V.L. Ginzburg, L. Pitaevskii, On the theory of superfluidity. Sov. Phys. JETP **7**, 858–861 (1958)
51. R.T. Glassey, On the blowing up of solutions to the Cauchy problem for nonlinear Schrödinger equations. J. Math. Phys. **18**(9), 1794–1797 (1977)
52. M. Grassin, D. Serre, Existence de solutions globales et régulières aux équations d'Euler pour un gaz parfait isentropique. C. R. Acad. Sci. Paris Sér. I Math. **325**, 721–726 (1997)
53. A. Griffin, T. Nikuni, E. Zaremba, *Bose-condensed Gases at Finite Temperatures* (Cambridge University Press, Cambridge, 2009)
54. E.P. Gross, Structure of a quantumzed vortex in boson systems. Il Nuovo Cimento (1995–1965) **20**, 454–477 (1961)
55. F. Haas. *Quantum Plasmas*. Springer Series on Atomic, Optical, and Plasma Physics, vol. 65 (Springer, New York, 2011)
56. N. Hayashi, K. Nakamitsu, M. Tsutsumi, On solutions of the initial value problem for the nonlinear Schrödinger equations in one space dimension. Math. Z. **192**, 637–650 (1986)
57. N. Hayashi, K. Nakamitsu, M. Tsutsumi, On solutions of the initial value problem for the nonlinear Schrödinger equations. J. Funct. Anal. **71**, 218–245 (1987)
58. L.E. Hientzsch, Non linear Schrödinger equations and quantum fluids non vanishing at infinity: incompressible limits and quantum vortices. Ph.D. Thesis, 2019. Available at https://iris.gssi.it/handle/20.500.12571/26564
59. F. Huang, H. Li, A. Matsumura, Existence and stability of steady-state of one-dimensional quantum hydrodynamic system for semiconductors. J. Differ. Equ. **225**, 1–25 (2006)
60. F. Huang, H. Li, A. Matsumura, S. Odanaka, Well-posedness and stability of quantum hydrodynamics for semiconductors in $\mathbf{R}^3$, in *Some Problems on Nonlinear Hyperbolic Equations and Applications* (World Scientific, 2010), pp. 131–160
61. R. Illner, G. Rein, Time decay of the solutions of the Vlasov-Poisson system in the plasma physical case. Math. Methods Appl. Sci. **19**, 1409–1413 (1996)
62. J. Jang, P.C. Jayanti, I. Kukavica, On the mass transfer on the 3D Pitaevskii model. J. Math. Fluid Mech. **26**, Paper no. 43 (2024)
63. A. Jüngel, *Transport Equations for Semiconductors*, vol. 773 (Springer, Berlin, 2009)
64. A. Jüngel, Dissipative quantum fluid models. Riv. Math. Univ. Parma (N.S.) 217–290 (2012)
65. A. Jüngel, M.C. Mariani, D. Rial, Local existence of solutions to the transient quantum hydrodynamic equations. Math. Models Meth. Appl. Sci. **12**, 485–495 (2002)

66. A. Jüngel, H.-L. Li, A. Matsumura, The relaxation-time limit in the quantum hydrodynamic equations for semiconductors. J. Differ. Equ. **225**, 440–464 (2006)
67. P. Kapitza, Viscosity of liquid helium below the $\lambda-$point. Nature **141**, 74 (1938)
68. T. Kato, On the Cauchy problem for the (generalized) Korteweg-de Vries equation. Stud. Appl. Math. Adv. Math. Suppl. Stud. **8**, 93–128 (1983)
69. T. Kato, Nonlinear Schrödinger equations, in *Schrödinger Operators (Sonderborg, 1988)*. Lecture Notes in Physics, vol. 345 (Springer, Berlin, 1989), pp. 218–263
70. M. Keel, T. Tao, Endpoint Strichartz estimates. Am. J. Math. **120**, 955–980 (1998)
71. I.M. Khalatnikov, *An Introduction to the Theory of Superfluidity* (CRC Press, Boca Raton, 2018)
72. B. Khesin, G. Misiolek, K. Modin, Geometry of the Madelung transform. Arch. Ration. Mech. Anal. **234**, 549–573 (2019)
73. S. Klainerman, Uniform decay estimates and the Lorentz invariance of the classical wave equation. Commun. Pure Appl. Math. **38**, 321–332 (1985)
74. D.J. Korteweg, Sur la forme que prennent les équations du mouvements des fluides si l'on tient compte des forces capillaires causées par des variations de densité considérables mais connues et sur la théorie de la capillarité dans l'hypothèse d'une variation continue de la densité. Archives Néerlandaises des Sciences Exactes et Naturelles **6**, 1–24 (1901)
75. M. Kostin, On the Schrödinger-Langevin equation. J. Chem. Phys. **57**, 3589–3591 (1972)
76. L.D. Landau, E.M. Lifshitz. *Fluid Mechanics*. Course of Theoretical Physics, vol. 6 (Elsevier, Amsterdam, 2013)
77. J. Li, Z. Xin, Global existence of weak solutions to the barotropic compressible Navier-Stokes flows with degenerate viscosities. Preprint, arXiv:1504.06826
78. H.-L. Li, P. Marcati, Existence and asymptotic behavior of multi-dimensional quantum hydrodynamic model for semiconductors. Commun. Math. Phys. **245**, 215–247 (2004)
79. H.-L. Li, J. Li, Z. Xin, Vanishing of vacuum states and blow-up phenomena of the compressible Navier-Stokes equations. Commun. Math. Phys. **281**, 401 (2008)
80. E.H. Lieb, M. Loss, *Analysis*. Graduate Studies in Mathematics, vol. 14, 2nd edn. (American Mathematical Society, Providence, 2001)
81. E. Madelung, Quantentheorie in hydrodynamischer Form. Zeit. Phys. **40**, 322–326 (1927)
82. P. Markowich, J. Sierra, Non-uniqueness of weak solutions of the quantum-hydrodynamic system. Kinet. Rel. Models **12**, 347–356 (2019)
83. C.S. Morawetz, Time decay for the nonlinear Klein-Gordon equations. Proc. R. Soc. Lond. Ser. A **306**, 291–296 (1968)
84. V. Naibo, A. Stefanov, On some Schrödinger and wave equations with time dependent potentials. Math. Ann. **334**, 325–338 (2006)
85. P. Nozieres, *The Theory of Quantum Liquids*, vol. II (CRC Press, Boca Raton, 1990)
86. W. Pauli, Handbuch der Physik. Geiger and Scheel **2**, 83–272 (1933)
87. B. Perthame, Time decay, propagation of low moments and dispersive effects for kinetic equations. Commun. Part. Differ. Equ. **21**, 659–686 (1996)
88. L. Pitaevskii, Vortex lines in an imperfect Bose gas. Sov. Phys. JETP **13**, 451–454 (1961)
89. L. Pitaevskii, S. Stringari, *Bose-Einstein Condensation and Superfluidity*, vol. 164 (Oxford University Press, Oxford, 2016)
90. F. Planchon, L. Vega, Bilinear virial identities and applications. Ann. Sci. École Normale Sup. **42**, 261–290 (2009)
91. J.M. Rakotoson, R. Temam, An optimal compactness theorem and application to elliptic-parabolic systems. Appl. Math. Lett. **14**, 303–306 (2001)
92. P.H. Roberts, R.J. Donnelly, Superfluid Mechanics. Ann. Rev. Fluid Mech. **6**, 179–225 (1974)
93. J.L. Roberts, N.R. Claussen, S.L. Cornish, E.A. Donley, E.A. Cornell, C.E. Wieman, Controlled collapse of a Bose-Einstein condensate. Phys. Rev. Lett. **86**, 4211 (2001)
94. D. Serre, Solutions classiques globales des équations d'Euler pour un fluide parfait compressible. Ann. Inst. Fourier (Grenoble) **47**, 139–153 (1997)
95. T.C. Sideris, Formation of singularities in three-dimensional compressible fluids. Commun. Math. Phys. **101**, 475–485 (1985)

96. W. Strauss, Nonlinear invariant wave equations, in *Invariant Wave Equations (Proc. "Ettore Majorana" Internat. School of Math. Phys. Erice, 1977)*. Lecture Notes in Ohysics, vol. 73 (Springer, New-York, 1978), pp. 197–249
97. R.S. Strichartz, Restrictions of Fourier transforms to quadratic surfaces and decay of solutions of wave equations. Duke Math. J. **44**, 705–714 (1977)
98. C. Sulem, P.-L. Sulem, *The Nonlinear Schrödinger Equation. Self-focusing and Wave Collapse*. Applied Mathematical Sciences, vol. 139 (Springer-Verlag, New York, 1999)
99. T. Tao, *Nonlinear Dispersive Equations: Local and Global Analysis*. CBMS Regional Conference Series in Mathematics, vol. 106 (American Mathematical Society, Providence, 2006)
100. M. Tsubota, M. Kobayashi, H. Takeuchi, Quantum hydrodynamics. Phys. Rep. **522**, 191–238 (2013)
101. Y. Tsutsumi, L^2-solutions for nonlinear Schrödinger equations and nonlinear groups. Funkcial. Ekvac. **30**, 115–125 (1987)
102. T.C. Wallstrom, Inequivalence between the Schrödinger equation and the Madelung hydrodynamic equations. Phys. Rev. A **49**, 1613 (1994)
103. S. Wiegert, Pauli problem for a spin of arbitrary length: a simple method to determine its wave function. Phys. Rev. A **45**, 7688 (1992)
104. Z. Xin, Blowup of smooth solutions to the compressible Navier-Stokes equation with compact density. Commun. Pure Appl. Math. **51**, 229–240 (1998)
105. K. Yajima, Existence of solutions for Schrödinger evolution equations. Commun. Math. Phys. **110**, 415–426 (1987)

Chapter 2
Singular Limits Arising in Mechanical Models of Tissue Growth

Noemi David

Abstract Based on the mechanical viewpoint that living tissues present a fluid-like behaviour, PDE models inspired by fluid dynamics are nowadays well established as one of the main mathematical tools for the macroscopic description of tissue growth. Depending on the type of tissue, these models link the pressure to the velocity field using either Brinkman's law (visco-elastic models) or Darcy's law (porous-medium equations (PME)). Moreover, the stiffness of the pressure law plays a crucial role in distinguishing density-based (compressible) models from free boundary (incompressible) problems where saturation of the density holds.

This course aims to analyse how to relate different mechanical models of living tissues through singular limits. In particular, we will address the inviscid limit towards the PME and the incompressible limit from the PME to free boundary problems of Hele-Shaw type for equations including convective effects as well as for systems of coupled nonlinear equations.

2.1 Introduction

Mechanical models of tumor growth borrow concepts from fluid dynamics to describe the mechanical properties of cell movement and proliferation. We consider the following continuity equation

$$\frac{\partial \varrho}{\partial t} + \nabla \cdot (\varrho \mathrm{v}) = \varrho G(p), \qquad x \in \mathbb{R}^d, t > 0, \tag{2.1}$$

that describes the evolution of the cell population density $\varrho = \varrho(x, t)$. The velocity field v is related to the pressure p which is a function of the density ϱ. Different choices of pressure-velocity and pressure-density relations lead to a variety of

N. David (✉)
Laboratoire de Mathématiques Raphaël Salem Avenue de l'Université, Saint-Étienne-du-Rouvray, France
e-mail: noemi.david@univ-rouen.fr

M. Bulíček et al. (eds.), *New Perspectives in Mathematical Fluid Mechanics*, Nečas Center Series, https://doi.org/10.1007/978-3-032-02319-3_2

models with very different mathematical properties. We will focus on the following laws:

- **Darcy's law**: $\mathrm{v} = -\nabla p$,
- **Brinkman's law**: $-\nu\Delta\mathrm{v} + \mathrm{v} = -\nabla p$.

Darcy's law describes the fact that cells move down pressure gradients towards regions with lower congestion. Brinkman's law also accounts for the viscosity of the tumor, $\nu > 0$, and can be seen as a non-local perturbation of Darcy's law.

The pressure is usually defined as a convex function of the density. Two classic examples are

$$p = \frac{m}{m-1}\varrho^{m-1}, \quad m > 1, \qquad p = \varepsilon\frac{\varrho}{\varrho - 1}, \quad \varepsilon > 0. \tag{2.2}$$

We will focus mainly on the power law, for which Eq. (2.1) is a reaction-*porous medium* equation (PME). As $m \to \infty$ (or $\varepsilon \to 0$) the limit density and pressure satisfy the following relation

$$p_\infty(1 - \varrho_\infty) = 0. \tag{2.3}$$

This means that, in the limit, the density is completely saturated, $\varrho_\infty = 1$, in the region where p_∞ is positive. This generates a *free boundary* problem describing the evolution of the set $\Omega(t) := \{x;\, p_\infty(x,t) > 0\}$—in our framework, the region occupied by the tumor.

In this course, we will focus on the theory of the *incompressible limit*, $m \to \infty$ ($\varepsilon \to 0$), for different kinds of tissue growth models. In particular, we will see a compendium of techniques apt to show the convergence of solutions to equations or systems of the form (2.1) to solutions of free boundary problems of Hele-Shaw type.

We will also address the *inviscid limit*, namely $\nu \to 0$ in Brinkman's law, which leads to a Darcy description. This asymptotics is another example of *singular limit* since the nature of the equation changes drastically—from hyperbolic to degenerate parabolic.

Outline

In this introductive section, we recall some basic properties of the porous medium equation (PME) and the original definition of the Hele-Shaw problem. Finally, we introduce the incompressible limit for the PME. Section 2.2 is devoted to the theory of the incompressible limit for tumor growth models. We review the seminal work by Perthame et al. and we present some more recent techniques applied to this problem, such as Aronson-Bénilan estimates in L^p spaces and energy methods. We also present some results on the rate of convergence of this asymptotics. In

Sect. 2.3, we discuss the inviscid limit from the Brinkman model to the PME. We conclude with a brief review of multi-species systems and structured equations for the description of heterogeneous tissues, Sect. 2.4.

2.1.1 The Porous Medium Equation

The porous medium equation is the simplest example of nonlinear (degenerate) diffusion equation

$$\frac{\partial \varrho}{\partial t} = \Delta \varrho^m, \qquad m > 1. \tag{2.4}$$

The range $m > 1$ is also known as *slow diffusion*, while for $m < 1$ Eq. (2.4) is known as *fast diffusion equation*. The original motivation of Eq. (2.4) is given by the modelling of a fluid's flow through a porous medium. This phenomenon is described by Darcy's law, which is usually written as

$$\mathrm{v} = -\frac{k}{\mu}\nabla p, \tag{2.5}$$

where k is the permeability of the medium and μ is the viscosity of the fluid. When the pressure is given by the following law

$$p = \frac{m}{m-1}\varrho^{m-1}, \; m > 1,$$

one recovers (up to rescaling the constants in Darcy's law) Eq. (2.4). The pressure satisfies the following equation

$$\frac{\partial p}{\partial t} = (m-1)p\Delta p + |\nabla p|^2. \tag{2.6}$$

It appears clear that the equation has a mixed character: it is strictly parabolic in the set where $p > 0$, and it has a hyperbolic flavour where $p \sim 0$. At the moving free boundary the gradient term $|\nabla p|^2$ "wins" over the degenerate diffusion part, and the equation resembles the eikonal equation. An essential feature of solutions to the PME is indeed the presence of a *free boundary*, $\partial\{\varrho > 0\}$, since—being degenerate parabolic—the equation admits solutions with compact support.

The definition of a class of weak solutions for Eq. (2.4) was first given in the one-dimensional case in [33], where the authors prove the existence and uniqueness of solutions in that class. From now on, we will consider the following definition of weak solutions.

Definition 2.1 (Weak Solutions of the PME) A locally integrable function ϱ defined in $\mathbb{R}^d \times (0, T)$ is said to be a weak solution of (2.4) with initial condition given by $\varrho_0 \in L^1(\mathbb{R}^d)$ if

(i) $\varrho^m \in L^2(0, T; H^1(\mathbb{R}^d))$,
(ii) ϱ satisfies

$$\int_0^T \int_{\mathbb{R}^d} \left(\nabla \varrho^m \cdot \nabla \varphi - \varrho \frac{\partial \varphi}{\partial t} \right) \mathrm{d}x\, \mathrm{d}t = \int_{\mathbb{R}^d} \varrho_0(x) \varphi(x, 0)\, \mathrm{d}x,$$

for any $\varphi \in C_c^1(\mathbb{R}^d \times [0, T))$.

Compact Support Property

If there exists a compact set $K \subset \mathbb{R}^d$, such that $\operatorname{supp} \varrho_0 \subset K$, then for all $T > 0$ there exists a bounded open set $\Omega \subset \mathbb{R}^d$ depending on the size of K, and on the final time T such that

$$\operatorname{supp} \varrho(t) \subset \Omega \text{ for all } t \in [0, T].$$

Self-similar Profile

An explicit formula is available for *source solutions* of Eq. (2.4), namely $\varrho(x, t)$ such that $\varrho(x, t) \to M\delta(x)$ as $t \to 0$, where $M := \int \varrho_0(x)\, \mathrm{d}x$. The self-similar profile, usually called Barenblatt solution, is

$$B(x, t; M) := t^{-\alpha} F(x t^{-\alpha/d}), \qquad F(\xi) := (C - \kappa |\xi|^2)_+^{\frac{1}{m-1}},$$

where

$$\alpha = \frac{d}{2 + d(m-1)}, \qquad \kappa = \frac{\alpha(m-1)}{2md},$$

and $C > 0$ is a constant determined by the mass M. This profile gives the simplest example of a solution that does not satisfy the equation in the classical sense since the gradient is not continuous across the free boundary, see Fig. 2.1.

Another important property of the PME, for which the Barenblatt profile represents a classic example, is that solutions with initial data bounded in $L^1(\mathbb{R}^d)$ but not in $L^\infty(\mathbb{R}^d)$ experience an instantaneous smoothness effect, in that the solution $\varrho(t)$ is always $L^\infty(\mathbb{R}^d)$ for $t > 0$. In particular,

$$\|\varrho(t)\|_\infty \leq C(m, d) \|\varrho_0\|_1^{2\alpha/d} t^{-\alpha}.$$

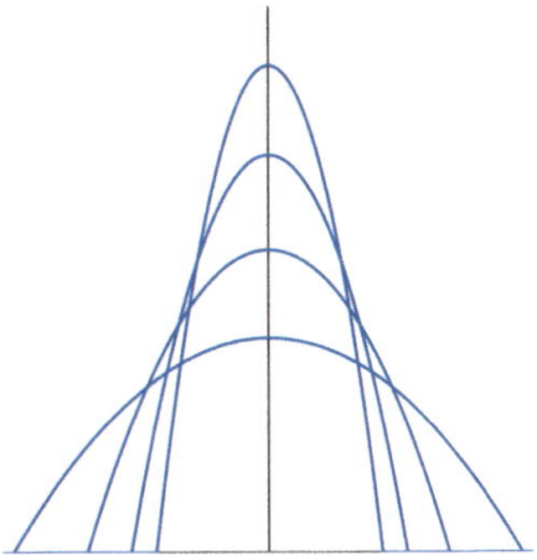

Fig. 2.1 Barenblatt profile

This property is a consequence of the fundamental estimate of the PME, also known as the Aronson-Bénilan estimate, and the following lemma.

Lemma 2.1 *Let $f(t,x)$ be any nonnegative, smooth, bounded and integrable function in $\mathbb{R}^d$ such that*

$$\Delta(f^{m-1}) \geq -K,$$

for some $m > 1$ and $K > 0$. Then $f \in L^\infty(\mathbb{R}^d)$ and

$$\|f\|_\infty \leq C(m,d)\|f\|_1^{2\alpha/d} K^\alpha.$$

For more details, we refer the reader to [40, Proposition 9.8].

Aronson-Bénilan Estimate

The estimate is named after Aronson and Bénilan, [5], who established the result in $\mathbb{R}^d$—while for $d = 1$ it was proven in [4]. The result states that the pressure of the solution to (2.4) satisfies

$$\Delta p = \Delta\left(\frac{m}{m-1}\varrho^{m-1}\right) \geq -\frac{1}{kt}, \qquad \text{with } k = m - 1 + \frac{2}{d}. \tag{2.7}$$

Let us mention that this lower bound on the Laplacian of the pressure is used by the authors to prove that there exists a unique strong and continuous solution to the Cauchy problem with L^1-bounded initial data. The Aronson-Bénilan estimate (AB in short) is referred to as the fundamental estimate in the theory of the porous medium equation.

Let us present here the proof of the AB estimate from [5].

Proof Let us recall the pressure equation, Eq. (2.6). We denote $w := \Delta p$. We have

$$\frac{\partial p}{\partial t} = (m-1)pw + |\nabla p|^2.$$

We compute the time derivative of w

$$
\begin{aligned}
\frac{\partial w}{\partial t} &= \Delta\left(\frac{\partial p}{\partial t}\right) \\
&= \Delta\left((m-1)pw + |\nabla p|^2\right) \\
&= (m-1)w^2 + 2(m-1)\nabla p\cdot\nabla w + (m-1)p\Delta w + 2\nabla p\cdot\nabla\Delta p \\
&\quad + 2\sum_{i,j}\left(\frac{\partial^2 p}{\partial x_i\partial x_j}\right)^2 \\
&\geq (m-1)w^2 + 2m\nabla p\cdot\nabla w + (m-1)p\Delta w + \frac{2}{d}(\Delta p)^2,
\end{aligned}
$$

and thus, we have

$$
\frac{\partial w}{\partial t} \geq \left(m-1+\frac{2}{d}\right)w^2 + 2m\nabla p\cdot\nabla w + (m-1)p\Delta w.
$$

Assuming p smooth and bounded away from zero, the above inequality implies that w is a supersolution of a quasi-linear parabolic operator. Let $\tau > 0$. The function

$$
W(t) := -\frac{1}{k(t+\tau)}, \quad \text{with } k := m-1+\frac{2}{d},
$$

is a (sub)solution of the same operator, since

$$
W'(t) = \frac{1}{k(t+\tau)^2} = kW(t)^2.
$$

Let us assume that, for τ small enough, $w(x,0) = \Delta p(x,0) \geq -\frac{1}{k\tau} = W(0)$. Therefore, by the comparison principle for uniformly parabolic operators we find $w(x,t) \geq -\frac{1}{k(t+\tau)}$ for $t > 0$ and for any $\tau > 0$. Letting $\tau \to 0$ we finally find

$$
w(x,t) \geq -\frac{1}{\left(m-1+\frac{2}{d}\right)t}.
$$

Let us notice that the above estimate is independent of any regularisation argument. Therefore, it can be understood in the sense of distributions for any solution obtained as the limit of regular solutions, see [5]. Here we give a gist of the argument. In order to apply the comparison principle for quasi-linear parabolic operators we need to approximate the solution of the PME taking $\varrho_{0,\varepsilon}(x) = \varrho_0(x) + \varepsilon$. For such initial data, the PME is no longer degenerate and there exists a unique solution that satisfies $\varrho_\varepsilon \in C^\infty(\mathbb{R}^d \times (0,\infty))$ and $\varrho_\varepsilon(x,t) \geq \varepsilon$. Consequently, the pressure $p_\varepsilon = \frac{m}{m-1}\varrho_\varepsilon^{m-1} \in C^\infty(\mathbb{R}^d \times (0,\infty))$ is bounded away from zero as well. Then, we can apply the above argument and the AB estimate (2.7) holds for p_ε uniformly in ε. One can show that $p_\varepsilon \to p$ in $L^1_{loc}(\mathbb{R}^d \times (0,\infty))$, *cf.* [40, Lemma 9.5]. Hence, for any $\varphi \in C_c^\infty(\mathbb{R}^d \times (0,\infty))$, $\varphi \geq 0$, we have

$$\int_0^\infty \int_{\mathbb{R}^d} \left(\Delta p_\varepsilon + \frac{1}{kt} \right) \varphi \, \mathrm{d}x \, \mathrm{d}t \geq 0,$$

and, thus

$$\int_0^\infty \int_{\mathbb{R}^d} \left(p_\varepsilon \Delta\varphi + \frac{1}{kt}\varphi \right) \mathrm{d}x \, \mathrm{d}t \geq 0,$$

$$\downarrow$$

$$\int_0^\infty \int_{\mathbb{R}^d} \left(p \Delta\varphi + \frac{1}{kt}\varphi \right) \mathrm{d}x \, \mathrm{d}t \geq 0,$$

and this completes the proof, namely (2.7) holds in the sense of distributions. □

A question that has fascinated many researchers in the last few decades is to understand the behavior of the porous medium equation as the exponent m becomes larger and larger. In the limit $m \to \infty$, the density has to satisfy the constraint $\varrho_\infty \leq 1$ (also called *hard congestion constraint* in the context of crowd motion modelling [2, 31]), and compactly supported solutions develop sharp interfaces. In particular, the appearance of a saturated region occurs, namely a zone where $\varrho_\infty = 1$, which is closely related to the non-coincidence set of a free boundary problem known as the *Hele-Shaw problem*. Before introducing the underlying principles behind the incompressible limit $m \to \infty$, we briefly present the original Hele-Shaw problem.

2.1.2 The Hele-Shaw Problem

The Hele-Shaw problem was introduced to model the injection of a fluid into a laminar cell [37]. The fluid surface expands in the small gap between two parallel flat plates that form the cell, also called *Hele-Shaw cell*. Originally the model was motivated by applications to the plastic industry, in particular to injection moulding. In [23], the authors approach the problem from a variational point of view. They consider a finite cell and a finite source; the injected fluid is assumed to be incompressible and the pressure variations perpendicular to the cell surface are neglected since the space between the plates is infinitesimally small. The fluid "blows" from the injection point with a constant rate Q, thus, at time $t > 0$, the increment of the fluid blob is Qt volume units. The movement of the fluid is

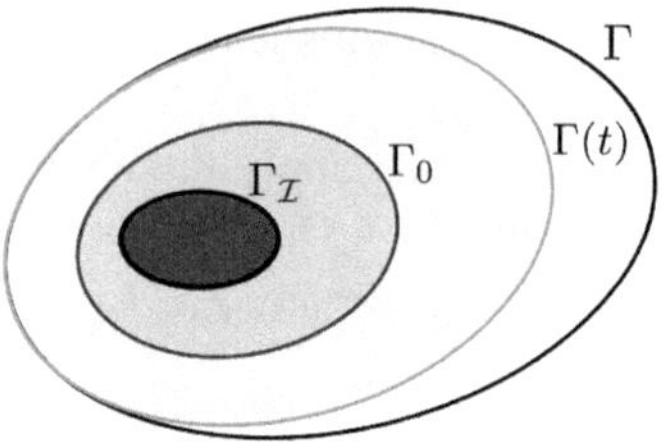

Fig. 2.2 Hele-Shaw problem in a bounded domain

governed by Darcy's law (2.5) (where without loss of generality we assume $k = \mu$). The problem is set as displayed in Fig. 2.2.
The fluid is injected into the cell along the curve $\Gamma_{\mathcal{I}}$. At time $t = 0$, the fluid occupies the region between $\Gamma_{\mathcal{I}}$ and Γ_0 denoted Ω_0, which expands into $\Omega(t)$ at $t > 0$, with free boundary $\Gamma(t)$.

The fluid incompressibility implies that the velocity field is divergence-free in the occupied region

$$\Delta p = 0, \text{ in } \Omega(t).$$

The pressure on $\Gamma(t)$ is assumed to be equal to zero, the velocity of the free boundary is opposite to the normal derivative of the pressure, $v = -\partial_n p$, and the flow is tangential to the outer boundary Γ. Let us assume that for some function $l = l(x)$, the moving boundary and the fluid surface can be defined as

$$\Gamma(t) = \{x;\ l(x) = t\}, \text{ and } \Omega(t) = \{x;\ l(x) < t\}.$$

The Hele-Shaw problem in the sense of Elliot-Janovský [23] can be stated as follows.

Problem 2.1 (Original Hele-Shaw Problem) Find $l(x)$ and $p(x, t)$ such that

$$\begin{cases} l(x) = 0, & \text{for } x \in \Omega_0, \\ \Delta p(x,t) = 0, & \text{for } x \in \Omega(t), \end{cases} \qquad \begin{cases} p = 0, & \text{on } \Gamma(t), \\ \partial_n p = -v, & \text{on } \Gamma(t), \\ \partial_n p = Q, & \text{on } \Gamma_{\mathcal{I}}, \\ \partial_n p = 0, & \text{on } \Gamma. \end{cases}$$

2.1.3 *The Mesa-limit*

We here present in a formal and heuristic way, the leading mechanisms behind the asymptotics $m \to \infty$ of the porous medium equation.

2.1.3.1 Incompressible Limit of the Cauchy Problem

We are interested in the asymptotic behavior of the following Cauchy problem as $m \to \infty$

$$\begin{cases} \dfrac{\partial \varrho}{\partial t} = \Delta \varrho^m, & x \in \mathbb{R}^d,\ t > 0, \\ \varrho(x,t) = \varrho_0(x) \geq 0, & \varrho_0 \in L^1(\mathbb{R}^d) \cap L^\infty(\mathbb{R}^d). \end{cases} \tag{2.8}$$

The PME can be rewritten as a diffusion equation

$$\frac{\partial \varrho_m}{\partial t} = \nabla \cdot (D(\varrho_m) \nabla \varrho_m), \qquad D(\varrho_m) = m \varrho_m^{m-1},$$

where the subscript highlights the dependency upon the parameter m. In the limit the nonlinear density-dependent diffusivity coefficient $D(\varrho_m)$ behaves like

$$D(\varrho) \xrightarrow{m \to \infty} \begin{cases} 0, & \text{when } \varrho < 1, \\ \infty, & \text{when } \varrho \geq 1. \end{cases}$$

Since the diffusivity coefficient blows up where the initial data is larger than 1, the solution instantaneously collapses to the level 1 as $m \to \infty$. On the other hand, where $\varrho_0 < 1$ the diffusivity vanishes as $m \to \infty$. Therefore, in those regions, the solution "stays still" as $m \to \infty$. This heuristic argument suggests that in the limit the solution of the Cauchy problem (2.8) converges to a stationary profile $\varrho_\infty = \varrho_\infty(x)$. The plateau-like region where $\varrho_\infty = 1$ is referred to as *mesa* (from Spanish, mesa are mountain formations that have a flat region on top) is associated with the variational inequality formulation of the Hele-Shaw problem, see [6, 10, 24, 38].

The "stationarity" of the limit density, namely $\varrho_\infty = \varrho_\infty(x)$, is deduced upon combining three tools: uniform bounds, the AB estimate, and the conservation of mass [10]. In fact, the AB estimate (2.7) implies monotonicity, $\partial_t \varrho_\infty \geq 0$, while the equation preserves the mass. In [10] the authors also show that $0 \leq \varrho_\infty \leq 1$. In particular

$$\varrho_\infty(x) = \begin{cases} 1, & \text{for } x \in A, \\ \varrho_0(x), & \text{for } x \notin A, \end{cases}$$

where A is defined as the non-coincidence set of the solution to a variational inequality problem, equivalent to a "stationary" Hele-Shaw free boundary model.

2.1.3.2 The Boundary Value Problem

The asymptotic behavior of the PME changes significantly for the Cauchy-Dirichlet problem with non-homogeneous boundary conditions

$$\begin{cases} \dfrac{\partial \varrho_m}{\partial t} = \Delta \varrho_m^m, & x \in \Omega,\ t > 0, \\ \varrho_m(x,0) = \varrho_0(x) \geq 0, & x \in \Omega, \\ \varrho_m(x,t)^m = g(x,t) \geq 0, & x \in \partial\Omega,\ t > 0. \end{cases} \tag{2.9}$$

Let us point out that as $m \approx \infty$, $\varrho_m^m \approx p_m$. Hence, the boundary condition in (2.9) is equivalent to fixing the value of the pressure on the boundary. Usually, the quantity ϱ_m^m is referred to as the *generalized pressure*.

For vanishing Dirichlet boundary data, $g \equiv 0$, the stationarity of the limit Hele-Shaw problem can also be directly seen by analysing the asymptotic behavior of p_m: from the pressure Eq. (2.6) we have

$$\int_0^T \int_\Omega |\nabla p_m|^2 \,\mathrm{d}x\,\mathrm{d}t = \frac{1}{m-2} \int_\Omega (p_m(0) - p_m(T)) \,\mathrm{d}x \xrightarrow{m\to\infty} 0.$$

Therefore, the boundary of the limit HS problem is actually motionless, and $p_\infty = 0$ everywhere.

On the other hand, for non-trivial boundary data, p_m does not vanish as $m \to \infty$ since it is forced to be positive near the outer boundary $\partial\Omega$. Hence, the motion of the free boundary $\partial\{p_\infty > 0\}$ is governed by Darcy's law $v = -\partial_{\mathrm{n}} p_\infty$. Notice that the mass conservation property no longer holds, therefore the stationarity argument mentioned before fails.

2.1.3.3 The Graph Relation

The relation between p_m and ϱ_m implies

$$\left(\frac{m}{m-1} p_m\right)^{\frac{m}{m-1}} = \varrho_m^m = \frac{m-1}{m} p_m \varrho_m,$$

from which we can formally recover $p_\infty = p_\infty \varrho_\infty$, namely $p_\infty(1 - \varrho_\infty) = 0$, which is equivalent to $p_\infty \in P_\infty(\varrho_\infty)$ where P_∞ is the monotone graph

$$P_\infty(\varrho) = \begin{cases} \{0\}, & \text{for } \varrho < 1, \\ [0, +\infty), & \text{for } \varrho = 1, \\ \emptyset, & \text{for } \varrho > 1. \end{cases} \tag{2.10}$$

We will refer to this relation as *saturation constraint* since it implies that in regions of positive pressure the density is totally saturated, $\varrho_\infty \equiv 1$. Moreover, from (2.6), one can easily see that p_∞ satisfies

$$p_\infty \Delta p_\infty = 0,$$

which implies the incompressibility condition in $\Omega(t) = \{p_\infty > 0\} \subset \{\varrho_\infty = 1\}$ already seen in the definition of the Hele-Shaw problem.

2.2 The Incompressible Limit

Emanating from the early works on the mesa problem for the porous medium equation, research began branching out in different directions. In this section, we address the incompressible limit for porous medium equations including reaction and/or advection terms and their applications to tissue growth modelling. The first generalisation concerns the inclusion of a pressure-dependent growth term and was proposed in [35].

2.2.1 A Purely Mechanical Model of Tumor Growth

Let us consider the following equation

$$\frac{\partial \varrho_m}{\partial t} - \nabla \cdot (\varrho_m \nabla p_m) = \varrho_m G(p_m), \quad x \in \mathbb{R}^d,\ t > 0, \tag{2.11}$$

with

$$p_m = P_m(\varrho_m) := \frac{m}{m-1} \varrho_m^{m-1}. \tag{2.12}$$

The equation for the pressure now reads

$$\frac{\partial p_m}{\partial t} = (m-1) p_m (\Delta p_m + G(p_m)) + |\nabla p_m|^2. \tag{2.13}$$

The pressure-dependent reaction rate, G, models the birth and death of cells. It is assumed to decrease with increasing pressure, as higher pressure limits the available space for cell division. Therefore, a standard assumption is that there exists a value $p_H > 0$, called *homeostatic pressure*, such that

$$G \in C^1([0, +\infty)),\ G(0) > 0,\ G' \le -\alpha,\ G(p_H) = 0, \tag{2.14}$$

where $\alpha > 0$ is a fixed positive constant. This model was introduced in [35] where the authors study the incompressible limit $m \to \infty$. As mentioned in the introduction, this limit bridges the gap between the density-based model (2.11) and a free boundary problem of Hele-Shaw type. We recall that, in the limit, the pressure and density satisfy the graph relation (2.10), hence

$$0 \le \varrho_\infty \le 1, \quad p_\infty(1 - \varrho_\infty) = 0. \tag{2.15}$$

Let us state the assumptions on the initial data.

Assumption 2.1 We assume that $\varrho_{m,0} \ge 0$, $\varrho_{m,0} \in L^1(\mathbb{R}^d)$, and $p_{m,0} = P_m(\varrho_m)$ defined in (2.12) satisfy

- $0 \le p_{m,0} \le p_H$,
- $\|\partial_{x_i} \varrho_{m,0}\|_{L^1(\mathbb{R}^d)} \le C$ for all $i = 1, ..., d$,

and there exists $0 \le \varrho_0 \le 1$, $\varrho_0 \in L^1(\mathbb{R}^d)$, such that $\|\varrho_{m,0} - \varrho_0\|_{L^1(\mathbb{R}^d)} \to 0$ as $m \to \infty$.

For ease of exposition, from now on we will assume that the initial data have compact support. As will be stated in the next section, this implies that the solution has compact support for all finite times. Specifically, we assume that there exists a compact set $K \subset \mathbb{R}^d$ such that $\text{supp}(\varrho_{m,0}) \subset K$ for all $m > 1$.

Theorem 2.1 (Incompressible Limit [35]) *Let ϱ_m be the weak solution of* (2.11) *with compactly supported initial data $\varrho_{m,0}$ satisfying Assumption 2.1.*

For all $T > 0$, there exist $\varrho_\infty \in L^\infty(0, T; L^\infty(\mathbb{R}^d) \cap L^1(\mathbb{R}^d))$ and $p_\infty \in L^\infty(0, T; L^\infty(\mathbb{R}^d) \cap L^1(\mathbb{R}^d)) \cap L^2(0, T; H^1(\mathbb{R}^d))$ that satisfy (2.15), *such that*

$$\varrho_m \to \varrho_\infty, \ p_m \to p_\infty, \ \textit{strongly in } L^p(0, T; L^p(\mathbb{R}^d)) \textit{ for all } 1 \le p < \infty,$$

and $(\varrho_\infty, p_\infty)$ satisfies the following equation (in the weak sense)

$$\frac{\partial \varrho_\infty}{\partial t} = \Delta p_\infty + \varrho_\infty G(p_\infty), \quad \varrho_\infty(0) = \varrho_0 \textit{ in } L^1(\mathbb{R}^d), \tag{2.16}$$

as well as

$$p_\infty(\Delta p_\infty + G(p_\infty)) = 0 \textit{ in } \mathcal{D}'(\mathbb{R}^d \times (0, \infty)). \tag{2.17}$$

Equation (2.17) is known as *complementarity relation*, and it links the limit problem to its free boundary description. Defining $\Omega(t) := \{x; \ p_\infty(x, t) > 0\}$, we would have

$$\begin{cases} -\Delta p_\infty = G(p_\infty), & \text{in } \Omega(t), \\ p_\infty = 0, & \text{on } \partial\Omega(t). \end{cases}$$

The velocity of the free boundary is given by Darcy's law. However, the law is influenced by the presence of outer density. It follows from (2.15) that $\varrho_\infty = 1$ in $\Omega(t)$, namely there is total saturation in the set representing the tumor bulk. Characteristic functions are solutions of the equation, in particular, if $\varrho_0 = \mathbb{1}_{\Omega(0)}$ then $\varrho_\infty(t) = \mathbb{1}_{\Omega(t)}$, and in this case the normal velocity of $\partial\Omega(t)$ is given by

$$v = -\nabla p_\infty \cdot \mathrm{n},$$

where n is the outer normal to the boundary. However, it is also possible that outside of $\Omega(t)$ the density, ϱ_∞, does not vanish, hence there is a positive measure set where $0 < \varrho_\infty < 1$. At the boundary, $\partial\Omega$, the density jumps from level 1 to some outer value $0 \leq \varrho_\infty^E(x,t) < 1$ and the velocity law now reads

$$(1 - \varrho_\infty^E)v = -\nabla p_\infty \cdot \mathrm{n}.$$

Throughout these notes, we will not study the rigorous derivation of such law—if interested the reader may look at [2, 28].

Our main goal will be to show how to prove the complementarity relation (2.17). This is clearly more involved than proving that the limit satisfies Eq. (2.16). While for the latter weak compactness of the pressure gradient ∇p_m in $L^2(0,T;L^2(\mathbb{R}^d))$ is enough, the former requires **strong compactness** of ∇p_m. Equation (2.17) can be rewritten as

$$\iint \left(-\varphi|\nabla p_\infty|^2 - p_\infty \nabla\varphi \cdot \nabla p_\infty + p_\infty G(p_\infty)\varphi\right) = 0, \tag{2.18}$$

for all $\varphi \in C_c^\infty(\mathbb{R}^d \times (0,\infty))$. Since the pressure sequence, p_m, satisfies

$$\iint \left(p_m \frac{\partial\varphi}{\partial t} - (m-2)\varphi|\nabla p_m|^2 - (m-1)(p_m \nabla\varphi \cdot \nabla p_m + p_m G(p_m)\varphi)\right) = 0,$$

proving (2.18) is equivalent to showing strong compactness of the pressure gradient as $m \to \infty$, see [35, Lemma 2.5]. In order to show that, up to a subsequence,

$$\nabla p_m \to \nabla p_\infty \text{ strongly in } L^2(0,T;L^2(\mathbb{R}^d)),$$

two methods are now well-established in the literature. The first one involves proving higher order stability estimates on the solution, in particular inferring uniform bounds on the Laplacian of the pressure Δp_m. The second method exploits the dissipation structure of the equation employing energy equalities.

2.2.2 *Aronson-Bénilan Estimates in L^p Spaces*

We present some new versions of the fundamental estimate of the PME, which have been first developed in [35] and applied to the incompressible limit for tumor growth models. These estimates have proven to be a useful tool to obtain the strong compactness of the pressure gradient.

Before discussing the Aronson-Bénilan estimates in L^p spaces, we recall some standard properties of Eq. (2.11)—the compact support property, the propagation of BV-bounds and the L^2-integrability of ∇p_m.

2.2.2.1 A Priori Estimates and Compact Support Property

For the comprehensive proofs of the following properties, we refer the reader to [14, 35].

Lemma 2.2 (Compact Support Property) *Let $K \subset \mathbb{R}^d$ be a compact set such that $\varrho_{m,0} \subset K$ for all $m > 1$. For all $T > 0$ there exist $\Omega \subset \mathbb{R}^d$ such that*

$$\operatorname{supp}(\varrho_m(t)) \subset \Omega, \textit{ for all } 0 \le t \le T.$$

Proof This property follows from the comparison principle and the fact that it is always possible to construct a supersolution with compact support. Let us take the function

$$\Pi(x,t) := G(0)\left(S(t) - \frac{|x|^2}{2}\right)_+,$$

with $S'(t) \ge 2G(0)S(t)$, with $S(0) = S_0$ large enough such that $\Pi(x,0) \ge p_{m,0}$. Then, Π is a supersolution to the equation

$$\frac{\partial p}{\partial t} = (m-1)p(\Delta p + G(0)) + |\nabla p|^2,$$

for which p_m is a subsolution. Therefore $p_m \le \Pi$ almost everywhere, and p_m and ϱ_m are compactly supported. □

Lemma 2.3 (Uniform Bounds) *Both ϱ_m and p_m are uniformly bounded*

$$0 \le p_m \le p_H, \quad 0 \le \varrho_m \le \varrho_H := P_m^{-1}(p_H),$$

almost everywhere in $\mathbb{R}^d \times (0,T)$. Moreover, $\varrho_m, p_m \in L^\infty(0,T;L^1(\mathbb{R}^d))$ uniformly in m.

Proof Using Kato's inequality, we compute

$$\frac{\mathrm{d}}{\mathrm{d}t}\int_{\mathbb{R}^d}(\varrho_m-\varrho_H)_+\,\mathrm{d}x \leq \int_{\mathbb{R}^d}\Delta(\varrho_m^m-\varrho_H^m)_+\,\mathrm{d}x+\int_{\mathbb{R}^d}(\varrho_m-\varrho_H)_+G(p_m)\,\mathrm{d}x$$
$$+\int_{\mathbb{R}^d}\varrho_H G(p_m)\mathrm{sign}_+(\varrho_m-\varrho_H)\,\mathrm{d}x$$
$$\leq G(0)\int_{\mathbb{R}^d}(\varrho_m-\varrho_H)_+\,\mathrm{d}x,$$

where we used the fact that $G(p)\leq 0$ for $p\geq p_H$. By Gronwall's lemma and the assumption $p_{m,0}\leq p_H$, we get $p_m\leq p_H$.

Arguing in the same way and computing the time derivative of $\|\varrho_m(t)\|_{L^1(\mathbb{R}^d)}$ we obtain

$$\|\varrho_m(t)\|_{L^1(\mathbb{R}^d)}\leq e^{tG(0)}\|\varrho_{m,0}\|_{L^1(\mathbb{R}^d)}.$$

Since $p_m=\frac{m}{m-1}\varrho_m^{m-2}\varrho_m$, combining the uniform bound $p_m\leq p_H$ and the L^1-bound of ϱ_m we conclude. □

Lemma 2.4 (Propagation of BV Regularity) *There exists $C>0$ independent of m such that*

$$\|\partial_{x_i}\varrho_m\|_{L^\infty(0,T;L^1(\mathbb{R}^d))}+\|\partial_{x_i}p_m\|_{L^1(0,T;L^1(\mathbb{R}^d))}\leq C.$$

Proof Equation (2.11) propagates BV regularity along time. This can be seen by computing the time derivative of $|\partial_{x_i}\varrho|$ and applying Gronwall's lemma. Using Kato's inequality and the fact that $\mathrm{sign}(\partial_{x_i}\varrho_m)=\mathrm{sign}(\partial_{x_i}p_m)$, we have

$$\frac{\partial|\partial_{x_i}\varrho_m|}{\partial t}\leq\Delta(m\varrho_m^{m-1}|\partial_{x_i}\varrho_m|)+|\partial_{x_i}\varrho_m|G(p_m)+G'(p_m)|\partial_{x_i}p_m|\varrho_m$$
$$\leq\Delta(m\varrho_m^{m-1}|\partial_{x_i}\varrho_m|)+|\partial_{x_i}\varrho_m|G(0)-\alpha|\partial_{x_i}p_m|\varrho_m.$$

Thus, integrating in space and time we have

$$\int_{\mathbb{R}^d}|\partial_{x_i}\varrho_m|(t)\,\mathrm{d}x+\int_0^T\int_{\mathbb{R}^d}\alpha|\partial_{x_i}p_m|\varrho_m\,\mathrm{d}x\,\mathrm{d}t\leq G(0)\int_0^T\int_{\mathbb{R}^d}|\partial_{x_i}\varrho_m|\,\mathrm{d}x\,\mathrm{d}t$$
$$+\|\partial_{x_i}\varrho_m(0)\|_{L^1(\mathbb{R}^d)},$$

and Gronwall's lemma yields

$$\|\partial_{x_i}\varrho_m(t)\|_{L^1(\mathbb{R}^d)}+\int_0^T\int_{\mathbb{R}^d}\alpha|\partial_{x_i}p_m|\varrho_m\,\mathrm{d}x\,\mathrm{d}t\leq C.$$

Finally, this implies

$$\begin{aligned}\|\partial_{x_i} p_m\|_{L^1(0,T;L^1(\mathbb{R}^d))} &= \int_0^T \int_{\mathbb{R}^d} m\varrho_m^{m-2} \partial_{x_i}\varrho_m \,\mathrm{d}x\,\mathrm{d}t \\ &\le \iint_{\{\varrho_m \le 1/2\}} m\varrho_m^{m-2} \partial_{x_i}\varrho_m \,\mathrm{d}x\,\mathrm{d}t \\ &\quad + 2 \iint_{\{\varrho_m \ge 1/2\}} |\partial_{x_i} p_m| \varrho_m \,\mathrm{d}x\,\mathrm{d}t \\ &\le C,\end{aligned}$$

and the proof is complete. □

Lemma 2.5 (Pressure Gradient in L^2) *There exists $C > 0$ independent of m such that*

$$\|p_m\|_{L^2(0,T;H^1(\mathbb{R}^d))} \le C.$$

Proof This property simply follows by integrating the equation for p_m

$$\frac{\mathrm{d}}{\mathrm{d}t} \int_{\mathbb{R}^d} p_m \,\mathrm{d}x + (m-2) \int_{\mathbb{R}^d} |\nabla p_m|^2 \,\mathrm{d}x\,\mathrm{d}t \le G(0)(m-1) \int_{\mathbb{R}^d} p_m \,\mathrm{d}x.$$

We conclude by integrating in time and using the boundedness of p_m in L^1. □

2.2.2.2 Aronson-Bénilan Estimate in L^∞

We now state the result from [35] which gives a lower bound on the Laplacian of the pressure in the same fashion as (2.7). However, the estimate is revisited in order to take into account the reaction term.

Proposition 2.1 *Let ϱ_m, p_m be the weak solution to* (2.11)*, and let $r_G := \min_p G(p) - pG'(p) > 0$. The following inequality holds in the sense of distributions*

$$\Delta p_m + G(p_m) \ge -r_G \frac{e^{-(m-1)r_G t}}{1 - e^{-(m-1)r_G t}}. \tag{2.19}$$

Proof The proof follows a similar argument as the one used for (2.7). Let us denote $w := \Delta p + G(p)$. The pressure Eq. (2.13) reads

$$\frac{\partial p}{\partial t} = (m-1)pw + |\nabla p|^2. \tag{2.20}$$

Computing $\partial_t w$, we find

$$\frac{\partial w}{\partial t} = (m-1)\Delta(pw) + 2\nabla p \cdot \nabla \Delta p + 2\sum_{i,j}\left(\frac{\partial^2 p}{\partial x_i \partial x_j}\right)^2 + G'(p)\frac{\partial p}{\partial t}$$

$$\geq (m-1)p\Delta w + (m-1)w\Delta p + 2(m-1)\nabla p \cdot \nabla w + 2\nabla p \cdot \nabla(w - G(p))$$

$$+ G'(p)(m-1)pw + G'(p)|\nabla p|^2$$

$$= (m-1)p\Delta w + (m-1)w^2 - (m-1)wG(p) + 2m\nabla p \cdot \nabla w$$

$$- G'(p)|\nabla p|^2 + G'(p)(m-1)pw$$

$$\geq (m-1)p\Delta w + (m-1)w^2 + 2m\nabla p \cdot \nabla w - (m-1)(G(p) - pG'(p))w,$$

where in the last inequality we used the fact that $-G'(p)|\nabla p|^2 \geq 0$. Since $m > 1$ and $-(m-1)(G(p) - pG'(p)) \leq 0$ we can again apply the comparison principle. Treating p as a known function with enough regularity, w is a supersolution of a quasi-linear parabolic operator. Let us define

$$W_G(t) := -r_G \frac{e^{-(m-1)r_G t}}{1 - e^{-(m-1)r_G t}}.$$

Then, we have $W_G' = (m-1)W_G^2 - (m-1)r_G W_G$. In particular, W_G is a subsolution of the same parabolic operator. Therefore, the comparison principle yields (2.19). As in the proof for $G = 0$, the above estimate holds independently of any regularizing argument. Consequently, it is possible to prove that this lower bound holds in the sense of distributions for a larger class of functions obtained as the limit of solutions with enough regularity. Let us notice that for $t \sim 0$, the estimate turns out to be $\Delta p + G(p) \gtrsim -\frac{C}{(m-1)t}$. □

Let us remark that the lower bound (2.19) implies

$$\frac{\partial p_m}{\partial t} \geq -(m-1)p_m r_G \frac{e^{-(m-1)r_G t}}{1 - e^{-(m-1)r_G t}}, \tag{2.21}$$

and since $m\varrho_m^{m-2}\partial_t \varrho_m = \partial_t p_m$, we get

$$\frac{\partial \varrho_m}{\partial t} \geq -\varrho_m r_G \frac{e^{-(m-1)r_G t}}{1 - e^{-(m-1)r_G t}}. \tag{2.22}$$

Letting $m \to \infty$ we obtain the following monotonicity properties at the limit

$$\frac{\partial \varrho_\infty}{\partial t} \geq 0, \qquad \frac{\partial p_\infty}{\partial t} \geq 0. \tag{2.23}$$

Let us also remark that (2.21) and (2.22) imply the boundedness of the L^1-norms of both $\partial_t \varrho_m$ and $\partial_t p_m$—it is sufficient to split the L^1-norm using $|\partial_t \varrho_m| = \partial_t \varrho_m + 2(\partial_t \varrho_m)_-$ and bound each term. Therefore, space-time BV-regularity yields strong compactness of both pressure and density, while the boundedness of ∇p_m implies its weak compactness in L^2. This is enough to pass to the limit in (2.11).

From (2.19) and the positivity of the pressure, it is immediate to see

$$p_\infty(\Delta p_\infty + G(p_\infty)) \geq 0. \tag{2.24}$$

The converse inequality follows from a regularizing argument *à la Steklov*. Although we shall not discuss all the details here—we refer the reader to [35, Proof of Theorem 2.3.]—let us give an idea of the proof. Since from the bound on the Laplacian (2.19) follows compactness in space for the gradient of the pressure—as will be also discussed in Sect. 2.2.3—the only issue to conclude that the gradient converges strongly regards compactness in time. To this end, the authors take a time-dependent regularizing kernel $\omega_\varepsilon(t)$ with support in $(-\varepsilon, 0)$, take the convolution of Eq. (2.11) with ω_ε, and test it against p_m. Having now additional regularity in time provided by the convolution with the kernel, they pass to the limit $m \to \infty$ to obtain

$$p_\infty(\Delta(p_\infty \star \omega_\varepsilon) + \varrho_\infty G(p_\infty) \star \omega_\varepsilon) = \lim_{m\to\infty} p_m \partial_t \varrho_m \star \omega_\varepsilon.$$

Then, the rest of the proof consists of showing that the right-hand side is bounded from above by $o(1)$ as $\varepsilon \to 0$, which can be achieved by using the lower bounds on the time derivative provided by the Aronson-Bénilan estimate, see (2.21) and (2.22). Therefore, one has

$$p_\infty(\Delta p_\infty + G(p_\infty)) \leq 0,$$

which together with (2.24) yields the result and concludes the proof of Theorem 2.1.

We will now discuss why the lower bound (2.19) is not always available for models that include additional effects other than a pressure-dependent growth rate. In particular, one cannot expect the lower bound to converge to zero as $m \to \infty$ when monotonicity does not hold, which is the case for equations including drifts or coupled systems for which G is allowed to take negative values—which was not the case so far since $0 \leq p_m \leq p_H$ almost everywhere. For this reason, new versions of the Aronson-Bénilan estimate are needed.

2.2.2.3 Aronson-Bénilan Estimate in L^3

Let us consider the second tumor growth model studied in [35]. The cell population density, ϱ_m, satisfies a reaction-porous medium equation, while the concentration of nutrients (for instance, oxygen or glucose), c_m, diffuses and is consumed at rate $H(\varrho_m)$

$$\begin{cases} \dfrac{\partial \varrho_m}{\partial t} - \Delta \varrho_m^m = \varrho_m G(p_m, c_m), \\ \dfrac{\partial c_m}{\partial t} - \Delta c_m = -H(\varrho_m) c_m, \\ \qquad c_m(x,t) \to c_B, \quad \text{as } |x| \to \infty, \end{cases} \tag{2.25}$$

where $c_B > 0$ denotes the level of nutrients in the blood vessels. The equation on ϱ_m is the same as (2.11), where the pressure is given by (2.13), but the reaction term depends on the nutrient concentration as well. The assumptions on G are now the following

$$\partial_c G > 0, \ \partial_p G < 0, \ G(p,c) \le 0 \text{ for all } p \ge p_H, \ G(0,c) \le 0 \text{ for all } c \le \bar{c},$$

where $\bar{c} > 0$ represents the survival threshold below which the level of nutrients is not sufficient and cells die by necrosis. This happens at the centre of the tumor since as time passes the tumor bulk expands and it becomes more difficult for nutrients to reach the core. This immediately implies that the monotonicity property (2.23) cannot hold. Indeed, the Aronson-Bénilan estimate proved in Proposition 2.1 cannot hold either. There will be a region in the centre of the bulk in which the pressure will become zero (the necrotic core) and where $G(p_m, c_m) < 0$. This implies that it is not possible to find a lower bound on $\Delta p_m + G(p_m, c_m)$ that holds everywhere and vanishes as $m \to \infty$, such as (2.19).

The same issue concerns the following equation

$$\frac{\partial \varrho_m}{\partial t} - \Delta \varrho_m^m + \nabla \cdot (\varrho_m \nabla V) = \varrho_m G(p_m), \tag{2.26}$$

where $V : \mathbb{R}^d \times (0,\infty) \to \mathbb{R}$ is a potential that could represent, for instance, the concentration of a nutrient or a chemical towards which the tumor cells are attracted. In [29] the authors show that the following lower bound holds

$$\partial p_m \ge \frac{C_1}{mt} - C_2,$$

for some positive constants C_1, C_2. This version of the Aronson-Bénilan for Eq. (2.26) is again obtained by working with a subsolution. However, the lower bound does not converge to zero as $m \to \infty$. Here C_2 depends on the norm of the derivatives of the potential V.

For this reason, a different argument is necessary to establish the complementarity relation for (2.25) and (2.26). The trick is to consider AB estimates in “weaker" spaces, which means working in L^2 rather than with a subsolution. This argument was initiated in [25] for two-species systems, and further developed in [7, 13, 14] for the models with nutrients and drifts.

For the ease of exposition, we only prove the result for a simpler case, Eq. (2.11).

Proposition 2.2 *Let* ϱ_m *be the solution to Eq.* (2.11) *and* $m > 3-4/d$. *Assume that* $(\Delta p_{m,0} + G(p_{m,0}))_- \in L^2(\mathbb{R}^d)$ *uniformly. Then, the pressure* p_m *in* (2.12) *satisfies*

$$(\Delta p_m + G(p_m))_- \in L^3(0,T;L^3(\mathbb{R}^d)),$$

uniformly in m.

Proof For the sake of readability, we will remove the subscript m. We consider again Eq. (2.20) and multiply by $-(w)_-$, where we denote again $w := \Delta p + G$. Hence, from

$$\begin{aligned}\frac{\partial w}{\partial t} &= (m-1)\Delta(pw) + \Delta\left(|\nabla p|^2\right) + \frac{\partial G}{\partial t}\\ &= (m-1)\Delta(pw) + 2\nabla p\cdot\nabla(w-G) + 2\sum_{i,j}\left|\frac{\partial^2 p}{\partial x_i\partial x_j}\right|^2 + \frac{\partial G}{\partial t}\\ &\geq (m-1)\Delta(pw) + 2\nabla p\cdot\nabla(w-G) + \frac{2}{d}|\Delta p|^2 + \frac{\partial G}{\partial t},\end{aligned}$$

we obtain

$$\begin{aligned}\frac{1}{2}\frac{\partial (w)_-^2}{\partial t} \leq &(m-1)\Delta(p(w)_-)(w)_- + 2(w)_-\nabla p\cdot\nabla(w)_- + 2(w)_-\nabla p\cdot\nabla G\\ &-\frac{2}{d}(w)_-|w-G|^2 - \frac{\partial G}{\partial t}(w)_-\\ =&(m-1)\Delta(p(w)_-)(w)_- + \nabla p\cdot\nabla(w)_-^2 + 2(w)_-G'(p)|\nabla p|^2\\ &-\frac{2}{d}(w)_-|w-G|^2 - \frac{\partial G}{\partial t}(w)_-.\end{aligned}$$

The last term gives

$$\begin{aligned}-\frac{\partial G}{\partial t}(w)_- &= -\frac{\partial p}{\partial t}G'(p)(w)_-\\ &= (m-1)p(w)_-^2G'(p) - |\nabla p|^2G'(p)(w)_-\\ &\leq -|\nabla p|^2G'(p)(w)_-,\end{aligned}$$

and we finally have

$$\frac{1}{2}\frac{\partial (w)_-^2}{\partial t} \leq \underbrace{(m-1)\Delta(p(w)_-)(w)_- + \nabla p \cdot \nabla (w)_-^2}_{\mathcal{D}} + \underbrace{(w)_-|\nabla p|^2 G'(p)}_{\leq 0} - \frac{2}{d}(w)_-^3 + C(w)_-^2 + C(w)_-, \tag{2.27}$$

where C depends on the L^∞ norm of $G(p)$. We recall that the compact support property still holds, hence $\text{supp}(\varrho_m) = \text{supp}(p_m) \subset \Omega$. This is true also for (2.25) and (2.26) under the assumption that $\Delta V \in L^\infty$. Therefore we can integrate in space and use integration by parts. The same argument works for non-compactly supported solutions imposing a bound on the second moment and testing against a smooth test function with compact support, see [25] for details. After integration by parts, the term $\mathcal{D}$ gives

$$\begin{aligned}
\int_{\mathbb{R}^d} \mathcal{D}\,\mathrm{d}x &= \int_\Omega (1-m)\nabla(p(w)_-)\cdot\nabla(w)_- + \nabla p\cdot\nabla(w)_-^2\,\mathrm{d}x \\
&= \int_{\mathbb{R}^d} \frac{1-m}{2}\nabla p\cdot\nabla(w)_-^2 + (1-m)p|\nabla(w)_-|^2 + \nabla p\cdot\nabla(w)_-^2\,\mathrm{d}x \\
&= \int_{\mathbb{R}^d} \frac{3-m}{2}\nabla p\cdot\nabla(w)_-^2 + (1-m)p|\nabla(w)_-|^2\,\mathrm{d}x \\
&\leq \frac{m-3}{2}\int_{\mathbb{R}^d} \Delta p(w)_-^2\,\mathrm{d}x \\
&= \frac{3-m}{2}\int_{\mathbb{R}^d} (w)_-^3 + G(w)_-^2\,\mathrm{d}x,
\end{aligned}$$

where we used again $\Delta p = w - G$. Finally, combining these estimates with (2.27), we find

$$\frac{\mathrm{d}}{\mathrm{d}t}\int_\Omega \frac{(w)_-^2}{2}\,\mathrm{d}x + \left(\frac{2}{d} - \frac{3-m}{2}\right)\int_\Omega (w)_-^3\,\mathrm{d}x \leq C\int_\Omega (w)_-^2 + C(w)_-\,\mathrm{d}x.$$

For $m > 3 - 4/d$ the coefficient in front of the higher-order term is positive. Thus, we can integrate in time to obtain subsequently

$$\int_\Omega \frac{(w(T))_-^2}{2}\,\mathrm{d}x + \left(\frac{2}{d} - \frac{3-m}{2}\right)\int_0^T\int_\Omega (w)_-^3\,\mathrm{d}x\,\mathrm{d}t \leq C\int_0^T\int_\Omega (w)_-^2 + (w)_-\,\mathrm{d}x\,\mathrm{d}t + \int_\Omega \frac{(w_0)_-^2}{2}\,\mathrm{d}x$$

and

$$\left(\frac{4+d(m-3)}{2d}\right)\int_0^T\int_\Omega (w)_-^3 \,\mathrm{d}x\,\mathrm{d}t \le C\int_0^T\int_\Omega (w)_-^2 + (w)_- \,\mathrm{d}x\,\mathrm{d}t + C,$$

hence, we conclude. Notice that even if C is proportional to m, this still yields the uniform bound of $(w)_-$ in L^3. This is indeed the case for (2.25) and (2.26). □

Remark 2.1 Let us stress again that, as seen in the previous chapter, solutions to the porous medium equation which are compactly supported exhibit a corner at the free boundary. Hence the fact that $\Delta p_m(t) \in \mathcal{M}(\mathbb{R}^d)$ is sharp.

Proposition 2.3 (Uniform Bound on Δp_m) *There exists $C > 0$ independent of m such that*

$$\int_0^T\int_{\mathbb{R}^d} |\Delta p_m| \,\mathrm{d}x\,\mathrm{d}t \le C. \tag{2.28}$$

Proof Since the integral of $w = \Delta p_m + G$ is controlled, we can also control the positive part of w, using $(w)_+ = w + (w)_-$

$$\int_0^T\int_\Omega (w)_+ \,\mathrm{d}x\,\mathrm{d}t \le C + \int_0^T\int_\Omega (w)_- \,\mathrm{d}x\,\mathrm{d}t \le C + C\left(\int_0^T\int_\Omega (w)_-^3 \,\mathrm{d}x\,\mathrm{d}t\right)^{\frac{1}{3}}.$$

Thus it holds

$$\int_0^T\int_\Omega |\Delta p_m + G| \,\mathrm{d}x\,\mathrm{d}t \le C,$$

and the boundedness of G yields the result. □

2.2.3 *Strong Compactness of the Pressure Gradient*

Now we show how the a priori estimates proven so far yield the strong compactness of ∇p_m in $L^2(0, T; L^2(\mathbb{R}^d))$.

2.2.3.1 Strong Compactness of the Pressure Gradient

The bound (2.28) provides compactness in space of ∇p_m in any L^q with $1 \le q < d/(d-1)$, see [8]. For compactness in time, we notice that since $\partial_t p_m$ is uniformly bounded in L^1, the time derivative of ∇p_m, which is equal to $\nabla(\partial_t p_m)$, is a gradient of a L^1-bounded function. This is enough to prove strong compactness, by means of Aubin-Lions' lemma. We refer to [14] for the details and to [9] for the simple version of the lemma (which actually relies on the Fréchet-Kolmogorov Theorem) that yields compactness. Therefore, up to a subsequence, we have

$$\nabla p_m \to \nabla p_\infty, \text{ strongly in } L^1(0,T;L^q(\mathbb{R}^d)), \text{ for all } 1 \le q < d/(d-1).$$

The last step consists in uplifting the strong compactness up to (at least) L^2. To this end, we need a higher (uniform) order of integrability.

2.2.3.2 L^4-bound on the Pressure Gradient

We show that the gradient can be controlled uniformly in L^4 (thus the compactness proven above will be uplifted up to $L^q(0,T;L^q(\mathbb{R}^d))$ for any $1 \le q < 4$). This estimate follows from a weighted control on the Hessian, namely, we will prove that $\sqrt{p_m} D^2 p_m$ is uniformly controlled in L^2. Let us remark that it is not possible to establish an L^p-bound on $D^2 p_m$ since, as we have seen, the Laplacian of the pressure can be a singular measure on the free boundary of the positivity set. It is then reasonable to expect that weighting the Hessian with a power of the pressure (which is zero on the free boundary) would "kill" the singularity. Similar bounds have been established for multi-species counterparts of the same model, see [12, 26], as well as for fluid dynamics models such as the thin film equation, see [1].

To prove the result, the following lemma is needed.

Lemma 2.6 *Let $0 \le p \in L^\infty(\mathbb{R}^d)$ with $\sqrt{p} D^2 p \in L^2(\mathbb{R}^d)$ be a compactly supported function. Then, the following holds*

$$\int_{\mathbb{R}^d} |\nabla p|^4 \, dx \le 8 \int_{\mathbb{R}^d} p^2 |\Delta p|^2 \, dx + 4 \int_{\mathbb{R}^d} p^2 |D^2 p|^2 \, dx.$$

Proof We integrate by parts to obtain

$$\begin{aligned}
\int_{\mathbb{R}^d} |\nabla p|^4 \, dx &= -\int_{\mathbb{R}^d} p \nabla \cdot (\nabla p |\nabla p|^2) \, dx \\
&= -\int_{\mathbb{R}^d} p \Delta p |\nabla p|^2 \, dx - \frac{1}{2} \int_{\mathbb{R}^d} \nabla p^2 \cdot \nabla |\nabla p|^2 \, dx \\
&= -\int_{\mathbb{R}^d} p \Delta p |\nabla p|^2 \, dx + \frac{1}{2} \int_{\mathbb{R}^d} p^2 \Delta |\nabla p|^2 \, dx \\
&= -\int_{\mathbb{R}^d} p \Delta p |\nabla p|^2 \, dx + \int_{\mathbb{R}^d} p^2 \nabla p \cdot \nabla \Delta p \, dx + \int_{\mathbb{R}^d} p^2 |D^2 p|^2 \, dx \\
&= -3 \int_{\mathbb{R}^d} p \Delta p |\nabla p|^2 \, dx - \int_{\mathbb{R}^d} p^2 |\Delta p|^2 + \int_{\mathbb{R}^d} p^2 |D^2 p|^2 \, dx.
\end{aligned}$$

Upon using Young's inequality we obtain

$$\int_{\mathbb{R}^d} |\nabla p|^4 \,\mathrm{d}x \leq \left(\frac{3\alpha}{2} - 1\right) \int_{\mathbb{R}^d} p^2 |\Delta p|^2 \,\mathrm{d}x + \frac{3}{2\alpha} \int_{\mathbb{R}^d} |\nabla p|^4 \,\mathrm{d}x + \int_{\mathbb{R}^d} p^2 |D^2 p|^2 \,\mathrm{d}x.$$

Setting $\alpha = 2$ and rearranging, we find

$$\frac{1}{4} \int_{\mathbb{R}^d} |\nabla p|^4 \,\mathrm{d}x \leq 2 \int_{\mathbb{R}^d} p^2 |\Delta p|^2 \,\mathrm{d}x + \int_{\mathbb{R}^d} p^2 |D^2 p|^2 \,\mathrm{d}x,$$

and this concludes the proof of the lemma. □

Proposition 2.4 *Let p_m be the pressure related to the solution to* (2.11)*, with $m > 2$. There exists a positive constant $C > 0$ independent of m such that*

$$\int_0^T \int_{\mathbb{R}^d} (D^2 p_m)^2 p_m + |\nabla p_m|^4 \,\mathrm{d}x \,\mathrm{d}t \leq C. \tag{2.29}$$

As done for the Aronson-Bénilan estimate, also the proof of this bound will be shown in the simple setting, namely Eq. (2.11). For the proofs in the case with nutrients and drifts we refer the reader to [13, 14, 18].

Proof Let us test the pressure equation by $-(\Delta p + G(p))$ (for readability, let us only write the integration in space for now, and lose the subscript m). We obtain

$$-\int_{\mathbb{R}^d} \frac{\partial p}{\partial t}(\Delta p + G(p)) + (m-1)p|\Delta p + G(p)|^2 \,\mathrm{d}x = -\int_{\mathbb{R}^d} |\nabla p|^2 \Delta p \,\mathrm{d}x + \int_{\mathbb{R}^d} |\nabla p|^2 G(p) \,\mathrm{d}x. \tag{2.30}$$

By integration by parts, the first term of the right-hand side becomes

$$\begin{aligned} \int_{\mathbb{R}^d} |\nabla p|^2 \Delta p \,\mathrm{d}x &= \int_{\mathbb{R}^d} \Delta(|\nabla p|^2) p \,\mathrm{d}x \\ &= \int_{\mathbb{R}^d} 2p \nabla p \cdot \nabla \Delta p + 2(D^2 p)^2 p \,\mathrm{d}x \\ &= -\int_{\mathbb{R}^d} 2|\nabla p|^2 \Delta p - 2|\Delta p|^2 p + 2(D^2 p)^2 p \,\mathrm{d}x, \end{aligned}$$

where $(D^2 p)^2 = \sum_{i,j} (\partial^2_{i,j} p)^2$. Therefore, we have

$$\int_{\mathbb{R}^d} |\nabla p|^2 \Delta p \, \mathrm{d}x = -\frac{2}{3} \int_{\mathbb{R}^d} |\Delta p|^2 p \, \mathrm{d}x + \frac{2}{3} \int_{\mathbb{R}^d} (D^2 p)^2 p \, \mathrm{d}x.$$

From (2.30) we get

$$\begin{aligned} \frac{\mathrm{d}}{\mathrm{d}t} \int_{\mathbb{R}^d} \frac{|\nabla p|^2}{2} \, \mathrm{d}x &+ (m-1) \int_{\mathbb{R}^d} p |\Delta p + G(p)|^2 \, \mathrm{d}x \\ &+ \frac{2}{3} \int_{\mathbb{R}^d} (-|\Delta p|^2 p + (D^2 p)^2 p) \, \mathrm{d}x \\ &= \int_{\mathbb{R}^d} \frac{\partial p}{\partial t} G(p) \, \mathrm{d}x - \int_{\mathbb{R}^d} |\nabla p|^2 G(p) \, \mathrm{d}x. \end{aligned}$$

Let us call $R(p) = \int_0^p G(q) \, \mathrm{d}q$. Then $\partial_t R(p) = G(p) \partial_t p$ and integrating in time on $[0, T]$ we obtain

$$\begin{aligned} \frac{1}{2} \|\nabla p(T)\|_{L^2(\mathbb{R}^d)} &+ (m-1) \int_0^T \int_{\mathbb{R}^d} p(|\Delta p + G(p)|^2 - \frac{2}{3} |\Delta p|^2) \, \mathrm{d}x \, \mathrm{d}t \\ &+ \frac{2}{3} \int_0^T \int_{\mathbb{R}^d} (D^2 p)^2 p \, \mathrm{d}x \, \mathrm{d}t \\ \leq \int_{\mathbb{R}^d} & R(p(T)) - R(p_0) \, \mathrm{d}x + G(0) \|\nabla p\|^2_{L^2(0,T;L^2(\mathbb{R}^d))} + \frac{1}{2} \|\nabla p_0\|^2_{L^2(\mathbb{R}^d)} \\ \leq C&, \end{aligned}$$

where C is a positive constant independent of m. Now let us use the bound

$$|\Delta p|^2 \leq \frac{3}{2} |\Delta p + G(p)|^2 + 3 |G(p)|^2,$$

and the fact that $|G(p)| \leq G(0)$, to find

$$\frac{1}{2} \|\nabla p(T)\|_{L^2(\mathbb{R}^d)} + (m-2) \int_0^T \int_{\mathbb{R}^d} p |\Delta p + G(p)|^2 + \frac{2}{3} \int_0^T \int_{\mathbb{R}^d} (D^2 p)^2 p \leq C.$$

For $m > 2$, this estimate provides the following uniform bounds

$$(m-2) \int_0^T \int_{\mathbb{R}^d} p |\Delta p + G(p)|^2 \leq C, \qquad \int_0^T \int_{\mathbb{R}^d} (D^2 p)^2 p \leq C. \tag{2.31}$$

The uniform bound (2.29) follows from Lemma 2.6, (2.31) and $\|p_m\|_\infty \leq p_H$. □

2.2.4 Energy Methods

A different method that does not employ uniform estimates on the second derivatives of the pressure to establish the strong compactness of the pressure gradient has been recently developed in [30, 36] and subsequently extended in [12, 27]. This technique relies on energy equalities satisfied by p_m (and p_∞). Although this method can be employed to study the incompressible limit $m \to \infty$, it was first applied to prove the existence of solutions to two-species systems of cross-diffusion equations. In such systems, whose coupling is degenerate and nonlinear, it is essential to infer the strong compactness of the pressure gradient to show the existence of solutions, as will be further discussed in Sect. 2.4. The goal of this section is to review this method in the simple setting of one-species models. We will consider the equation with external drift given by (2.26).

The properties stated for the solution of Eq. (2.11) hold true also in the presence of a drift under suitable assumptions on the potential V—for instance $D^2 V \in L^\infty_{x,t}$, $\partial_t \nabla V \in L^1_t L^\infty_x$, $\partial_t \Delta V \in L^\infty_{x,t}$. Let us recall here those properties.

Proposition 2.5 *The solution to Eq.* (2.26) *satisfies the following a priori estimates uniformly in m*

(i) $\varrho_m, p_m \in L^\infty(0, T; L^\infty(\mathbb{R}^d) \cap L^1(\mathbb{R}^d))$,
(ii) $p_m \in L^2(0, T; H^1(\mathbb{R}^d))$,
(iii) $\varrho_m, p_m \in L^1(0, T; BV(\mathbb{R}^d))$, $\partial_t \varrho_m, \partial_t p_m \in L^1(0, T; L^1(\mathbb{R}^d))$.

The above bounds imply that

$$\begin{aligned} &\varrho_m \to \varrho_\infty, \quad \text{strongly in } L^q(0, T; L^q(\mathbb{R}^d)), \ \text{for } 1 \le q < \infty, \\ &p_m \to p_\infty, \quad \text{strongly in } L^q(0, T; L^q(\mathbb{R}^d)), \ \text{for } 1 \le q < \infty, \\ &\nabla p_m \rightharpoonup \nabla p_\infty, \quad \text{weakly in } L^2(0, T; L^2(\mathbb{R}^d)). \end{aligned}$$

As a consequence, we have $p_\infty(1 - \varrho_\infty) = 0$.

Remark 2.2 This also implies

$$\nabla \varrho_m^m \rightharpoonup \nabla p_\infty, \quad \text{weakly in } L^2(0, T; L^2(\mathbb{R}^d)),$$

which allows to pass to the limit in the porous medium part of the equation.

This compactness is sufficient to obtain

$$\frac{\partial \varrho_\infty}{\partial t} - \Delta p_\infty + \nabla \cdot (\varrho_\infty \nabla V) = \varrho_\infty G(p_\infty). \tag{2.32}$$

Exploiting the fact that the limit satisfies the above equation, it is possible to uplift the weak convergence of the pressure gradient into strong convergence. We already know by lower semicontinuity that

$$\|\nabla p_\infty\|_2^2 \leq \liminf_{m\to\infty} \int_0^T\int_{\mathbb{R}^d} |\nabla p_m|^2 \,\mathrm{d}x\,\mathrm{d}t.$$

The goal is then to show

$$\limsup_{m\to\infty} \int_0^T\int_{\mathbb{R}^d} |\nabla p_m|^2 \,\mathrm{d}x\,\mathrm{d}t \leq \|\nabla p_\infty\|_2^2.$$

Hence, convergence in norm and weak convergence imply strong convergence.

In order to do so, we will exploit the following identities (obtained by multiplying (2.26) and (2.32) by ϱ_m^m and p_∞, respectively)

$$\frac{\mathrm{d}}{\mathrm{d}t}\int_{\mathbb{R}^d} \frac{\varrho_m^{m+1}}{m+1}\,\mathrm{d}x + \int_{\mathbb{R}^d} |\nabla \varrho_m^m|^2\,\mathrm{d}x = \int_{\mathbb{R}^d} \nabla\varrho_m^m \cdot \nabla V \varrho_m \,\mathrm{d}x + \int_{\mathbb{R}^d} \varrho_m^{m+1} G(p_m)\,\mathrm{d}x,$$

and

$$\int_{\mathbb{R}^d} p_\infty \frac{\partial \varrho_\infty}{\partial t}\,\mathrm{d}x + \int_{\mathbb{R}^d} |\nabla p_\infty|^2\,\mathrm{d}x = \int_{\mathbb{R}^d} \nabla p_\infty \cdot \nabla V \varrho_\infty\,\mathrm{d}x + \int_{\mathbb{R}^d} p_\infty G(p_\infty)\,\mathrm{d}x,$$

where we used $p_\infty(1-\varrho_\infty) = 0$.

Proposition 2.6 (Strong Convergence of $\nabla\varrho_m^m$) *As $m \to \infty$, we have*

$$\nabla\varrho_m^m \to \nabla p_\infty, \text{ strongly in } L^2(0,T; L^2(\mathbb{R}^d)).$$

Proof From the two energy equalities, we have

$$\begin{aligned}
\int_0^T\int_{\mathbb{R}^d} |\nabla\varrho_m^m|^2\,\mathrm{d}x\,\mathrm{d}t &= \int_0^T\int_{\mathbb{R}^d} p_\infty \frac{\partial\varrho_\infty}{\partial t}\,\mathrm{d}x\,\mathrm{d}t - \frac{1}{m+1}\int_0^T\int_{\mathbb{R}^d} \frac{\partial \varrho_m^{m+1}}{\partial t}\,\mathrm{d}x\,\mathrm{d}t \\
&\quad - \int_0^T\int_{\mathbb{R}^d} \nabla p_\infty \cdot \nabla V\varrho_\infty\,\mathrm{d}x\,\mathrm{d}t + \int_0^T\int_{\mathbb{R}^d} \nabla\varrho_m^m \cdot \nabla V \varrho_m\,\mathrm{d}x\,\mathrm{d}t \\
&\quad - \int_0^T\int_{\mathbb{R}^d} p_\infty G(p_\infty)\,\mathrm{d}x\,\mathrm{d}t + \int_0^T\int_{\mathbb{R}^d} \varrho_m^{m+1} G(p_m)\,\mathrm{d}x\,\mathrm{d}t \\
&\quad + \int_0^T\int_{\mathbb{R}^d} |\nabla p_\infty|^2\,\mathrm{d}x\,\mathrm{d}t \\
&\to \int_0^T\int_{\mathbb{R}^d} p_\infty\frac{\partial\varrho_\infty}{\partial t}\,\mathrm{d}x\,\mathrm{d}t + \int_0^T\int_{\mathbb{R}^d} |\nabla p_\infty|^2\,\mathrm{d}x\,\mathrm{d}t,
\end{aligned}$$

thanks to the weak compactness of $\nabla \varrho_m^m$ and the strong compactness of ϱ_m and p_m. Since $p_\infty > 0$ implies $\varrho_\infty = 1$, while $\varrho_\infty \leq 1$ almost everywhere, we have

$$\int_0^T \int_{\mathbb{R}^d} p_\infty \frac{\partial \varrho_\infty}{\partial t} \,\mathrm{d}x\,\mathrm{d}t \leq 0,$$

and the proof is concluded. Let us remark that it can be proven that the last inequality is actually an equality. For the rigorous argument, we refer the reader to [12, 30]. □

2.2.4.1 Removing the Assumptions for Strong Compactness

Let us stress that actually strong compactness of neither ϱ_m or p_m is needed to pass to the limit in the density equation (except for the nonlinear reaction term $G(p)$). In fact, one would obtain the equation

$$\frac{\partial \varrho_\infty}{\partial t} - \Delta p_\infty + \nabla \cdot (\varrho_\infty \nabla V) = R_\infty,$$

where R_∞ is the weak limit of $\varrho_m G(p_m)$. The only things left to prove in order to apply the energy method are

(i) $p_\infty(1 - \varrho_\infty) = 0$,
(ii) $\limsup_{m\to\infty} \iint \varrho_m^{m+1} G(p_m) \leq \iint p_\infty R_\infty$.

It is interesting to notice that strong compactness is not needed to prove *(i)*. By the uniform bounds $p_m \in L^2(0, T; H^1(\mathbb{R}^d))$ and $\partial_t \varrho_m \in L^2(0, T; H^{-1}(\mathbb{R}^d))$, we know that, by compensated compactness, the product $\varrho_m p_m$ is converging to $\varrho_\infty p_\infty$ in the sense of distributions (and consequently in any L^q space, with $1 \leq q < \infty$ thanks to the uniform bounds). It remains to show that the product $\varrho_\infty p_\infty$ is indeed equal to p_∞. This holds by convexity arguments. Let $\psi_\delta = \psi_\delta(x)$ be a convex function such that $\psi_\delta(x) \to x$ as $\delta \to 0$. We denote $\Psi_m(x) = \frac{m}{m-1}\left(\frac{m-1}{m}x\right)^{m/(m-1)}$. Let us take $\delta > 0$ small enough such that $\psi_\delta(x) \leq \Psi_m(x)$. Therefore, we have

$$\psi_\delta(p_\infty) \leq \liminf_{m\to\infty} \psi_\delta(p_m) \leq \liminf_{m\to\infty} \Psi_m(p_m) = \liminf_{m\to\infty} \varrho_m p_m.$$

Since we chose $\delta > 0$ arbitrarily, we take $\delta \to 0$ to obtain $\varrho_\infty p_\infty \geq p_\infty$, which implies $\varrho_\infty p_\infty = p_\infty$ since $\varrho_\infty \leq 1$.

To prove *(ii)* we use the monotonicity of the reaction rate, $G' \leq 0$. We denote H a convex function such that $H' = -G$. Since R_∞ is the weak limit of $\varrho_m G(p_m)$, we need to show

$$\liminf_{m\to\infty} \iint -\varrho_m^{m+1} G(p_m) + p_\infty \varrho_m G(p_m) \,\mathrm{d}x\,\mathrm{d}t \geq 0.$$

The integral can be rewritten as

$$\begin{aligned}\iint \varrho_m H'(p_m)(\varrho_m^m - p_\infty)\,\mathrm{d}x\,\mathrm{d}t &= \iint \varrho_m H'(p_m)(p_m - p_\infty)\,\mathrm{d}x\,\mathrm{d}t \\ &\quad + \iint \varrho_m H'(p_m)(\varrho_m^m - p_m)\,\mathrm{d}x\,\mathrm{d}t \\ &\geq \iint \varrho_m (H(p_m) - H(p_\infty))\,\mathrm{d}x\,\mathrm{d}t \\ &\quad + \iint \varrho_m H'(p_m) p_m \left(\frac{m-1}{m}\varrho_m - 1\right)\mathrm{d}x\,\mathrm{d}t,\end{aligned}$$

where we used the convexity of H. We consider the first integral. Since the convex conjugate of H satisfies $H(p_m) \geq p_m \psi - H^*(\psi)$, we have

$$\begin{aligned}\iint \varrho_m (H(p_m) - H(p_\infty))\,\mathrm{d}x\,\mathrm{d}t &\geq \iint \varrho_m (p_m \psi - H^*(\psi) - H(p_\infty))\,\mathrm{d}x\,\mathrm{d}t \\ &\to \iint \varrho_\infty (p_\infty \psi - H^*(\psi) - H(p_\infty))\,\mathrm{d}x\,\mathrm{d}t,\end{aligned}$$

and taking the sup over ψ we infer nonnegativity of the limit. Finally, the limit of the second integral can be proven to be zero. We can split the set of integration between the region where $\varrho_m \leq 1 - \varepsilon$ and $\varrho_m > 1 - \varepsilon$. Hence, using the uniform bounds and a Taylor expansion of $\varrho_m \sim p_m^{m-1}$ at $m \to \infty$, we have

$$\begin{aligned}\iint \varrho_m H'(p_m) p_m \left(\frac{m-1}{m}\varrho_m - 1\right)\mathrm{d}x\,\mathrm{d}t &\leq \varrho_H \|G\|_\infty (1-\varepsilon)^{m-1} \\ &\quad + \varrho_H \|G\|_\infty p_H \max\left(\varepsilon, \frac{|\ln p_H| + o(1)}{m-1}\right).\end{aligned}$$

Choosing $\varepsilon = 1/\sqrt{m}$ and passing to the limit $m \to \infty$, we conclude the proof of *(ii)*.

Finally, the strong compactness of $\nabla \varrho_m^m$ implies the strong compactness of ϱ_m^m which, in turn, implies the strong compactness of the pressure itself and we can identify R_∞ as $\varrho_\infty G(p_\infty)$.

This is particularly relevant when the potential V does not satisfies the assumptions required to propagate BV-regularity of ϱ_m and p_m. This has implications, for instance, in mean field games for pedestrian motion—we refer the reader to [14, Theorem 1.1.] for the weaker assumptions needed on V.

2.2.5 Rate of Convergence

Although the incompressible limit for tumor growth and crowd motion models such as (2.26) has attracted a lot of attention from different communities, little was known about quantitative rates of convergence for this asymptotics. One result was known in the 2-Wasserstein distance and provided a polynomial rate of $1/24$, see [2]. This question was addressed in [15] in the $\dot{H}^{-1}$ norm.

Let us consider the following equation

$$\frac{\partial \varrho_m}{\partial t} = \nabla \cdot (\varrho_m(\nabla p_m + \nabla V)), \qquad t > 0,\ x \in \mathbb{R}^d, d \geq 3, \tag{2.33}$$

where

$$p_m = f'(\varrho_m).$$

In [15] reaction terms of the form $\varrho_m G(x,t)$ are also considered, but, for the sake of simplicity, in these notes, we will present the argument only for drift terms, $V = V(x)$. The same equation was also considered in [2] since the conservation of mass is essential to be able to compute the 2-Wasserstein distance. These kinds of equations, as well as their incompressible counterparts for $m = \infty$, are also frequently used in crowd motion modelling, see for instance [31, 32].

Remark 2.3 It is well known, see for instance [3, Theorem 11.2.8], that Eq. (2.33) is a gradient flow in the 2-Wasserstein distance, associated with the energy

$$\mathcal{E}(\varrho_m) = \int_{\mathbb{R}^d} f(\varrho_m)\,\mathrm{d}x + \int_{\mathbb{R}^d} \varrho_m V\,\mathrm{d}x.$$

Usually, in optimal transport problems, the potential is assumed to be bounded from below (or even convex), hence we can take $V \geq 0$ without loss of generality. Both internal and potential energy are then positive, and the dissipation is given by

$$\frac{\mathrm{d}}{\mathrm{d}t}\mathcal{E}(\varrho_m) = -\int_{\mathbb{R}^d} \varrho_m(t)|\nabla f'(\varrho_m(t)) + \nabla V|^2\,\mathrm{d}x, \tag{2.34}$$

and it holds

$$\int_0^t \int_{\mathbb{R}^d} \varrho_m(s)|\nabla f'(\varrho_m(s)) + \nabla V|^2\,\mathrm{d}x\,\mathrm{d}s \leq \mathcal{E}(\varrho_m(0)). \tag{2.35}$$

Since it is standard to assume that the initial energy is uniformly bounded, this gives a uniform L^1 bound on the dissipation.

Let us consider the following two choices for the internal energy

$$f_m(\varrho) = \frac{1}{m-1}\varrho^m, \quad \text{or} \quad f_m(\varrho) = -\frac{1}{m}(\varrho + \ln(1-\varrho)),$$

so that the pressure is given by either the barotropic or the singular law already mentioned in the introduction

$$p_m = \frac{m}{m-1}\varrho^{m-1}, \quad \text{or} \quad p_m = \frac{\varrho}{m(1-\varrho)}. \tag{2.36}$$

We first present the result achieved in the $\dot{H}^{-1}$ norm (which guarantees to take into account also reaction terms), and we will then comment on improvements recently obtained in the W_2 distance.

2.2.5.1 Results in $\dot{H}^{-1}$

To work in $\dot{H}^{-1}$ we do not need to impose any lower bound on the potential. On the other hand, the assumption on the second derivatives of V look less "natural" than the assumptions imposed for the result in the 2-Wasserstein distance. This is due to the fact that, unlike the porous medium part, the drift term in the equation is not a gradient flow in the H^{-1}-norm.

We consider a potential V satisfying the following assumption.

Assumption 2.2 Let $V : \mathbb{R}^d \to \mathbb{R}$ and $\alpha \in \mathbb{R}$ be such that

$$D^2V - \frac{\Delta V}{2} I_d \geq \alpha I_d. \tag{2.37}$$

Theorem 2.2 (Convergence Rate in $\dot{H}^{-1}$) *Let $\varrho_{m,0}$ be a sequence of non-negative initial data uniformly bounded in $L^1(\mathbb{R}^d)$ and $\varrho_{m,0} \to \varrho_{\infty,0} \leq 1$. Let the pressure be defined as in* (2.36) *with $p_{m,0} \in L^\infty(\mathbb{R}^d)$. In the power law case, we additionally assume that $\varrho_{m,0}$ is compactly supported.*

Let ϱ_m be the solution to (2.33) *with V satisfying* (2.37)*. Then, for all $T > 0$, there exists a unique $\varrho_\infty \in C([0,T]; H^{-1}(\mathbb{R}^d))$ such that the sequence ϱ_m converges to ϱ_∞, as $m \to \infty$, with the following rate*

$$\sup_{t\in[0,T]} \|\varrho_m(t) - \varrho_\infty(t)\|_{\dot{H}^{-1}(\mathbb{R}^d)} \leq \frac{C(T)}{m^{1/2}} + \|\varrho_{m,0} - \varrho_{\infty,0}\|_{\dot{H}^{-1}(\mathbb{R}^d)}.$$

Proof of Theorem 2.2 We provide different proofs for the singular law and the power law.

Singular Pressure

If the pressure is given by the singular law, the equation can be written as

$$\frac{\partial \varrho_m}{\partial t} = \Delta H_m(\varrho_m) + \nabla(\varrho_m \nabla V),$$

where

$$H_m(\varrho) := \frac{1}{m}\left(\frac{\varrho}{1-\varrho} + \ln(1-\varrho)\right).$$

Recall that we have the uniform bound $\varrho < 1$, so that the right-hand side above is well-defined with $\ln(1-\varrho) \leq 0$.

Let us take $m_2 > m_1 > 1$. Let ϱ_i be solutions to (2.33) with respect to m_1, m_2, and $\varphi_i = \mathcal{N} \star \varrho_i$ (with $\mathcal{N}$ the Newtonian potential) hence $-\Delta\varphi_i = \varrho_i$, for $i = 1, 2$. We call $\psi := \varphi_1 - \varphi_2$ and we compute

$$\begin{aligned}\frac{1}{2}\frac{\mathrm{d}}{\mathrm{d}t}\int_{\mathbb{R}^d} |\nabla\psi|^2 \,\mathrm{d}x &= \int_{\mathbb{R}^d} \psi\Delta(H_1(\varrho_1) - H_2(\varrho_2))\,\mathrm{d}x - \int_{\mathbb{R}^d}(\varrho_1 - \varrho_2)\nabla V \cdot \nabla\psi \,\mathrm{d}x\\ &= \int_{\mathbb{R}^d}(\varrho_1 - \varrho_2)(H_2(\varrho_2) - H_1(\varrho_1))\,\mathrm{d}x\\ &\quad - \int_{\mathbb{R}^d}(\varrho_1 - \varrho_2)\nabla V \cdot \nabla\psi \,\mathrm{d}x.\end{aligned}$$

Let us compute the two terms separately, starting from the one with the potential. By integration by parts, we obtain

$$\begin{aligned}-\int_{\mathbb{R}^d}(\varrho_1 - \varrho_2)\nabla V \cdot \nabla\psi \,\mathrm{d}x &= \int_{\mathbb{R}^d} \Delta\psi\nabla V \cdot \nabla\psi \,\mathrm{d}x\\ &= -\int_{\mathbb{R}^d} \nabla\psi \cdot D^2 V \nabla\psi \,\mathrm{d}x - \int_{\mathbb{R}^d} \nabla\psi \cdot D^2\psi\nabla V \,\mathrm{d}x\\ &= -\int_{\mathbb{R}^d} \nabla\psi \cdot D^2 V \nabla\psi \,\mathrm{d}x\\ &\quad - \int_{\mathbb{R}^d} \nabla\left(\frac{|\nabla\psi|^2}{2}\right) \cdot \nabla V \,\mathrm{d}x\\ &= -\int_{\mathbb{R}^d} \nabla\psi \cdot D^2 V \nabla\psi \,\mathrm{d}x + \int_{\mathbb{R}^d} \frac{|\nabla\psi|^2}{2}\Delta V \,\mathrm{d}x\\ &\leq -\alpha\int_{\mathbb{R}^d} |\nabla\psi|^2 \,\mathrm{d}x,\end{aligned}$$

where the last inequality follows from (2.37).

We now estimate the part coming from nonlinear diffusion. On the set $\{\varrho_1 > \varrho_2\}$ we make use of the non-negativity of $H_1(\varrho_1)$ and non-positivity of the logarithmic term in $H_2(\varrho_2)$ to write

$$\int_{\{\varrho_1 > \varrho_2\}} (\varrho_1 - \varrho_2)(H_2(\varrho_2) - H_1(\varrho_1))\,\mathrm{d}x \leq \frac{1}{m_2}\int_{\{\varrho_1 > \varrho_2\}} (\varrho_1 - \varrho_2)\frac{\varrho_2}{1-\varrho_2}\,\mathrm{d}x$$

$$\leq \frac{1}{m_2} \int_{\{\varrho_1 > \varrho_2\}} \varrho_2 \, \mathrm{d}x,$$

where we used $\varrho_1 - \varrho_2 \leq 1 - \varrho_2$. Similarly, on the complementary set $\{\varrho_1 \leq \varrho_2\}$ we have

$$\int_{\{\varrho_1 \leq \varrho_2\}} (\varrho_1 - \varrho_2)(H_2(\varrho_2) - H_1(\varrho_1)) \, \mathrm{d}x \leq \frac{1}{m_1} \int_{\{\varrho_1 \leq \varrho_2\}} \varrho_1 \, \mathrm{d}x.$$

Therefore we have

$$\begin{aligned}
\frac{1}{2} \frac{\mathrm{d}}{\mathrm{d}t} \int_{\mathbb{R}^d} |\nabla \psi|^2 \, \mathrm{d}x &\leq \frac{1}{m_1} \int_{\{\varrho_1 \leq \varrho_2\}} \varrho_1 \, \mathrm{d}x + \frac{1}{m_2} \int_{\{\varrho_1 \geq \varrho_2\}} \varrho_2 \, \mathrm{d}x - \alpha \int_{\mathbb{R}^d} |\nabla \psi|^2 \mathrm{d}x \\
&\leq \frac{1}{m_1} \|\varrho_1(t)\|_{L^1(\mathbb{R}^d)} + \frac{1}{m_2} \|\varrho_2(t)\|_{L^1(\mathbb{R}^d)} - \alpha \int_{\mathbb{R}^d} |\nabla \psi|^2 \mathrm{d}x,
\end{aligned}$$

and since ϱ_1 and ϱ_2 are uniformly bounded in $L^\infty(0, T; L^1(\mathbb{R}^d))$ (we can assume without loss of generality $\|\varrho_m(t)\|_1 = 1$), we find

$$\frac{\mathrm{d}}{\mathrm{d}t} \int_{\mathbb{R}^d} \frac{|\nabla \psi|^2}{2} \, \mathrm{d}x \leq \frac{1}{m_1} + \frac{1}{m_2} - \alpha \int_{\mathbb{R}^d} |\nabla \psi|^2 \, \mathrm{d}x.$$

Taking $m_2 \to \infty$ (and calling m_1 simply m) we obtain

$$\frac{1}{2} \frac{\mathrm{d}}{\mathrm{d}t} \|\varrho_m(t) - \varrho_\infty(t)\|^2_{\dot{H}^{-1}} \leq \frac{1}{m} - \alpha \|\varrho_m(t) - \varrho_\infty(t)\|^2_{\dot{H}^{-1}},$$

and by Gronwall's lemma we conclude.

Power Law: The Compact Support Assumption

In the power law case, we assume that the initial data $p_{m,0}$ is L^∞ and compactly supported uniformly in m. This implies that there exists $\bar{p} = \bar{p}(T) > 0$ such that $p_m \leq \bar{p}$ almost everywhere. This bound will be essential throughout the proof. Consequently, $\varrho_m \leq (\bar{p})^{1/(m-1)} \leq 1$ for m large enough. We can therefore assume without loss of generality that $\varrho_m \leq 1$ almost everywhere. If this does not hold, we can rescale the equation by a factor $(\bar{p})^{1/(m-1)}$ and proceed in the same way, see [15] for the details.

Let us remark that, in the singular law case, the compact support assumption is not needed. Indeed, the uniform bound $\varrho_m < 1$ follows directly from the singularity of the pressure law.

We argue as before and we compute

$$\frac{\mathrm{d}}{\mathrm{d}t}\int_{\mathbb{R}^d}\frac{|\nabla\psi|^2}{2}\,\mathrm{d}x=\int_{\mathbb{R}^d}\psi\Delta(\varrho_1^{m_1}-\varrho_2^{m_2})\,\mathrm{d}x-\int_{\mathbb{R}^d}(\varrho_1-\varrho_2)\nabla V\cdot\nabla\psi\,\mathrm{d}x.$$

By definition of ψ, we have

$$\begin{aligned}\int_{\mathbb{R}^d}\psi\Delta(\varrho_1^{m_1}-\varrho_2^{m_2})\,\mathrm{d}x&=\int_{\mathbb{R}^d}(\varrho_2-\varrho_1)(\varrho_1^{m_1}-\varrho_2^{m_2})\,\mathrm{d}x\\&=\int_{\mathbb{R}^d}(\varrho_2-1+1-\varrho_1)(\varrho_1^{m_1}-\varrho_2^{m_2})\,\mathrm{d}x\\&\le\int_{\mathbb{R}^d}(1-\varrho_1)\varrho_1^{m_1}\,\mathrm{d}x+\int_{\mathbb{R}^d}(1-\varrho_2)\varrho_2^{m_2}\,\mathrm{d}x\\&\le\frac{1}{m_1}+\frac{1}{m_2},\end{aligned}$$

where first we use that $\varrho_i\le 1$, and then the inequality

$$s^\gamma(1-s)\le\frac{s}{\gamma},\quad \text{for } 0\le s\le 1.$$

Let us recall that we can assume without loss of generality $\|\varrho_m\|_1=1$. Finally, we have

$$\frac{\mathrm{d}}{\mathrm{d}t}\int_{\mathbb{R}^d}\frac{|\nabla\psi|^2}{2}\,\mathrm{d}x\le\frac{1}{m_1}+\frac{1}{m_2}-\alpha\int_{\mathbb{R}^d}|\nabla\psi|^2\,\mathrm{d}x,$$

and we conclude as before. □

2.2.5.2 Global-in-time Result

The same quantitative rate can be also found in the 2-Wasserstein distance, and it can actually be shown to hold globally in time for strictly convex potentials [17]. Furthermore, the result holds without requiring the initial data to be compactly supported. For the sake of simplicity, we consider as initial data a probability measure with bounded second moment, $\varrho_0\le 1$, that is independent of m. This implies that, if $|\nabla V|^2\lesssim|x|^2$, the initial energy is uniformly bounded for m large enough

$$\mathcal{E}(\varrho_0)=\frac{1}{m-1}\int_{\mathbb{R}^d}\varrho_0^m+\int_{\mathbb{R}^d}\varrho_0 V\le C.$$

The result can be generalized for a sequence of initial data $\varrho_{0,m}$ that converges to ϱ_0 in W_2.

Proposition 2.7 *Let* $\varrho_0 \leq 1$*,* $\varrho_0 \in \mathcal{P}_2(\mathbb{R}^d)$ *be the initial data for Eq.* (2.33)*. If the potential satisfies*

$$\alpha I \leq D^2 V \leq \beta I,$$

with $\alpha, \beta > 0$*, then we have*

$$\sup_{t\in[0,\infty)} W_2(\varrho_m(t), \varrho_\infty(t)) \leq \frac{C}{m^{1/2}}.$$

Before proceeding with the proof, let us recall some basic definitions and tools from the theory of optimal transport, see, for instance, [39].

2.2.5.3 The 2-Wasserstein Distance

Given two probability measures $\mu, \nu \in \mathcal{P}_2(\mathbb{R}^d)$, the 2-Wasserstein distance is defined as

$$W_2(\mu, \nu) = \inf_{\gamma\in\Pi(\mu,\nu)} \left(\iint_{\mathbb{R}^d\times\mathbb{R}^d} |x-y|^2 \, \mathrm{d}\gamma(x,y) \right)^{\frac{1}{2}}, \tag{2.38}$$

where $\Pi(\mu, \nu) := \left\{\gamma \in \mathcal{P}_2(\mathbb{R}^d \times \mathbb{R}^d) : (\pi^x)_\# \gamma = \mu, (\pi^y)_\# \gamma = \nu\right\}$ and the maps $\pi^x, \pi^y : \mathbb{R}^d \times \mathbb{R}^d \to \mathbb{R}^d$ are the canonical projections. Moreover, under the assumption that $\mu \ll \mathcal{L}^d$ (where $\mathcal{L}^d$ denotes the Lebesgue measure supported on $\mathbb{R}^d$), the optimal transport plan $\bar{\gamma}$ is induced by a map $T : \mathbb{R}^d \to \mathbb{R}^d$ which is the gradient of a convex function, $T = \nabla u$. This means $\bar{\gamma} = (\mathrm{id}, T)_\# \mu$. The function $u : \mathbb{R}^d \to \mathbb{R}$ is given by $u(x) = |x|^2/2 - \varphi(x)$, where φ is the so-called Kantorovich potential for the transport between μ and ν. Let us recall also that the map T is in fact the solution to $\inf\left\{\int_{\mathbb{R}^d} |x - T(x)|^2 \, \mathrm{d}\mu : T_\#\mu = \nu\right\}^{\frac{1}{2}}$, and it can be written as $T = \mathrm{id} - \nabla\varphi$. In particular, we have

$$W_2(\mu, \nu) = \left(\int_{\mathbb{R}^d} |T(x) - x|^2 \, \mathrm{d}\mu(x) \right)^{\frac{1}{2}}.$$

Equivalently, there exists ψ such that $T^{-1} = \mathrm{id} - \nabla\psi$, and, we can summarize these properties as follows

$$(\mathrm{id} - \nabla\varphi)_\#\mu = \nu, \quad (\mathrm{id} - \nabla\psi)_\#\nu = \mu.$$

Proof of Proposition 2.7 The proof relies on computing the time derivative of the distance between two solutions of the equation for $1 < m_1 < m_2$, rather than working in a discrete setting like in [2]. A classic result in optimal transport is the following equality, see for instance [39, Theorem 5.24]

$$\frac{\mathrm{d}}{\mathrm{d}t}\left(\frac{1}{2}W_2^2(\varrho_1(t),\varrho_2(t))\right)=\int_{\mathbb{R}^d}\nabla\varphi^t\cdot v_1^t\,\mathrm{d}\varrho_1(t)+\int_{\mathbb{R}^d}\nabla\psi^t\cdot v_2^t\,\mathrm{d}\varrho_2(t),$$

where (φ^t,ψ^t) is the pair of Kantorovich potentials in the transport between $\varrho_1(t)$ and $\varrho_2(t)$ with quadratic cost and $v_i^t=-\nabla p_i(t)-\nabla V$ for $i=1,2$. Thus, we obtain

$$\begin{aligned}\frac{\mathrm{d}}{\mathrm{d}t}\left(\frac{1}{2}W_2^2(\varrho_1(t),\varrho_2(t))\right)&=\int_{\mathbb{R}^d}\left(\Delta\varphi^t\varrho_1^{m_1}(t)+\Delta\psi^t\varrho_2^{m_2}(t)\right)\mathrm{d}x\\&\quad-\int_{\mathbb{R}^d}(\nabla\varphi^t\cdot\nabla V\varrho_1(t)+\nabla\psi^t\cdot\nabla V\varrho_2(t))\,\mathrm{d}x.\end{aligned}\tag{2.39}$$

From the contractivity of the 2-Wasserstein distance for strictly convex potentials, it is easy to see

$$-\int_{\mathbb{R}^d}(\nabla\varphi^t\cdot\nabla V\varrho_1(t)+\nabla\psi^t\cdot\nabla V\varrho_2(t))\,\mathrm{d}x\le-\alpha W_2^2(\varrho_1(t),\varrho_2(t)).$$

Indeed, using the fact that $\nabla\varphi^t(x)=x-T(x)=-\nabla\psi^t(T(x))$, and $\varrho_2(t)=T_\#\varrho_1(t)$, we have

$$\begin{aligned}&-\int_{\mathbb{R}^d}(\nabla\varphi^t\cdot\nabla V\varrho_1(t)+\nabla\psi^t\cdot\nabla V\varrho_2(t))\,\mathrm{d}x\\&=\int_{\mathbb{R}^d}(T(x)-x)\cdot(\nabla V(x)-\nabla V(T(x))\varrho_1(t)\,\mathrm{d}x,\\&\le-\alpha\int_{\mathbb{R}^d}|T(x)-x|^2\,\mathrm{d}\varrho_1(t),\end{aligned}$$

where we used $D^2V\ge\alpha I$, see [39, Corollary 5.25].

It remains now to control the porous medium term, which will give the rate. Let us recall the Monge–Ampère equations satisfied by the Kantorovich potentials

$$\det(I-D^2\varphi^t(x))=\frac{\varrho_1(t,x)}{\varrho_2(t,T(x))},\ \varrho_1(t)-a.e.,$$
$$\det(I-D^2\psi^t(y))=\frac{\varrho_2(t,y)}{\varrho_1(t,T^{-1}(y))},\ \varrho_2(t)-a.e.,$$

where T is the optimal transport map in the transport between $\varrho_1(t)$ and $\varrho_2(t)$. By the inequality between the arithmetic and geometric means for symmetric positive semi-definite matrices, we have

$$\Delta\varphi^t(x) \le d\left(1 - \left(\frac{\varrho_1(t,x)}{\varrho_2(t,T(x))}\right)^{\frac{1}{d}}\right), \quad \Delta\psi^t(y) \le d\left(1 - \left(\frac{\varrho_2(t,y)}{\varrho_1(t,T^{-1}(y))}\right)^{\frac{1}{d}}\right),$$

pointwise a.e. on the supports of the corresponding measures. Using these bounds, one can show that there exists C only depending on the dimension d (see [17] for details) such that

$$\begin{aligned}
&\int_{\mathbb{R}^d} (\Delta\varphi^t \varrho_1^{m_1}(t) + \Delta\psi^t \varrho_2^{m_2}(t))\,\mathrm{d}x \\
&\quad \le \frac{C}{d\min\{m_1,m_2\}+1}\int_{\mathbb{R}^d} \varrho_1(t,x)\left(\varrho_1^{m_2-1}(t,x) + \varrho_2^{m_1-1}(t,T(x))\right)\mathrm{d}x \\
&\quad \le \frac{C}{dm_1+1}\int_{\mathbb{R}^d}\left(\varrho_1^{m_2}(t,x) + \varrho_1(t,x)\varrho_2^{m_1-1}(t,T(x))\right)\mathrm{d}x.
\end{aligned}$$

Therefore, the argument boils down to estimating the two integrals

$$\int_{\mathbb{R}^d} \varrho_1^{m_2}(t,x)\,\mathrm{d}x, \quad \text{and} \quad \int_{\mathbb{R}^d} \varrho_1(t,x)\varrho_2^{m_1-1}(t,T(x))\,\mathrm{d}x = \int_{\mathbb{R}^d} \varrho_2^{m_1}(t,x)\,\mathrm{d}x.$$

Assuming $m_2 = 2m_1 - 1$, the bound is given by [17, Proposition 3.5], which we report here for sake of completeness.

Proposition 2.8 *Let ϱ_1, ϱ_2 be solutions of Eq.* (2.33) *with power pressure law with exponents m_1, m_2, respectively. Let us take $m_2 = 2m_1 - 1$. There exists a uniform positive constant C such that*

$$\int_{\mathbb{R}^d} \varrho_1^{m_2}(t)\,\mathrm{d}x + \int_{\mathbb{R}^d} \varrho_2^{m_1}(t)\,\mathrm{d}x \le C + g_1(t) + g_2(t), \tag{2.40}$$

where $g_i := \frac{\mathrm{d}}{\mathrm{d}t}\mathcal{E}(\varrho_i)$ is the dissipation given by (2.34).

The idea of the proof of (2.40) is to use Sobolev's and Hölder's inequalities to control the integral of ϱ_m^{2m-1} by

$$\int_{\mathbb{R}^d} \left|\nabla\varrho_m^{m-1/2}(t)\right|^2 \mathrm{d}x = \int_{\mathbb{R}^d} \varrho_m(t)\,|\nabla p_m(t)|^2\,\mathrm{d}x,$$

that is part of the energy dissipation (2.34).

Let us stress that this argument allows to remove the compact support assumption needed for $\dot{H}^{-1}$, because the L^∞ uniform bound on the pressure is no longer necessary.

Finally, we get

$$\frac{\mathrm{d}}{\mathrm{d}t} W_2^2(\varrho_m(t), \varrho_{2m-1}(t)) \le \frac{C + g(t)}{m} - \alpha W_2^2(\varrho_m(t), \varrho_{2m-1}(t)),$$

where $g(t) := g_1(t)+g_2(t)$ is uniformly bounded in $L^1(0,\infty)$ thanks to (2.35). The argument is then completed by using Gronwall's lemma and writing the distance between ϱ_m and ϱ_∞ as the series

$$W_2(\varrho_m, \varrho_\infty) = \sum_{k=0}^{\infty} W_2(\varrho_{2^k m}, \varrho_{2^{k+1} m-1}).$$

The fact that the rate is global in time holds since the potential is strictly convex, namely $\alpha > 0$. For potentials whose second derivatives are simply uniformly bounded the result holds locally in time. □

2.3 The Inviscid Limit

Let us consider the following Brinkman model of tissue growth

$$\begin{aligned} &\frac{\partial \varrho_\nu}{\partial t} - \nabla \cdot (\varrho_\nu \nabla W_\nu) = \varrho_\nu G(p_\nu), \\ &-\nu \Delta W_\nu + W_\nu = p_\nu, \end{aligned} \tag{2.41}$$

where the pressure p_ν is given by the usual power law

$$p_\nu = \frac{m}{m-1} \varrho_\nu^{m-1}, \quad m > 1. \tag{2.42}$$

This model was first studied in [34], where the authors address the incompressible limit $m \to \infty$ and obtain a viscous free boundary problem, see also [19, 20]. Brinkman's equation allows to account not only for cell-to-environment friction but also for the friction between cells themselves through the viscosity coefficient $\nu > 0$.

Here we want to address the inviscid limit $\nu \to 0$, which relates the above nonlocal diffusion equation to the model that was the main topic of the previous chapters, namely the reaction-porous medium equation

$$\frac{\partial \varrho}{\partial t} - \Delta \varrho^m = \varrho G(p). \tag{2.43}$$

The potential W_ν in (2.41) has the form

$$W_\nu = \mathcal{K}_\nu \star p_\nu, \text{ with } -\nu \Delta \mathcal{K}_\nu + \mathcal{K}_\nu = \delta_0.$$

This means that (2.41) is a nonlocal diffusion equation, and by passing to $\nu \to 0$ we are considering a nonlocal-to-local approximation towards Darcy model (2.43), since the kernel (or interaction potential) satisfies $\mathcal{K}_\nu \to \delta_0$, as $\nu \to 0$, and thus $W_\nu \to p$. The assumptions on G are the same as the one stated in the previous section, see (2.14).

Let us introduce the definition of solutions we will be working with.

Definition 2.2 (Weak Solutions: Brinkman) A nonnegative function $\varrho_\nu \geq 0$ is a weak solution to (2.41) with nonnegative initial data $\varrho_0 \in L^1(\mathbb{R}^d) \cap L^\infty(\mathbb{R}^d)$, if $\varrho_\nu \in L^\infty(0,T; L^1(\mathbb{R}^d) \cap L^\infty(\mathbb{R}^d))$, and there holds

$$\int_0^T \int_{\mathbb{R}^d} \varrho_\nu \frac{\partial \varphi}{\partial t} \,\mathrm{d}x\,\mathrm{d}t - \int_0^T \int_{\mathbb{R}^d} \varrho_\nu \nabla W_\nu \cdot \nabla \varphi \,\mathrm{d}x\,\mathrm{d}t = -\int_0^T \int_{\mathbb{R}^d} \varphi \varrho_\nu G(p_\nu) \,\mathrm{d}x\,\mathrm{d}t - \int_{\mathbb{R}^d} \varphi(x,0)\varrho_0(x)\,\mathrm{d}x,$$

for any test function $\varphi \in C_c^\infty(\mathbb{R}^d \times [0,T))$, as well as

$$-\nu \Delta W_\nu + W_\nu = p_\nu,$$

almost everywhere in $\mathbb{R}^d \times (0,T)$, where p_ν is given by (2.42).

Definition 2.3 (Weak Solutions: Darcy) A nonnegative function ϱ is a weak solution to (2.43) with nonnegative initial data $\varrho_0 \in L^1(\mathbb{R}^d) \cap L^\infty(\mathbb{R}^d)$, if $\varrho \in L^\infty(0,T; L^1(\mathbb{R}^d) \cap L^\infty(\mathbb{R}^d))$, and $p \in L^2(0,T; H^1(\mathbb{R}^d))$ and there holds

$$\int_0^T \int_{\mathbb{R}^d} \varrho \frac{\partial \varphi}{\partial t} \,\mathrm{d}x\,\mathrm{d}t - \int_0^T \int_{\mathbb{R}^d} \varrho \nabla p \cdot \nabla \varphi \,\mathrm{d}x\,\mathrm{d}t = -\int_0^T \int_{\mathbb{R}^d} \varphi \varrho G(p) \,\mathrm{d}x\,\mathrm{d}t - \int_{\mathbb{R}^d} \varphi(x,0)\varrho_0(x)\,\mathrm{d}x,$$

for any test function $\varphi \in C_c^\infty(\mathbb{R}^d \times [0,T))$.

2.3.1 Brinkman Model with Linear Pressure

This section is focused on the inviscid limit $\nu \to 0$, for model (2.41) with linear pressure (for simplicity we rescaled the pressure law such that $p_\nu = \varrho_\nu$)

$$\begin{aligned} &\frac{\partial \varrho_\nu}{\partial t} - \nabla \cdot (\varrho_\nu \nabla W_\nu) = \varrho_\nu G(\varrho_\nu), \\ &-\nu \Delta W_\nu + W_\nu = \varrho_\nu, \end{aligned} \tag{2.44}$$

which leads to the following porous medium (with exponent $m = 2$)

$$\frac{\partial \varrho}{\partial t} - \frac{1}{2}\Delta \varrho^2 = \varrho G(\varrho),$$

or equivalently

$$\frac{\partial \varrho}{\partial t} - \nabla \cdot (\varrho \nabla \varrho) = \varrho G(\varrho). \tag{2.45}$$

In our analysis of the inviscid limit, we make use of the entropy structure of Eqs. (2.44) and (2.45). We define the usual entropy functional

$$\mathcal{H}[f](t) := \int_{\mathbb{R}^d} f(x,t)(\ln f(x,t) - 1)\,\mathrm{d}x.$$

Let us now state the main result.

Theorem 2.3 (Inviscid Limit) *Let $\varrho_0 \in L^1(\mathbb{R}^d) \cap L^\infty(\mathbb{R}^d)$ satisfy*

$$\mathcal{H}[\varrho_0] < \infty, \quad \int_{\mathbb{R}^d} |x|^2 \varrho_0 \,\mathrm{d}x < \infty, \quad \int_{\mathbb{R}^d} \varrho_0 \mathcal{K}_\nu \star \varrho_0 \,\mathrm{d}x < \infty.$$

Let $(\varrho_\nu)_{\nu>0}$ be a sequence of solutions to (2.44) *in the sense of Definition 2.2. There exists a function $\varrho \in L^\infty(0,T; L^1(\mathbb{R}^d) \cap L^\infty(\mathbb{R}^d))$, such that, up to a subsequence,*

$$\varrho_\nu \to \varrho, \text{ strongly in } L^2(0,T; L^2(\mathbb{R}^d)),$$

and

$$W_\nu \to \varrho, \text{ strongly in } L^2(0,T; H^1(\mathbb{R}^d)).$$

Furthermore, ϱ is a weak solution to (2.45) *in the sense of Definition 2.3, with initial data ϱ_0.*

Remark 2.4 The result remains true for "well-prepared" sequences of initial data $\varrho_{\nu,0}$ which are weakly compact in $L^1(\mathbb{R}^d)$ and satisfy

$$\sup_{\nu>0} \mathcal{H}[\varrho_{\nu,0}] < \infty, \quad \sup_{\nu>0} \int_{\mathbb{R}^d} |x|^2 \varrho_{\nu,0} \,\mathrm{d}x < \infty, \quad \sup_{\nu>0} \int_{\mathbb{R}^d} \varrho_{\nu,0} \mathcal{K}_\nu \star \varrho_{\nu,0} \,\mathrm{d}x < \infty,$$

as well as

$$\liminf_{\nu\to 0} \mathcal{H}[\varrho_{\nu,0}] \le \mathcal{H}[\varrho_0],$$

where ϱ_0 is the weak limit of $\varrho_{\nu,0}$.

Let us point out that the *strong* compactness of ∇W_ν is not needed to pass to the limit in (2.44)—weak convergence is sufficient. However, we shall prove how to upgrade weak to strong compactness for the gradient. This will come in handy to show the convergence of solutions to systems of coupled equations, which will be addressed in the next section.

Here we outline the proof of the main result

- we prove the *entropy inequality* for solutions to (2.44),
- we infer the strong compactness of W_ν by means of the Aubin-Lions lemma,
- we pass to $\nu \to 0$ in (2.44), hence showing that the limit of ϱ_ν satisfies (2.45),
- we use the entropy structure of (2.45)—in particular, the entropy *equality*—to uplift the weak compactness of ∇W_ν to strong compactness.

Let us summarise some fundamental regularity results for (2.44) that were obtained recently in a series of works on Brinkman models, see [19, 20, 34]. For any $\nu > 0$, there is a unique solution ϱ_ν to (2.44) satisfying (uniformly in $\nu > 0$)

(i) $\varrho_\nu \in L^\infty(0, T; L^1(\mathbb{R}^d))$,
(ii) $0 \le \varrho_\nu(x,t) \le \varrho_H$, almost everywhere,
(iii) $W_\nu \in L^\infty(0, T; L^1(\mathbb{R}^d) \cap L^\infty(\mathbb{R}^d))$.

From these a priori bounds we deduce that, up to a subsequence, there exists ϱ and W such that $\varrho_\nu \rightharpoonup^* \varrho$, $W_\nu \rightharpoonup^* \varrho$ in $L^\infty(0, T; L^\infty(\mathbb{R}^d))$. We now proceed to prove Theorem 2.3. We will not give here all the details of the proof—we refer the reader to [16].

2.3.1.1 Entropy Dissipation Inequality

The key result of this section is the following entropy dissipation inequality.

Proposition 2.9 (Entropy Inequality) *The solution ϱ_ν to* (2.44) *satisfies the entropy inequality*

$$\mathcal{H}[\varrho_\nu](T) - \mathcal{H}[\varrho_0] - \int_0^T\int_{\mathbb{R}^d} \varrho_\nu \Delta W_\nu \, \mathrm{d}x \, \mathrm{d}t \le \int_0^T\int_{\mathbb{R}^d} \varrho_\nu \ln \varrho_\nu G(\varrho_\nu) \, \mathrm{d}x \, \mathrm{d}t. \tag{2.46}$$

The idea of the proof consists of considering the entropy functional along the flow, namely, computing

$$\frac{\mathrm{d}}{\mathrm{d}t}\mathcal{H}[\varrho_\nu(t)].$$

However, this is not rigorously possible because the solution ϱ_ν does not have enough regularity to be able to test Eq. (2.44) by $\ln(\varrho_\nu)$. Therefore we need to introduce a regularized equation. Let $\epsilon > 0$, we consider

$$\frac{\partial \varrho_{\nu,\epsilon}}{\partial t} - \epsilon \Delta \varrho_{\nu,\epsilon} - \nabla \cdot (\varrho_{\nu,\epsilon} \nabla W_\nu) = \varrho_{\nu,\epsilon} G(\varrho_\nu), \tag{2.47}$$

where ϱ_ν, W_ν is the unique solution to (2.44). The above equation has a unique bounded solution $\varrho_{\nu,\epsilon} \in L^\infty(0,T;L^1(\mathbb{R}^d)) \cap L^2(0,T;H^1(\mathbb{R}^d)) \cap C(0,T;L^2(\mathbb{R}^d))$. This sequence of solutions is compact, and as $\epsilon \to 0$ convergences to the solution of the original Eq. (2.44)—for the full compactness argument we refer the reader to [16].

Now we show that $\varrho_{\nu,\epsilon}$ satisfies an entropy equality. This result relies on the boundedness of the second moment. In fact, assuming that the density second moment is initially bounded, this is preserved by Eq. (2.47) over time.

Lemma 2.7 *Let $\varrho_{\nu,\epsilon}$ be the solution to Eq.* (2.47) *with initial data ϱ_0. Then, there holds*

$$\sup_{t\in[0,T]} \int_{\mathbb{R}^d} \varrho_{\nu,\epsilon}(x,t)|x|^2 \,\mathrm{d}x \le C + C \int_0^T \int_{\mathbb{R}^d} \varrho_{\nu,\epsilon} |\nabla W_\nu|^2 \,\mathrm{d}x \,\mathrm{d}t =: C(\nu),$$

for some constant $C(\nu) > 0$, independent of ϵ.

Proof We compute

$$\begin{aligned}
\frac{1}{2}\frac{\mathrm{d}}{\mathrm{d}t} \int_{\mathbb{R}^d} \varrho_{\nu,\epsilon}|x|^2 \,\mathrm{d}x &= \epsilon d \int_{\mathbb{R}^d} \varrho_{\nu,\epsilon} \,\mathrm{d}x + \int_{\mathbb{R}^d} \varrho_{\nu,\epsilon} x \cdot \nabla W_\nu \,\mathrm{d}x \\
&\quad + \frac{1}{2} \int_{\mathbb{R}^d} |x|^2 \varrho_{\nu,\epsilon} G(\varrho_{\nu,\epsilon}) \,\mathrm{d}x \\
&\le \epsilon d \|\varrho_{\nu,\epsilon}\|_{L^\infty(0,T;L^1(\mathbb{R}^d))} + \left(\frac{1}{2} + \|G\|_{L^\infty}\right) \int_{\mathbb{R}^d} \varrho_{\nu,\epsilon}|x|^2 \,\mathrm{d}x \\
&\quad + \frac{1}{2} \int_{\mathbb{R}^d} \varrho_{\nu,\epsilon} |\nabla W_\nu|^2 \,\mathrm{d}x \\
&\le C_1 + \frac{1}{2} \int_{\mathbb{R}^d} \varrho_{\nu,\epsilon} |\nabla W_\nu|^2 \,\mathrm{d}x + \frac{C_2}{2} \int_{\mathbb{R}^d} \varrho_{\nu,\epsilon} |x|^2 \,\mathrm{d}x.
\end{aligned}$$

Then, by Gronwall's lemma, we get

$$\begin{aligned}
&\frac{1}{2} \int_{\mathbb{R}^d} |x|^2 \varrho_{\nu,\epsilon}(x,t) \,\mathrm{d}x \\
&\quad \le \left[\frac{1}{2} \int_{\mathbb{R}^d} |x|^2 \varrho_0(x) \,\mathrm{d}x + C_1 T + \frac{1}{2} \int_0^T \int_{\mathbb{R}^d} \varrho_{\nu,\epsilon} |\nabla W_\nu|^2 \,\mathrm{d}x \,\mathrm{d}t \right] e^{C_2 T},
\end{aligned}$$

and the proof is complete. □

A priori, the constant $C(\nu)$ depends on ν. We will later show that this is indeed not the case.

The control on the second moment also implies the control of $\varrho_{\nu,\epsilon} \ln \varrho_{\nu,\epsilon}$ in $L^\infty(0,T;L^1(\mathbb{R}^d))$.

Lemma 2.8 (Entropy Bounds) *Let $\varrho_{\nu,\epsilon}$ be the solution to Eq.* (2.47). *There exists C independent of ϵ such that*

$$\sup_{\epsilon>0}\sup_{t\in[0,T]}\int_{\mathbb{R}^d}\varrho_{\nu,\epsilon}(t)|\ln\varrho_{\nu,\epsilon}(t)|\,\mathrm{d}x \le C.$$

Proof We use the Legendre transform of $f(x) = x\ln x$, which is $f^*(y) = e^{y-1}$. We have

$$f(x) \ge xy - f^*(y),$$

hence, by the second moment bound

$$\int_{\mathbb{R}^d}\varrho_{\nu,\epsilon}\ln\varrho_{\nu,\epsilon}\,\mathrm{d}x \ge \int -\varrho_{\nu,\epsilon}|x|^2 - e^{-|x|^2-1}\,\mathrm{d}x \ge -C.$$

Together with

$$\int_{\mathbb{R}^d}\varrho_{\nu,\epsilon}\ln\varrho_{\nu,\epsilon}\,\mathrm{d}x \le \int_{\{\varrho_{\nu,\epsilon}>1\}}\varrho_{\nu,\epsilon}\ln\varrho_{\nu,\epsilon}\,\mathrm{d}x \le \|\varrho_{\nu,\epsilon}\|_1\|\varrho_{\nu,\epsilon}\|_\infty,$$

we conclude. □

Now we show that the second moment and entropy bounds actually hold *uniformly* in ν for the limit ϱ_ν.

Lemma 2.9 *The limit ϱ_ν of the sequence $(\varrho_{\nu,\epsilon})_{\epsilon>0}$ satisfies*

$$\sup_{\nu>0}\sup_{t\in[0,T]}\int_{\mathbb{R}^d}\varrho_\nu|x|^2\,\mathrm{d}x < \infty, \qquad \sup_{\nu>0}\sup_{t\in[0,T]}\int_{\mathbb{R}^d}\varrho_\nu|\ln\varrho_\nu|\,\mathrm{d}x < \infty.$$

Proof The main argument consists in multiplying (2.47) by $\mathcal{K}_\nu \star \varrho_{\nu,\epsilon}$ and use the strong compactness of $\varrho_{\nu,\epsilon}$ as $\epsilon \to 0$. Indeed, integrating in space, we obtain

$$\begin{aligned}&\frac{1}{2}\frac{\mathrm{d}}{\mathrm{d}t}\int_{\mathbb{R}^d}\varrho_{\nu,\epsilon}\mathcal{K}_\nu\star\varrho_{\nu,\epsilon}\,\mathrm{d}x + \epsilon\int_{\mathbb{R}^d}\nabla\mathcal{K}_\nu\star\varrho_{\nu,\epsilon}\cdot\nabla\varrho_{\nu,\epsilon}\,\mathrm{d}x\\ &\quad= -\int_{\mathbb{R}^d}\nabla\mathcal{K}_\nu\star\varrho_{\nu,\epsilon}\cdot\varrho_{\nu,\epsilon}\nabla W_\nu\,\mathrm{d}x + \int_{\mathbb{R}^d}\mathcal{K}_\nu\star\varrho_{\nu,\epsilon}\varrho_{\nu,\epsilon}G(\varrho_{\nu,\epsilon})\,\mathrm{d}x.\end{aligned} \tag{2.48}$$

The bound follows after integrating in time and passing to the limit $\epsilon \to 0$ using the compactness and the uniform bounds satisfied by $\varrho_{\nu,\epsilon}$. One can indeed show that

$$\int_0^T \int_{\mathbb{R}^d} \nabla \mathcal{K}_\nu \star \varrho_{\nu,\epsilon} \cdot \varrho_{\nu,\epsilon} \nabla W_\nu \, \mathrm{d}x \, \mathrm{d}t \to \int_0^T \int_{\mathbb{R}^d} \varrho_\nu |\nabla W_\nu|^2 \, \mathrm{d}x \, \mathrm{d}t,$$

and the remaining terms converge to uniformly bounded quantities. Hence, we get

$$\int_0^T \int_{\mathbb{R}^d} \varrho_\nu |\nabla W_\nu|^2 \, \mathrm{d}x \, \mathrm{d}t \le C + \frac{1}{2} \int_{\mathbb{R}^d} \varrho_0 \mathcal{K}_\nu \star \varrho_0 \, \mathrm{d}x.$$

Using Lemma 2.7 and taking $\epsilon \to 0$, we conclude. □

Proof of Proposition 2.9 We are now ready to prove the entropy dissipation inequality. Let $\delta > 0$ and $\phi \in C_c^\infty(\mathbb{R}^d)$, $\phi \ge 0$. We consider the following regularised form of the entropy functional

$$\mathcal{H}_\delta^\phi[\varrho_{\epsilon,\nu}](t) := \int_{\mathbb{R}^d} \left(\varrho_{\epsilon,\nu}(x,t) + \delta\right) (\ln(\varrho_{\epsilon,\nu}(x,t) + \delta) - 1)\phi(x) \, \mathrm{d}x,$$

and by $\mathcal{H}^\phi[f]$ we denote the above functional with $\delta = 0$. Given the regularity of $\varrho_{\nu,\epsilon}$, the weak form of Eq. (2.47) can be written as

$$\int_0^T \int_{\mathbb{R}^d} \frac{\partial \varrho_{\nu,\epsilon}}{\partial t} \varphi(x,t) + \varrho_{\nu,\epsilon} \nabla \varphi(x,t) \cdot \nabla W_\nu \, \mathrm{d}x \, \mathrm{d}t + \epsilon \int_0^T \int_{\mathbb{R}^d} \nabla \varrho_{\nu,\epsilon} \cdot \nabla \varphi(x,t) \, \mathrm{d}x \, \mathrm{d}t$$
$$= \int_0^T \int_{\mathbb{R}^d} \varrho_{\nu,\epsilon} G(\varrho_\nu) \varphi(x,t) \, \mathrm{d}x \, \mathrm{d}t,$$

for any $\varphi \in L^2(0,T; H^1(\mathbb{R}^d))$. We choose

$$\varphi(x,t) := \ln(\varrho_{\nu,\epsilon} + \delta)\phi(x),$$

and we obtain

$$\int_0^T \frac{\mathrm{d}}{\mathrm{d}t} \mathcal{H}_\delta^\phi[\varrho_{\nu,\epsilon}](t) \, \mathrm{d}t + I_1^{\epsilon,\delta} + I_2^{\epsilon,\delta} + I_3^{\epsilon,\delta} + I_4^{\epsilon,\delta} = I_5^{\epsilon,\delta},$$

where

$$
\begin{aligned}
I_1^{\epsilon,\delta} &= \int_0^T\int_{\mathbb{R}^d} \frac{\varrho_{\nu,\epsilon}}{\varrho_{\nu,\epsilon}+\delta}\nabla\varrho_{\nu,\epsilon}\cdot\nabla W_\nu\phi\,\mathrm{d}x\,\mathrm{d}t,\\
I_2^{\epsilon,\delta} &= \int_0^T\int_{\mathbb{R}^d} \varrho_{\nu,\epsilon}\ln(\varrho_{\nu,\epsilon}+\delta)\nabla\phi\cdot\nabla W_\nu\,\mathrm{d}x\,\mathrm{d}t,\\
I_3^{\epsilon,\delta} &= \epsilon\int_0^T\int_{\mathbb{R}^d} \frac{|\nabla\varrho_{\nu,\epsilon}|^2}{\varrho_{\nu,\epsilon}+\delta}\phi\,\mathrm{d}x\,\mathrm{d}t,\\
I_4^{\epsilon,\delta} &= \epsilon\int_0^T\int_{\mathbb{R}^d} \ln(\varrho_{\nu,\epsilon}+\delta)\nabla\phi\cdot\nabla\varrho_{\nu,\epsilon}\,\mathrm{d}x\,\mathrm{d}t,\\
I_5^{\epsilon,\delta} &= \int_0^T\int_{\mathbb{R}^d} \ln(\varrho_{\nu,\epsilon}+\delta)\left(\varrho_{\nu,\epsilon}G(\varrho_\nu)\right)\phi\,\mathrm{d}x\,\mathrm{d}t.
\end{aligned}
$$

Let us notice that, since $\varrho_{\nu,\epsilon}\in C([0,T];L^2(\mathbb{R}^d))$, the mapping $t\mapsto\mathcal{H}_\delta^\phi[\varrho_{\nu,\epsilon}(t)]$ is continuous in $[0,T]$, and therefore we obtain

$$
\mathcal{H}_\delta^\phi[\varrho_{\nu,\epsilon}](T)-\mathcal{H}_\delta^\phi[\varrho_{\nu,\epsilon}](0)+I_1^{\epsilon,\delta}+I_2^{\epsilon,\delta}+I_3^{\epsilon,\delta}+I_4^{\epsilon,\delta}=I_5^{\epsilon,\delta}. \tag{2.49}
$$

Now our goal is to pass to the limit first as $\epsilon\to 0$ and then as $\delta\to 0$. Let us notice that $I_3^{\epsilon,\delta}$ is nonnegative and can be dropped from the equality. This is the reason why for Eq. (2.44) we will only obtain an entropy dissipation *inequality* rather than an equality, but this is enough for our purposes. Moreover, $I_4^{\epsilon,\delta}\to 0$ as $\epsilon\to 0$. Using the fact that $\varrho_{\nu,\epsilon}\to\varrho_\nu$ strongly in $L^1(0,T;L^1(\mathbb{R}^d))$, the other terms in (2.49) converge to their counterparts as $\epsilon\to 0$, and $\delta\to 0$, so in the limit we obtain

$$
\begin{aligned}
\mathcal{H}^\phi[\varrho_\nu](T)-\mathcal{H}^\phi[\varrho_0]-\int_0^T\int_{\mathbb{R}^d}\varrho_\nu\Delta W_\nu\phi\,\mathrm{d}x\,\mathrm{d}t-\int_0^T\int_{\mathbb{R}^d}\varrho_\nu\nabla W_\nu\cdot\nabla\phi\,\mathrm{d}x\,\mathrm{d}t\\
+\int_0^T\int_{\mathbb{R}^d}\varrho_\nu\ln\varrho_\nu\nabla\phi\cdot\nabla W_\nu\,\mathrm{d}x\,\mathrm{d}t\le\int_0^T\int_{\mathbb{R}^d}\ln\varrho_\nu\varrho_\nu G(\varrho_\nu)\phi\,\mathrm{d}x\,\mathrm{d}t,
\end{aligned}
$$

for any $\phi\in C_c^\infty(\mathbb{R}^d)$, $\phi\ge 0$. Finally, we choose $\phi=\chi_R$ to be a sequence of smooth cut-off functions such that $|\nabla\chi_R|\lesssim R^{-1}$. Using the $L^\infty L^1$-control of $\varrho_\nu\ln\varrho_\nu$ from Lemma 2.9, we can pass to the limit $R\to\infty$ to obtain (2.46). □

2.3.1.2 Strong Compactness of W_ν and ϱ_ν

The entropy inequality just proven allows us to show strong convergence of both W_ν and ϱ_ν.

Proposition 2.10 (Regularity of W_ν**)** *Let* $(W_\nu)_{\nu>0}$ *be a sequence of potentials associated to solutions of* (2.44). *Then, there exists a constant* $C>0$ *independent* *of ν such that*

$$\|\nabla W_\nu\|_{L^2(0,T;L^2(\mathbb{R}^d))} \le C, \qquad \textit{and} \qquad \|\partial_t W_\nu\|_{L^2(0,T;H^{-1}(\mathbb{R}^d))} \le C.$$

Moreover,

$$W_\nu \to \varrho \ \textit{strongly in } L^2(0,T;L^2(\mathbb{R}^d)),$$

$$\varrho_\nu \to \varrho \ \textit{strongly in } L^2(0,T;L^2(\mathbb{R}^d)),$$

$$\nabla W_\nu \to \nabla\varrho \ \textit{weakly in } L^2(0,T;L^2(\mathbb{R}^d)).$$

Proof We begin by proving the bound on ∇W_ν. Integration by parts and Brinkman's law give

$$\begin{aligned}\int_0^T\int_{\mathbb{R}^d} |\nabla W_\nu|^2 \,\mathrm{d}x\,\mathrm{d}t &= -\int_0^T\int_{\mathbb{R}^d} W_\nu \Delta W_\nu \\ &= -\int_0^T\int_{\mathbb{R}^d} \varrho_\nu \Delta W_\nu \,\mathrm{d}x\,\mathrm{d}t - \nu\int_0^T\int_{\mathbb{R}^d} |\Delta W_\nu|^2 \,\mathrm{d}x\,\mathrm{d}t \\ &\le -\int_0^T\int_{\mathbb{R}^d} \varrho_\nu \Delta W_\nu \,\mathrm{d}x\,\mathrm{d}t.\end{aligned}$$

Rearranging the entropy dissipation inequality, (2.46), we find

$$-\int_0^T\int_{\mathbb{R}^d} \varrho_\nu \Delta W_\nu \,\mathrm{d}x\,\mathrm{d}t \le \int_0^T\int_{\mathbb{R}^d} \varrho_\nu \ln\varrho_\nu G(\varrho_\nu) \,\mathrm{d}x\,\mathrm{d}t + \mathcal{H}[\varrho_0] \le C.$$

Hence $\nabla W_\nu \in L^2(0,T;L^2(\mathbb{R}^d))$ uniformly in ν. Let us notice that as a byproduct, we also have

$$-\nu\int_0^T\int_{\mathbb{R}^d} |\Delta W_\nu|^2 \,\mathrm{d}x\,\mathrm{d}t \le C. \tag{2.50}$$

Let us recall that $W_\nu = \mathcal{K}_\nu \star \varrho_\nu$, and $\|\mathcal{K}_\nu\|_{L^1(\mathbb{R}^d)} = 1$. Then, for any $\varphi \in L^2(0,T;H^1(\mathbb{R}^d))$, we have

$$\begin{aligned}\int_0^T\int_{\mathbb{R}^d} \varphi\frac{\partial W_\nu}{\partial t} \,\mathrm{d}x\,\mathrm{d}t &= \int_0^T\int_{\mathbb{R}^d} \frac{\partial\varrho_\nu}{\partial t}\mathcal{K}_\nu \star \varphi \,\mathrm{d}x\,\mathrm{d}t \\ &= -\int_0^T\int_{\mathbb{R}^d} \varrho_\nu \nabla W_\nu \cdot \nabla\varphi \star \mathcal{K}_\nu \,\mathrm{d}x\,\mathrm{d}t + \int_0^T\int_{\mathbb{R}^d} \varrho_\nu G(\varrho_\nu)\varphi \star \mathcal{K}_\nu \,\mathrm{d}x\,\mathrm{d}t \\ &\le \|\varrho_\nu\|_{L^\infty}\|\nabla W_\nu\|_{L^2}\|\nabla\varphi\|_{L^2} + C\|\varphi\|_{L^2}.\end{aligned}$$

It follows that $\partial_t W_\nu \in L^2(0,T;H^{-1}(\mathbb{R}^d))$, uniformly in $\nu > 0$. We deduce the existence of a function $\chi \in L^2(0,T;H^1(\mathbb{R}^d))$ such that, up to a subsequence,

$$\begin{aligned} W_\nu &\to \chi, \quad \textit{strongly in } L^2(0,T;L^2(\mathbb{R}^d)),\\ \nabla W_\nu &\rightharpoonup \nabla\chi, \quad \textit{weakly in } L^2(0,T;L^2(\mathbb{R}^d)). \end{aligned}$$

It is easy to identify χ as ϱ, the weak limit of ϱ_ν. In fact, we have

$$\int_0^T\int_{\mathbb{R}^d} W_\nu\varphi \,\mathrm{d}x\,\mathrm{d}t = \int_0^T\int_{\mathbb{R}^d} \varrho_\nu \mathcal{K}_\nu \star \varphi \,\mathrm{d}x\,\mathrm{d}t \to \int_0^T\int_{\mathbb{R}^d} \varrho\varphi \,\mathrm{d}x\,\mathrm{d}t,$$

by weak-strong convergence. Finally, ϱ_ν converges strongly to ϱ since by Brinkman's law

$$\varrho_\nu = -\nu\Delta W_\nu + W_\nu,$$

and W_ν converges strongly to ϱ, while, from (2.50), we know that $-\nu\Delta W_\nu$ converges strongly to 0. □

2.3.1.3 Strong Compactness of ∇W_ν

Finally, we show how to uplift the weak compactness of ∇W_ν to strong compactness. From the previous estimates and Lemma 2.9, we know that the limit, ϱ, satisfies

$$\begin{aligned} &\varrho \in C([0,T];L^2(\mathbb{R}^d)) \cap L^2(0,T;H^1(\mathbb{R}^d)).\\ &\partial_t\varrho \in L^2(0,T;H^{-1}(\mathbb{R}^d)), \end{aligned}$$

as well as

$$\sup_{t\in[0,T]}\int_{\mathbb{R}^d} |x|^2\varrho \,\mathrm{d}x < \infty, \qquad \sup_{t\in[0,T]}\int_{\mathbb{R}^d} \varrho|\ln\varrho| \,\mathrm{d}x < \infty.$$

Passing to the limit in (2.44) we get

$$\frac{\partial\varrho}{\partial t} - \nabla\cdot(\varrho\nabla\varrho) = \varrho G(\varrho),$$

and the regularity of ϱ is enough to guarantee the entropy *equality*

$$\mathcal{H}[\varrho](T) - \mathcal{H}[\varrho_0] + \int_0^T\int_{\mathbb{R}^d} |\nabla\varrho|^2 \,\mathrm{d}x\,\mathrm{d}t = R(\varrho), \tag{2.51}$$

where $R(\varrho) := \iint \ln \varrho\, [\varrho G(\varrho)]$. Upon rearranging the entropy inequality (2.46) and (2.51), we get

$$-\int_0^T \int_{\mathbb{R}^d} \varrho_\nu \Delta W_\nu \,\mathrm{d}x\,\mathrm{d}t \leq \underbrace{\mathcal{H}[\varrho(T)] - \mathcal{H}[\varrho_\nu](T) + R(\varrho_\nu) - R(\varrho)}_{\mathcal{E}_\nu} + \int_0^T \int_{\mathbb{R}^d} |\nabla \varrho|^2 \,\mathrm{d}x\,\mathrm{d}t. \tag{2.52}$$

Recall that

$$\int_0^T \int_{\mathbb{R}^d} |\nabla W_\nu|^2 \,\mathrm{d}x\,\mathrm{d}t \leq -\int_0^T \int_{\mathbb{R}^d} \varrho_\nu \Delta W_\nu \,\mathrm{d}x\,\mathrm{d}t.$$

This, together with Eq. (2.52), gives

$$\int_0^T \int_{\mathbb{R}^d} |\nabla W_\nu|^2 \,\mathrm{d}x\,\mathrm{d}t \leq \int_0^T \int_{\mathbb{R}^d} |\nabla \varrho|^2 \,\mathrm{d}x\,\mathrm{d}t + \mathcal{E}_\nu.$$

To conclude it only remains to show that the term $\mathcal{E}_\nu$ vanishes in the limit $\nu \to 0$. Given the uniform $L^1 \cap L^\infty$-bound on $\varrho_\nu(T)$ and uniform control of the second moment, we infer that the sequence $\varrho_\nu(T)$ converges to $\varrho(T)$ weakly in $L^1(\mathbb{R}^d)$. Therefore, by the convexity of the entropy functional, we have

$$\mathcal{H}[\varrho](T) \leq \liminf_{\nu \to 0} \mathcal{H}[\varrho_\nu](T),$$

or equivalently

$$\limsup_{\nu \to 0} \left(\mathcal{H}[\varrho](T) - \mathcal{H}[\varrho_\nu](T)\right) \leq 0.$$

The reaction terms are more involved. We shall use the tightness of the sequence $(\varrho_\nu)_\nu$ to split the domain of integration between a subdomain on which the integrals involving ϱ_ν converge to their counterparts for ϱ and remainders which can be made arbitrarily small. The proof that $|R(\varrho_\nu) - R(\varrho)| \to 0$ can be found in Lemma 2.10 below.

Therefore, we deduce upper semicontinuity of the norm

$$\limsup_{\nu \to 0} \int_0^T \int_{\mathbb{R}^d} |\nabla W_\nu|^2 \,\mathrm{d}x\,\mathrm{d}t \leq \int_0^T \int_{\mathbb{R}^d} |\nabla \varrho|^2 \,\mathrm{d}x\,\mathrm{d}t.$$

On the other hand, weak convergence implies lower semicontinuity and thus

$$\lim_{\nu \to 0} \|\nabla W_\nu\|_2 = \|\nabla \varrho\|_2.$$

It follows that $\nabla W_\nu \to \nabla \varrho$ strongly in $L^2(0, T; L^2(\mathbb{R}^d))$.

Lemma 2.10 *Let* $R(\varrho) := \iint \ln \varrho \,[\varrho G(\varrho)]$ *, and* ϱ_ν *be the solution to* (2.44)*. Then, as* $\nu \to 0$*, we have*

$$|R(\varrho_\nu) - R(\varrho)| = \left| \int_0^T \int_{\mathbb{R}^d} \varrho_\nu \ln \varrho_\nu G(\varrho_\nu) - \varrho \ln \varrho G(\varrho) \,\mathrm{d}x \,\mathrm{d}t \right| \to 0.$$

Proof Let us remark the strong convergence of ϱ_ν is not sufficient since the logarithmic function is not Lipschitz near $\varrho = 0$.

By the bound on the second moment, Lemma 2.9, for any $\varepsilon > 0$, there exists a compact set $K \subset \mathbb{R}^d$ such that

$$\sup_{\nu \geq 0} \int_0^T \int_{K^c} \varrho_\nu |\ln \varrho_\nu| \,\mathrm{d}x \,\mathrm{d}t < \varepsilon.$$

Therefore, we write

$$\begin{aligned} \int_0^T \int_{\mathbb{R}^d} \varrho_\nu \ln \varrho_\nu G(\varrho_\nu) \,\mathrm{d}x \,\mathrm{d}t &= \int_0^T \int_{K^c} \varrho_\nu \ln \varrho_\nu G(\varrho_\nu) \,\mathrm{d}x \,\mathrm{d}t \\ &\quad + \int_0^T \int_K \varrho_\nu \ln \varrho_\nu G(\varrho_\nu) \,\mathrm{d}x \,\mathrm{d}t, \\ &\leq C\varepsilon + \int_0^T \int_K \varrho_\nu \ln \varrho_\nu G(\varrho_\nu) \,\mathrm{d}x \,\mathrm{d}t, \end{aligned}$$

with $C > 0$ only depending on the L^∞ norm of G.

We use Egorov's Theorem to further split K into the set where the sequence converges uniformly and the remainder with arbitrary small measure. Specifically, let $A \subset K$ with $\mathrm{meas}(A) < \varepsilon$ such that

$$\varrho_\nu \rightrightarrows \varrho \text{ in } K \setminus A.$$

Let $0 < \beta < e^{-1}$, then, for ν sufficiently small, we have

$$\varrho \geq \frac{\beta}{2} \implies \varrho_\nu \geq \frac{\beta}{4}, \quad \text{and} \quad \varrho < \frac{\beta}{2} \implies \varrho_\nu < \beta.$$

We split K into three disjoint subdomains as follows

$$K = (K \setminus A \cap \{\varrho \geq \beta/2\}) \cup (K \setminus A \cap \{\varrho < \beta/2\}) \cup A \equiv K_1 \cup K_2 \cup A.$$

For the integral over A, we notice that

$$\left|\int_0^T\int_A \varrho_\nu \ln \varrho_\nu G(\varrho_\nu)\,\mathrm{d}x\,\mathrm{d}t\right|$$
$$\leq \|G^{(i)}\|_{L^\infty(\mathbb{R})}\left(1+\|\varrho_\nu\|^2_{L^\infty(0,T;L^\infty(\mathbb{R}^d))}\right)\text{meas}(A)\leq C\varepsilon,$$

while the integral over K_2 is bounded as follows

$$\left|\int_0^T\int_{K_2} \varrho_\nu \ln \varrho_\nu G(\varrho_\nu)\,\mathrm{d}x\,\mathrm{d}t\right| \leq \|G\|_{L^\infty(\mathbb{R})}\text{meas}(K)\beta|\ln\beta|\leq C\beta|\ln\beta|.$$

Finally, on the set K_1, the sequence $\ln\varrho_\nu$ converges strongly to $\ln\varrho$ and therefore

$$\int_0^T\int_{K_1} \varrho_\nu \ln \varrho_\nu G(\varrho_\nu)\,\mathrm{d}x\,\mathrm{d}t \to \int_0^T\int_{K_1} \varrho \ln \varrho G(\varrho)\,\mathrm{d}x\,\mathrm{d}t.$$

Combining the estimates for $K^c\times[0,T]$ and $K\times[0,T]$, it follows that

$$|R(\varrho_\nu)-R(\varrho)|\leq\left|\int_0^T\int_{K_1} \varrho_\nu \ln \varrho_\nu G(\varrho_\nu)\,\mathrm{d}x\,\mathrm{d}t-\int_0^T\int_{K_1} \varrho \ln \varrho G(\varrho)\,\mathrm{d}x\,\mathrm{d}t\right|$$
$$+C\left(\varepsilon+\beta|\ln\beta|\right).$$

Since ε and β are arbitrarily small, we conclude. □

2.3.2 Nonlinear Pressure Laws

The inviscid limit has also been studied for nonlinear pressure laws. In particular, in [22] the authors consider the usual power law and study the limit $\nu\to 0$ for Eq. (2.41). In this section, we want to address the case of a general nonlinear pressure law of the form

$$p_\nu = f'(\varrho_\nu),$$

where f is a convex, continuous function, with $f(0)=0$. We consider the system

$$\begin{aligned}\frac{\partial\varrho_\nu}{\partial t}-\nabla\cdot(\varrho_\nu\nabla W_\nu)&=\varrho_\nu G(p_\nu),\\ -\nu\Delta W_\nu+W_\nu&=p_\nu,\end{aligned}\tag{2.53}$$

where G is decreasing and $G(p_H)=0$, for some $p_H>0$.

Theorem 2.4 (Inviscid Limit) *Let $(\varrho_\nu)_{\nu>0}$ be a sequence of weak solutions to Eq. (2.53) in the sense of Definition 2.2 with $p_\nu = f'(\varrho_\nu)$. There exist $\varrho, p \in L^\infty(0,T; L^1(\mathbb{R}^d) \cap L^\infty(\mathbb{R}^d))$ such that, up to a subsequence,*

$$\varrho_\nu \to \varrho, \text{ strongly in } L^2(0,T; L^2(\mathbb{R}^d)),$$
$$p_\nu \to p, \text{ strongly in } L^2(0,T; L^2(\mathbb{R}^d)),$$

and

$$W_\nu \to p, \text{ strongly in } L^2(0,T; H^1(\mathbb{R}^d)).$$

Furthermore, ϱ is a weak solution to

$$\frac{\partial \varrho}{\partial t} - \nabla \cdot (\varrho \nabla p) = \varrho G(p),$$
$$p = f'(\varrho),$$

with initial data ϱ_0.

We give here a gist of the proof, which follows the same main strategy as the argument for the linear pressure law presented in the previous section.

It is easy to see that weak solutions to (2.53) satisfy (uniformly in $\nu > 0$)

(i) $\varrho_\nu, p_\nu \in L^\infty(0,T; L^1(\mathbb{R}^d))$,
(ii) $0 \le p_\nu(x,t) \le p_H$, almost everywhere,
(iii) $W_\nu \in L^\infty(0,T; L^1(\mathbb{R}^d) \cap L^\infty(\mathbb{R}^d))$,

hence, $\varrho_\nu, p_\nu, W_\nu$ are weakly compact. Let us recall that since $p_\nu = f'(\varrho_\nu)$, we have

$$\varrho_\nu p_\nu = f(\varrho_\nu) + f^*(p_\nu), \quad \varrho_\nu = f^{*\prime}(p_\nu),$$

where f^* is the convex conjugate of f, namely

$$f^*(b) = \sup_a ab - f(a).$$

2.3.2.1 Energy Dissipation Inequality

Arguing in the same way as for the linear pressure case, one can show that weak solutions to (2.53) satisfy an energy dissipation inequality. Instead of testing the equation against $\ln \varrho_\nu$, we test against $z(p_\nu)$, where $z : [0,+\infty) \to \mathbb{R}$ is a positive and increasing function such that

$$\varrho_\nu z'(p_\nu) = 1.$$

Proposition 2.11 (Energy Inequality) *The solution ϱ_ν to (2.44) satisfies the energy inequality*

$$\int_{\mathbb{R}^d} p_\nu(T)\,\mathrm{d}x - \int_0^T\int_{\mathbb{R}^d} p_\nu \Delta W_\nu \,\mathrm{d}x\,\mathrm{d}t \le \int_0^T\int_{\mathbb{R}^d} z(p_\nu)\varrho_\nu G(p_\nu)\,\mathrm{d}x\,\mathrm{d}t + \int_{\mathbb{R}^d} p_0\,\mathrm{d}x. \tag{2.54}$$

Once again the proof follows from considering a parabolic regularization (by adding $+\,\varepsilon\Delta\varrho_{\nu,\varepsilon}$ to the equation), testing against $z(p_{\nu,\varepsilon})$ and then taking $\varepsilon \to 0$.

The boundedness of both pressure and density yields a uniform control on the integral on the right-hand side of (2.54). This implies the following result.

Lemma 2.11 *The following bounds hold uniformly in ν*

(i) $\sqrt{\nu}\Delta W_\nu \in L^2(0,T; L^2(\mathbb{R}^d))$,
(ii) $\nabla W_\nu \in L^2(0,T; L^2(\mathbb{R}^d))$,
(iii) $\partial_t \varrho_\nu \in L^2(0,T; H^{-1}(\mathbb{R}^d))$.

Proof From the energy dissipation inequality and Brinkman's law, we have

$$\int_0^T\int_{\mathbb{R}^d} (\nu\Delta W_\nu - W_\nu)\Delta W_\nu \,\mathrm{d}x\,\mathrm{d}t = -\int_0^T\int_{\mathbb{R}^d} p_\nu \Delta W_\nu\,\mathrm{d}x\,\mathrm{d}t \le C,$$

from which by integration by parts we get

$$\int_0^T\int_{\mathbb{R}^d} \nu|\Delta W_\nu|^2 + |\nabla W_\nu|^2\,\mathrm{d}x\,\mathrm{d}t \le C.$$

From the above bounds and the density equation *(iii)* follows and we conclude. □

Let us remark that as a consequence of *(i)* the difference between p_ν and W_ν converges strongly to zero in $L^2(0,T; L^2(\mathbb{R}^d))$, since $p_\nu - W_\nu = -\nu\Delta W_\nu$. By a compensated compactness argument we can show that the product $\varrho_\nu p_\nu$ converges to the product of the limits.

Lemma 2.12 *Up to a subsequence we have*

$$\varrho_\nu p_\nu \rightharpoonup \varrho p \text{ weakly in } L^2(0,T; L^2(\mathbb{R}^d)).$$

Proof Let φ be a smooth compactly supported test function. Thanks to the bounds *(ii)* and *(iii)* in Lemma 2.11 the product $\varrho_\nu W_\nu$ converges weakly to ϱp, while thanks to *(i)* $\nu\Delta W_\nu$ converges strongly to zero. Thus, the product $\varrho_\nu p_\nu$ converges in the sense of distributions

$$\int_0^T\int_{\mathbb{R}^d} \varrho_\nu p_\nu \varphi\,\mathrm{d}x\,\mathrm{d}t = \int_0^T\int_{\mathbb{R}^d} (\varrho_\nu W_\nu - \varrho_\nu \nu\Delta W_\nu)\varphi\,\mathrm{d}x\,\mathrm{d}t \to \int_0^T\int_{\mathbb{R}^d} \varrho p\varphi\,\mathrm{d}x\,\mathrm{d}t.$$

Since both pressure and density are uniformly bounded, they also converge up to a subsequence in any L^p space. □

Lemma 2.13 *Up to a subsequence, we have*

$$\varrho_\nu \to \varrho \text{ strongly in } L^2(0,T;L^2(\mathbb{R}^d)),$$

$$p_\nu \to p \text{ strongly in } L^2(0,T;L^2(\mathbb{R}^d)),$$

$$W_\nu \to p \text{ strongly in } L^2(0,T;L^2(\mathbb{R}^d)).$$

Proof We know that the product converges, hence

$$f(\varrho_\nu) + f^*(p_\nu) = \varrho_\nu p_\nu \rightharpoonup \varrho p \le f(\varrho) + f^*(p).$$

By lower semi-continuity of convex functions, we have

$$f(\varrho) \le \liminf_{\nu\to 0} f(\varrho_\nu), \qquad f^*(p) \le \liminf_{\nu\to 0} f^*(p_\nu).$$

This implies

$$f(\varrho_\nu) \rightharpoonup f(\varrho), \qquad f^*(p_\nu) \rightharpoonup f^*(p). \tag{2.55}$$

Since f and f^* are convex and increasing, (2.55) and weak compactness of ϱ_ν, p_ν imply strong compactness. Finally, W_ν converges strongly as well because, as we have already proven, $W_\nu - p_\nu \to 0$ strongly. □

Once again, the above result, together with the weak compactness of ∇W_ν, is enough to pass to the limit in the equation for ϱ_ν and obtain

$$\frac{\partial \varrho}{\partial t} - \nabla\cdot(\varrho \nabla p) = \varrho G(p), \tag{2.56}$$

with

$$p = f'(p).$$

The solution to the above equation has enough regularity to satisfy the following energy dissipation *equality*

$$\int_{\mathbb{R}^d} p(T)\,\mathrm{d}x + \int_0^T\!\!\int_{\mathbb{R}^d} |\nabla p|^2\,\mathrm{d}x\,\mathrm{d}t = \int_0^T\!\!\int_{\mathbb{R}^d} z(p)\varrho G(p)\,\mathrm{d}x\,\mathrm{d}t + \int_{\mathbb{R}^d} p_0\,\mathrm{d}x, \tag{2.57}$$

where z is such that $z'(p)\varrho = 1$. To uplift the weak compactness of ∇W_ν to strong compactness, we argue as in the previous section and we compare the above *equality* to the inequality (2.54) to show upper semi-continuity of $\|\nabla W_\nu\|_2$. Let us remark again that, thanks to Brinkman's law, we have

$$-\int_0^T\int_{\mathbb{R}^d} p_\nu \Delta W_\nu = \int_0^T\int_{\mathbb{R}^d} \nu|\Delta W_\nu|^2 + |\nabla W_\nu|^2 \, \mathrm{d}x \, \mathrm{d}t,$$

and therefore (2.54) implies

$$\int_{\mathbb{R}^d} p_\nu(T) \, \mathrm{d}x + \int_0^T\int_{\mathbb{R}^d} |\nabla W_\nu|^2 \, \mathrm{d}x \, \mathrm{d}t \le \int_0^T\int_{\mathbb{R}^d} z(p_\nu)\varrho_\nu G(p_\nu) \, \mathrm{d}x \, \mathrm{d}t + \int_{\mathbb{R}^d} p_0 \, \mathrm{d}x.$$

To conclude it is sufficient to show

$$\limsup_{\nu\to 0} \int_0^T\int_{\mathbb{R}^d} z(p_\nu)\varrho_\nu G(p_\nu) \, \mathrm{d}x \, \mathrm{d}t \le \int_0^T\int_{\mathbb{R}^d} z(p)\varrho G(p) \, \mathrm{d}x \, \mathrm{d}t,$$

which follows directly from the strong compactness of ϱ_ν and p_ν.

2.4 Cross-diffusion Systems

In the modelling of tissue growth, understanding the interactions between different species of cells (for instance, healthy tissue, cancer cells, quiescient cells, ...) and their impact on collective behavior is essential. A central role is played by systems of cross-diffusion equations with species-dependent reaction rates. The coupling between the two equations entails several analytical challenges, especially when both equations are nonlinear and degenerate. In this section, we give a brief overview of some questions that have attracted interest in recent years and the methods that have been developed to address them.

2.4.1 Two-species Models

We consider two populations of cells with densities $\varrho^{(1)}$ and $\varrho^{(2)}$, evolving under Darcy's law and with different pressure-dependent reaction rates

$$\begin{cases} \dfrac{\partial \varrho^{(1)}}{\partial t} = \nabla \cdot (\varrho^{(1)} \nabla p) + \varrho^{(1)} G^{(1)}(p), \\ \dfrac{\partial \varrho^{(2)}}{\partial t} = \nabla \cdot (\varrho^{(2)} \nabla p) + \varrho^{(2)} G^{(2)}(p), \end{cases} \qquad x \in \mathbb{R}^d, t > 0. \tag{2.58}$$

Both species are subject and contribute to the total pressure which is given by $p = f'(\varrho)$, where f is convex and ϱ is the sum of the two densities. A standard choice

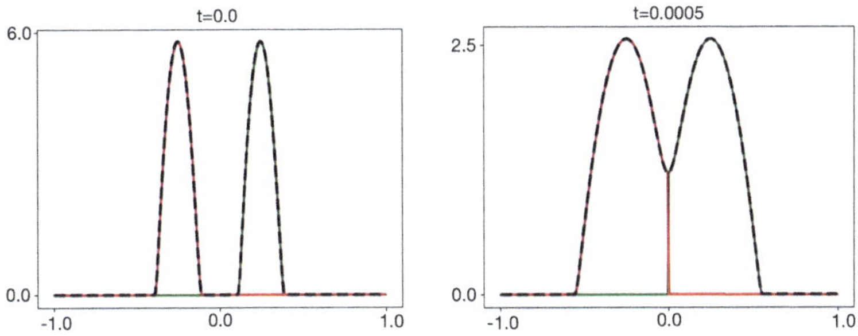

Fig. 2.3 Plot of $\varrho^{(1)}$, $\varrho^{(2)}$ *(red and green lines)*, and $\varrho = \varrho^{(1)} + \varrho^{(2)}$ *(black dashed line)*

is, for instance, the power law

$$p = \frac{m}{m-1}(\varrho^{(1)} + \varrho^{(2)})^{m-1}, \quad m > 1.$$

Degenerate cross-diffusion systems of this kind exhibit a common property: sharp interfaces may appear between the two phases. In fact, due to the degeneracy of the equations, if solutions are initially compactly supported they remain so since the speed of propagation is finite. Moreover, if the species are initially segregated, namely $\varrho^{(1)}(\cdot, 0)\varrho^{(2)}(\cdot, 0) = 0$ almost everywhere, they will remain so for all positive times, since both evolve under the same velocity field, see [9, 26]. This leads to the appearance of sharp interfaces between the two supports. As a consequence, no higher regularity than BV can be expected, see Fig. 2.3.

Actually, BV regularity (and hence strong compactness of the individual species) has been proven in [11] only for $d = 1$. For higher dimensions such regularity do not propagate. Moreover, it is not possible to apply the L^1 theory of DiPerna-Lions to find strong compactness of the single species $\varrho^{(i)}$ since the velocity field, $-\nabla p$, is not regular enough.

Therefore, in order to prove existence of weak solutions through approximation arguments it is essential to show the strong compactness of the pressure gradient. To this end, in [25] the authors use the L^3-version of the Aronson-Bénilan estimate discussed in Sect. 2.2.2. For the sake of completeness, we state here the result, see [25, Theorem 1].

Theorem 2.5 *Assuming that the initial data satisfy*

$$\varrho_0^i \geq 0, \ \textit{for}\ i = 1, 2, \quad 0 \leq p_0 \leq p_H,$$

$$\varrho_0(1 + |x|^2) \in L^1(\mathbb{R}^d), \quad \nabla p_0 \in L^2(\mathbb{R}^d), \quad (\Delta p_0)_- \in L^2(\mathbb{R}^d),$$

and additionally

$$m > 3 - \frac{4}{d}, \quad \sup_{0 \leq p \leq p_H} \frac{|G^{(1)}(p) - G^{(2)}(p)|}{p^{1/(m-1)}} \leq C, \tag{2.59}$$

then, for $t \in (0, T)$

$$\int_{\mathbb{R}^d} (\Delta p(t))_-^2 \, \mathrm{d}x \le C(T), \qquad \int_0^T \int_{\mathbb{R}^d} (\Delta p)_-^3 \, \mathrm{d}x \, \mathrm{d}t \le C(T). \tag{2.60}$$

The proof of existence of solutions then follows by an approximating argument with respect to which the above estimates are stable. It is then possible to infer the *strong* compactness of the pressure gradient, which is indeed crucial since only weak compactness is known for $\varrho^{(1)}$ and $\varrho^{(2)}$.

2.4.1.1 Incompressible Limit

We point out that besides the technicalities of the AB estimate which require strong assumptions on $G^{(1)}$ and $G^{(2)}$, see (2.59), this method does not allow to pass to the incompressible limit $m \to \infty$, since for two species systems such as (2.58) it is not clear how to prove a bound on $\partial_t p_m$ that is uniform in m. For this reason, energy methods discussed in Sect. 2.2.4 have been developed and successfully applied to systems such as (2.58). The main idea consist in considering the equation on the total density $\varrho = \varrho_1 + \varrho_2$, which is indeed a porous medium-type equation

$$\frac{\partial \varrho_m}{\partial t} = \Delta \varrho_m^m + \varrho_m^{(1)} G^{(1)}(p_m) + \varrho_m^{(2)} G^{(2)}(p_m).$$

Weak compactness of $\varrho_m^{(i)}$, for $i = 1, 2$, and $\nabla \varrho_m^m$ as $m \to \infty$ is enough to establish that the limit satisfies

$$\frac{\partial \varrho_\infty}{\partial t} = \Delta p_\infty + R_\infty, \tag{2.61}$$

where R_∞ is the weak limit of the sum of the reaction terms. In the same fashion as the proof in Sect. 2.2.4, by comparing the energy equalities for ϱ_m and ϱ_∞, one can show that $\nabla \varrho_m^m \to \nabla p_\infty$ strongly and identify R_∞ as $\varrho_\infty^{(1)} G^{(1)}(p_\infty) + \varrho_\infty^{(2)} G^{(2)}(p_\infty)$ using the monotonicity of $G^{(i)}$. This way, one obtains the following complementarity relation on the limit pressure

$$p_\infty(\Delta p_\infty + \varrho_\infty^{(1)} G^{(1)}(p_\infty) + \varrho_\infty^{(2)} G^{(2)}(p_\infty)) = 0.$$

For more details on these results we refer the reader to [12, 27, 30, 36], as well as to [26] where in order to construct Lagrangian solutions to system (2.58) and to pass to the incompressible limit the author exploits the estimate on the Hessian of the pressure presented in Sect. 2.2.3.

2.4.1.2 Inviscid Limit

The same strategy can be applied to the inviscid limit of Brinkman multi-species models such as

$$\begin{cases} \dfrac{\partial \varrho_\nu^{(1)}}{\partial t} = \nabla \cdot (\varrho_\nu^{(1)} \nabla W_\nu) + \varrho_\nu^{(1)} G^{(1)}(p_\nu), \\ \dfrac{\partial \varrho_\nu^{(2)}}{\partial t} = \nabla \cdot (\varrho_\nu^{(2)} \nabla W_\nu) + \varrho_\nu^{(2)} G^{(2)}(p_\nu), \end{cases} \tag{2.62}$$

where

$$-\nu \Delta W_\nu + W_\nu = p_\nu := \frac{m}{m-1} (\varrho_\nu^{(1)} + \varrho_\nu^{(2)})^{m-1}, \quad m > 1.$$

The limit $\nu \to 0$ was studied in [16] for $m = 2$ through entropy dissipation inequalities, and extended in [22] for $m > 2$, while the limit $m \to \infty$ can be found in [19, 20]. The idea is once again to consider the equation satisfied by $\varrho_\nu = \varrho_\nu^{(1)} + \varrho_\nu^{(2)}$, establish the entropy energy satisfied by ϱ_ν and then compare it with the energy equality of the limit density, to finally infer the strong compactness of ∇W_ν. The proof proceed in the same way as presented in Sects. 2.3.1 and 2.3.2, where the only technical difference is the convergence of the reaction terms.

Notice that the limit $\nu \to 0$ of (2.62) also yields a new proof of the existence of weak solutions to the Darcy system (2.58).

2.4.2 *Structured Equations for Phenotypic Heterogeneity*

As already mentioned, tissue are characterized by a large diversity of cells, and this heterogeneity plays a crucial role in their development. In particular, in tumors the variety of different phenotypes is so large that it becomes reasonable to consider a continuum of traits, rather than a finite number of species. Therefore, *structured models* have attracted major attention thanks to their ability of capturing the role of phenotypic heterogeneity in cancer growth.

In this section, we want to briefly review some results concerning structured models which, as the two-species models presented above, capture the fluid-dynamical aspects of tissue growth, namely they use either an elastic (Darcy's law) or viscoelastic (Brinkman's law) description.

The first example of such models for which a proof of existence of solutions was provided is the following Darcy model where the cell population density $n(y, x, t)$ is structured by a continuous phenotypic trait $y \in [0, 1]$, and the growth rate is phenotypic dependent and monotone decreasing with respect to the total pressure

$$\begin{cases} \dfrac{\partial n(y,x,t)}{\partial t} &= \nabla_x \cdot (n(y,x,t)\nabla_x p(x,t)) + n(y,x,t)G(y,p), \\ \varrho(x,t) &= \displaystyle\int_0^1 n(y,x,t)\,\mathrm{d}y, \\ p(x,t) &= \dfrac{m}{m-1}\left(\varrho(x,t)\right)^{m-1}, \quad m>1. \end{cases} \tag{2.63}$$

As for the two-species counterpart, system (2.58), when studying the existence and asymptotic behavior of solutions the main difficulty relies on finding the strong compactness of the pressure gradient. The existence of weak solutions of (2.63) can be shown by extending the method developed in [36] and based on energy equalities for the total population ϱ which satisfies

$$\frac{\partial \varrho}{\partial t} = \Delta \varrho^m + \int_0^1 n(y,x,t)G(y,p)\,\mathrm{d}y.$$

Let us remark that it is not possible to obtain an AB estimate for solutions to (2.63) following the same method used in [25] for the two-species model. This is due to the fact that it is not possible to find an analogous assumption on the reaction rates as the one needed in the two-species case, see (2.59).

2.4.2.1 Inviscid Limit of a Brinkman Model

Existence of solutions for (2.63) can be also obtained by taking the inviscid limit of the following structured equation coupled with Brinkman's law

$$\begin{cases} \dfrac{\partial n(y,x,t)}{\partial t} &= \nabla_x \cdot (n(y,x,t)\nabla_x W(x,t)) + n(y,x,t)G(y,p), \\ p(x,t) &= \dfrac{m}{m-1}\left(\displaystyle\int_0^1 n(y,x,t)\,\mathrm{d}y\right)^{m-1}, \\ -\nu\Delta W + W &= p, \end{cases} \tag{2.64}$$

with $m > 1$. The argument follows the one used in [16] for the two-species case, which consists in considering the total density equation

$$\frac{\partial \varrho}{\partial t} = \nabla \cdot (\varrho \nabla W) + \int_0^1 n(y,x,t)G(y,p)\,\mathrm{d}y,$$

and comparing energy dissipation (in)equalities to get strong compactness of ∇W_ν. A proof for the case $m = 2$ can be found in the recent work [21].

2.4.2.2 Incompressible Limit

Using again energy methods in order to pass to the limit $m \to \infty$, it can be shown that a sequence of solutions (n_m, p_m) to (2.63) converges to (n_∞, p_∞) in an appropriate functional space, and the limit satisfies

$$\frac{\partial n_\infty}{\partial t} = \nabla \cdot (n_\infty \nabla p_\infty) + n_\infty G(y, p_\infty), \qquad \text{in } \mathcal{D}'((0,1) \times \mathbb{R}^d \times (0,\infty)),$$

$$p_\infty \left(\Delta p_\infty + \int_0^1 n_\infty G(y, p_\infty)\,\mathrm{d}y \right) = 0, \qquad \text{in } \mathcal{D}'(\mathbb{R}^d \times (0,\infty)).$$

Once again, the main difficulty lies in the fact that it is not possible to prove strong compactness of the pressure before having proved strong compactness of its gradient. Nevertheless, using the monotonicity of $G = G(y, p)$ with respect to p and compensated compactness arguments the proof follows in a similar fashion as for the two-species case [12].

2.4.2.3 Many-species Limit

We conclude by reviewing a recent result that relate the multi-species models presented in the previous section to structured models such as (2.63) and (2.64), see [21]. The authors consider a N-species Brinkman model with linear pressure—here we present the general power law case, for which the method in [21] can still be applied. The goal is to show that as $N \to \infty$ solutions to the N-species system converge to solutions of the structured model (2.64). To this end, let us define the growth rate of the ith species, $G_N^{(i)}$, as

$$G_N^{(i)}(p) = G\left(\frac{i}{N}, p\right),$$

where $G : [0,1] \times \mathbb{R} \to \mathbb{R}$ is the trait-dependent rate of Eq. (2.64). The evolution of $n_N^{(i)}$, for $i = 1, \dots, N$, (representing the density of the ith population) is then described by the system

$$\begin{cases} \dfrac{\partial n_N^{(i)}}{\partial t} - \nabla \cdot (n_N^{(i)} \nabla W_N) = n_N^{(i)} G_N^{(i)}(\underline{p}_N), \\[2ex] -\nu \Delta W_N + W_N = \underline{p}_N. \end{cases} \tag{2.65}$$

where the *rescaled* total pressure is given by

$$\underline{p}_N(x,t) = \frac{m}{m-1}\left(\frac{1}{N}\sum_{i=1}^{N} n_N^{(i)}(x,t)\right)^{m-1}.$$

We now introduce the *interpolated* densities and reaction terms

$$n_N(y,x,t) = \sum_{i=1}^{N} n_N^{(i)}(x,t)\mathbb{1}_{\left(\frac{i-1}{N},\frac{i}{N}\right]}(y), \quad G_N(y,p) = \sum_{i=1}^{N} G_N^{(i)}(p)\mathbb{1}_{\left(\frac{i-1}{N},\frac{i}{N}\right]}(y).$$

These functions naturally embed into function spaces that are continuous in phenotype, space and time. Therefore, for all $y \in [0,1]$, system (2.65) can be rewritten as

$$\begin{cases} \dfrac{\partial n_N}{\partial t} - \nabla\cdot(n_N\nabla W_N) = n_N G_N(y,\underline{p}_N), \\ \\ -\nu\Delta W_N + W_N = \underline{p}_N. \end{cases} \tag{2.66}$$

As $N \to \infty$ solutions to (2.66) converge to solutions to (2.64), see [21, Theorem 2.5]. In particular, up to a subsequence, we have

$$n_N(y,\cdot,\cdot) \rightharpoonup^* n(y,\cdot,\cdot) \text{ in } L^\infty(0,T;L^\infty(\mathbb{R}^d)), \text{ for almost every } y \in [0,1],$$

$$\underline{p}_N \to p \text{ in } L^2(0,T;L^2(\mathbb{R}^d)),$$

$$W_N \to W \text{ in } L^2(0,T;H^1(\mathbb{R}^d)).$$

It has also been shown that the above compactness also holds when taking $\nu = \nu_N$ to be any sequence such that $\nu_N \to 0$ as $N \to \infty$. Therefore, in [21, Theorem 2.6], the authors show that a solution to (2.66) convergence to a solution of (2.63) as $\nu \to 0$ and $N \to \infty$, bridging the gap between the N-species Brinkman model and Darcy's phenotype model. The continuous phenotype limit $N \to \infty$ has also been shown for a discrete N-species Darcy's model (one can simply consider $\nu = 0$ in (2.66))—hence, the diagram commutes.

References

1. T. Alazard, D. Bresch, Functional inequalities and strong lyapunov functionals for free surface flows in fluid dynamics. Preprint (2020). arXiv: 2004.03440
2. D. Alexander, I. Kim, Y. Yao, Quasi-static evolution and congested crowd transport. Nonlinearity **27**(4), 823–858 (2014)

3. L. Ambrosio, G. Nicola, G. Savaré, *Gradient Flows in Metric Spaces and in the Space of Probability Measures* (Birkhäuser Verlag, Basel, 2008)
4. D.G. Aronson, Regularity properties of flows through porous media: The interface. Arch. Ration. Mech. Anal. **37**, 1–10 (1970)
5. D.G. Aronson, P. Bénilan, Régularité des solutions de l'équation des milieux poreux dans $\mathbf{R}^N$. C. R. Acad. Sci. Paris Sér. A-B **288**(2), A103–A105 (1979)
6. P. Benilan, L. Boccardo, M. Herrero, On the limit of solutions of $u_t = \delta u^m$ as $m \to \infty$. Rendiconti del Seminario Matematico Università e Politecnico di Torino. Fascicolo Speciale on Nonlinear PDE's (1991), pp. 1–13
7. G. Bevilacqua, B. Perthame, M. Schmidtchen, The Aronson-Bénilan Estimate in Lebesgue Spaces. Preprint (2020). arXiv: 2007.15267
8. L. Boccardo, T. Gallouët, Non linear elliptic and parabolic equations involving measure data. J. Funct. Anal. **87**, 149–169 (1989)
9. F. Bubba, B. Perthame, C. Pouchol, M. Schmidtchen, Hele-Shaw limit for a system of two reaction-(cross-)diffusion equations for living tissues. Arch. Ration. Mech. Anal. **236**, 735–766 (2020)
10. L.A. Caffarelli, A. Friedman, Asymptotic behavior of solutions of $u_t = \Delta u^m$ as $m \to \infty$. Indiana University Math. J. **36**(4), 711–728 (1987)
11. J.A. Carrillo, S. Fagioli, F. Santambrogio, M. Schmidtchen, Splitting schemes and segregation in reaction cross-diffusion systems. SIAM J. Math. Anal. **50**(5), 5695–5718 (2018)
12. N. David, Phenotypic heterogeneity in a model of tumor growth: existence of solutions and incompressible limit. Commun. Partial Differ. Equ. **48**(4), 678–710 (2023)
13. N. David, B. Perthame, Free boundary limit of a tumor growth model with nutrient. J. Math. Pures Appl. **155**, 62–82 (2021)
14. N. David, M. Schmidtchen, On the incompressible limit for a tumor growth model incorporating convective effects. Commun. Pure Appl. Math. **77**(5), 2613–2650 (2024)
15. N. David, T. Dębiec, B. Perthame, Convergence rate for the incompressible limit of nonlinear diffusion-advection equations. AIHP **40**(3), 511–529 (2022)
16. N. David, T. Dębiec, M. Mandal, M. Schmidtchen, A degenerate cross-diffusion system as the inviscid limit of a nonlocal tissue growth model. SIAM J. Math. Anal. **56**(2), 2090–2114 (2024)
17. N. David, A.R. Mészáros, F. Santambrogio, Improved convergence rates for the hele-shaw limit in the presence of confining potentials. Preprint (2024). arXiv:2405.07227
18. N. David, F. Santambrogio, M. Schmidtchen, Uniform regularity estimates for nonlinear diffusion-advection equations in the hard-congestion limit. Preprint (2024). arXiv:2405.06942
19. T. Dębiec, M. Schmidtchen, Incompressible limit for a two-species tumor model with coupling through brinkman's law in one dimension. Acta Appl. Math. **169**, 593–611 (2020)
20. T. Dębiec, B. Perthame, M. Schmidtchen, N. Vauchelet, Incompressible limit for a two-species model with coupling through brinkman's law in any dimension. J. Math. Pures Appl. **145**, 204–239 (2020)
21. T. Dębiec, M. Mandal, M. Schmidtchen, From finite to continuous phenotypes in (visco-)elastic tissue growth models. Preprint (2024). arXiv:2409.02904
22. C. Elbar, J. Skrzeczkowski, On the inviscid limit connecting brinkman's and darcy's models of tissue growth with nonlinear pressure. Preprint (2023). arXiv:2306.03752
23. C.M. Elliott, V. Janovský, A variational inequality approach to Hele-Shaw flow with a moving boundary. Proc. R. Soc. Edinb. **88A**, 93–107 (1981)
24. C.M. Elliott, M.A. Herrero, J.R. King, J.R. Ockendon, The mesa problem: Diffusion patterns for $u_t = \nabla \cdot (u^m \nabla u)$ as $m \to \infty$. IMA J. Appl. Math. **37**(2), 147–154 (1986)
25. P. Gwiazda, B. Perthame, A. Świerczewska Gwiazda, A two-species hyperbolic–parabolic model of tissue growth. Commun. Partial Differ. Equ. **44**(12), 1605–1618 (2019)
26. M. Jacobs, Lagrangian solutions to the Porous Media Equation and Reaction Diffusion Systems. Preprint (2022). arXiv:2208.01792
27. M. Jacobs, Existence of solutions to reaction cross diffusion systems. SIAM J. Math. Anal. **55**(6), 6991–7023 (2023)

28. I. Kim, N. Požár, Porous medium equation to Hele-Shaw flow with general initial density. Trans. Am. Math. Soc. **370**(2), 873–909 (2018)
29. I. Kim, Y.P. Zhang, Porous medium equation with a drift: Free boundary regularity. Arch. Ration. Mech. Anal. **242**, 1177–1228 (2021)
30. J.-G. Liu, X. Xu, Existence and incompressible limit of a tissue growth model with autophagy. SIAM J. Math. Anal. **53**(5), 5215–5242 (2021)
31. B. Maury, A. Roudneff-Chupin, F. Santambrogio, A macroscopic crowd motion model of gradient flow type. Math. Models Methods Appl. Sci. **20**(10), 1787–1821 (2010)
32. B. Maury, A. Roudneff-Chupin, F. Santambrogio, J. Venel, Handling congestion in crowd motion modeling. Netw. Heterogeneous Media **6**(3), 485 (2011)
33. O.A. Oleinik, A.S. Kalashnikov, C. Jui-lin, The Cauchy problem and boundary problems for equations of the type of non-stationary filtration. Izv. Akad. Nauk SSSR Ser. Mat. **22**(5), 667–704 (1958)
34. B. Perthame, N. Vauchelet, Incompressible limit of a mechanical model of tumor growth with viscosity. Philos. Trans. R. Soc. A **373**(2050), 20140283 (2015)
35. B. Perthame, F. Quirós, J.L. Vázquez, The Hele-Shaw asymptotics for mechanical models of tumor growth. Arch. Ration. Mech. Anal. **212**(1), 93–127 (2014)
36. B.C. Price, X. Xu, Global existence theorem for a model governing the motion of two cell populations. Kinetic Relat. Models **13**(6), 1175–1191 (2020)
37. S. Richardson, Hele-Shaw flows with a free boundary produced by the injection of fluid into a narrow channel. J. Fluid Mech. **56**(4), 609–618 (1972)
38. P.E. Sacks, A singular limit problem for the porous medium equation. J. Math. Anal. Appl. **140**(2), 456–466 (1989)
39. F. Santambrogio, Optimal transport for applied mathematicians. *Progress in Nonlinear Differential Equations and their Applications*, vol. 87 (Birkhäuser, Basel, 2015)
40. J.L. Vazquez, *The Porous Medium Equation. Mathematical Theory*. Oxford Mathematical Monographs (The Clarendon Press/Oxford University Press, Oxford, 2007)

Chapter 3
Mathematical Modeling of Fluids from the Kinetic Theory

Vincent Giovangigli

Abstract A synthesis of mathematical fluid models derived from the kinetic theory of gases is presented. Kinetic theory yields well structured set of partial equations with entropies compatible with convection, diffusion and sources. It also yields properly structured nonlinear source terms as well as the transport coefficients. We first address multicomponent flow models including detailed transport and chemical reactions. Symmetrization, hyperbolic-parabolic structure and asymptotic stability of constant equilibrium states are established. We next investigate relaxation issues for a two temperature model that leads to symmetrized hyperbolic-parabolic systems with stiff source terms. Local existence of solutions is established and the apparition of the bulk viscosity coefficient is justified in the fast relaxation limit. We finally consider nonideal fluids and their cohesive properties that lead to diffuse interfaces models with capillary effects. Using an augmented formulation—with density gradient added as an extra unknown—a new type of hyperbolic-parabolic-dispersive structure is obtained with antisymmetric second order coupling terms arising from capillarity. Local existence of solutions is established as well as asymptotic stability of constant states.

3.1 Introduction

Two theoretical frameworks may essentially be used in order to derive systems of equations modeling fluids in close form. The first method is based on thermodynamic arguments allowing to identify the transport fluxes appearing in the conservation equations [1–3]. The alternative method is to derive the full set of equations from molecular physics by using the kinetic theory of gases [4–7].

The thermodynamic of irreversible processes is well described in the book of De Groot and Mazur [2]. Starting from conservation equations for mass, momentum, and energy—including unknown fluxes—a balance of entropy is formulated

V. Giovangigli (✉)
CMAP, CNRS, École Polytechnique, Palaiseau, France
e-mail: vincent.giovangigli@polytechnique.edu

M. Bulíček et al. (eds.), *New Perspectives in Mathematical Fluid Mechanics*,
Nečas Center Series, https://doi.org/10.1007/978-3-032-02319-3_3

with an entropy production rate written as a sum of scalar products between unknown fluxes and fluid variables gradients—the thermodynamic forces and affinities. The unknown transport fluxes appearing in the conservation equations are then expressed as linear combinations of the fluid variable gradients—thereby closing the system of conservation laws—that insure a nonnegative rate of entropy production. Such a thermodynamic of irreversible processes presents many advantages. It is first a low-cost procedure—since it mainly requires to derive an entropy balance—and it provides the transport fluxes that are required to close the conservation equations. Various important properties may further be imposed on the proportionality coefficients between fluxes and affinities, namely symmetry, isotropy and positivity. From a mathematical point of view such fluxes yield strong parabolicity conditions with respect to some fluid variables. Such derivations may also easily be included in mathematical courses on fluid systems of partial differential equations for pedagogical purposes. Thermodynamic methods have thus been highly successful in mathematical and physical modeling of fluid systems [2]. However, they unfortunately suffer from several weaknesses. The proportionality coefficients between fluxes and affinities—the transport coefficients—indeed remain unknown. Such derivations also assume that classic thermodynamics obtained from thermostatics locally holds in a fluid even in the presence of spatial gradients. When applied to reactive fluids, they further yield production rates that are proportional to affinities although they should depend exponentially [2, 7]. Finally, for nonlinear problems like diffuse interface fluids, thermodynamic methods remain ambiguous since they yield multiple solutions, including unphysical solutions.

The alternative method, the kinetic theory of gases, does not suffer from these weaknesses. Starting from a Boltzmann equation, the complete fluid equations are obtained by using asymptotic expansions—more specifically the Chapman-Enskog procedure—assuming that collisions are dominant. Such a molecular derivation yields the conservation equations, the transport fluxes, the thermodynamic properties as well as the transport coefficients—the proportionality coefficients between the transport fluxes and the fluid variable gradients. The kinetic theory derivations also yield a thermodynamic framework valid in the presence of macroscopic gradients and provide entropies compatible with convection, diffusion and chemical sources. The proper exponential dependence of production rates in terms of affinities is also obtained for reactive fluids. Finally, starting from a BBGKY hierarchy—rather than Boltzmann equation—nonideal thermodynamics is obtained as well as diffuse interface fluid models without any ambiguity. The molecular derivations therefore yield the complete set of equations without any of the weaknesses associated with thermodynamic methods but they are of course much more complex. Existence theorem may subsequently be obtained for such systems of equations and, in order to illustrate these ideas, three different fluid models are presented and their mathematical structure analyzed. Such a synthesis reveals the high mathematical quality of the models derived from the kinetic theory and illustrates various hyperbolic-parabolic structures modeling fluids. This paper being mainly a synthesis of previous results, we often rely on already published papers for detailed proofs or comments. New results will still be obtained incidentally for each of the three models discussed successively.

We first consider multicomponent reactive gas mixtures that play a role in many scientific and industrial applications like combustion [8–10], crystal growth [11–13] and reentry into the earth's atmosphere [14, 15]. The governing equations are presented including multicomponent fluxes and thermochemistry as derived from the kinetic theory of gases and this notably includes the transport coefficients [5–7]. The fluid model is embedded in a mathematical framework and recast into a quasilinear system of partial differential equations [7]. Symmetrizability properties are then addressed by using entropic and normal variables and the system recast into a symmetric hyperbolic-parabolic form [7, 16–24]. We study the Cauchy problem and discuss the local existence of strong solutions [7, 24–26, 28, 30]. New results are obtained incidentally in the situation of small transport coefficients. Global existence results and asymptotic stability of constant equilibrium states are finally obtained using the strict dissipativity of linearized systems at equilibrium [18, 19, 27, 29, 31–39].

We next analyze the fast relaxation of translational and internal temperatures in a nonequilibrium gas [40–46]. The system of partial differential equations derived from the kinetic theory of gases is recast in a convenient hyperbolic-parabolic symmetric form with stiff source terms [47–54]. We discuss the Chapman-Enskog expansion in the fast relaxation limit and establish that the temperature difference becomes asymptotically proportional to the divergence of the velocity field. This leads to the apparition of the volume viscosity coefficient whose impact in fluid mechanics has been established [55–57]. We prove local in time error estimates between the out of equilibrium solution and the one-temperature equilibrium fluid solution and justify the apparition of volume viscosity terms [53]. The situation of ill prepared data is also addressed with multiscale expansions taking into account initial layers [58–60]. A previously established result valid for a family of second order systems is shown to apply to the situation of two-temperature fluids [60].

We finally investigate compressible diffuse interface fluids also termed cohesive or capillary fluids [61–74]. Such fluid models involve van der Waals' gradient energy [61], Korteweg's tensor [62], Dunn and Serrin's heat flux [63] as well as diffusive fluxes [64–67]. Such diffuse interface fluids may be used to investigate multiphase flows with complex liquid-vapor interfaces [66–74]. Diffuse interface fluids have been the object of intensive mathematical research over the last decades [75–90]. The density gradient is added as an extra variable and the convective and capillary fluxes of the augmented system are identified by using the Legendre transform of entropy. The augmented system of equations is recast into a new type of hyperbolic-parabolic systems involving symmetric hyperbolic first order terms, symmetric dissipative second order terms and antisymmetric cohesive second order terms [76, 90]. A priori estimates are obtained for such augmented system of equations in normal form. Using the augmented system and a priori estimates, local existence of strong solutions is established in an Hilbertian framework [90]. We then investigate the asymptotic stability of thermodynamically stable states using a new characterization of strict dissipativity for augmented systems replacing conditions for non augmented third order systems [91–98].

The situation of multicomponent fluxes is investigated in Sect. 3.2, the relaxation of internal energy in Sect. 3.3, and diffuse interface fluids are addressed in Sect. 3.4.

3.2 Multicomponent Flows

Multicomponent reactive flows with complex chemistry and detailed transport phenomena arise in various scientific and engineering applications like combustion [8–10], crystal growth [11–13] and reentry into the earth's atmosphere [14, 15]. This is a strong motivation for investigating the corresponding governing equations and analyzing their mathematical structure and properties [7].

The governing equations for multicomponent reactive flows obtained from the kinetic theory of gases are first presented [5–7]. These equations may be split between conservation equations, transport fluxes, transport coefficients, thermodynamic properties and chemical production rates. The thermodynamics of ideal mixtures is naturally obtained from the kinetic theory of dilute gases and the situation of nonideal gases will be addressed in Sect. 3.4. Symmetrization of the resulting quasilinear system of equation is discussed by using entropic as well as normal variables and this yields the hyperbolic-parabolic structure. We investigate the Cauchy problem and local existence of solutions is obtained by using the hyperbolic-parabolic system in normal form [7, 28, 30]. New results are obtained in the situation of small transport coefficients. Global existence theorems around constant equilibrium states are obtained as well as asymptotic stability [7, 28–31].

A similar mathematical structure as well as existence results have also been obtained for ambipolar plasmas, where Poisson Equation is replaced by the zero current limit, and the smooth dependence on parameters has also been established [32]. The situations of partial equilibrium flows where a subset of chemical reactions are assumed to be infinitely fast [36], Saint-Venant equations modeling thin viscous layers over fluid substrates [99], and fluids in vibrational nonequilibrium [30] have also been investigated. The convergence towards equilibrium flows in the fast chemistry limit has further been established [54] as well as the existence of travelling wave solutions with complex chemistry and detailed transport [100]. Many important advances have also been made recently concerning the existence of weak solutions with large data and we refer to [33–35, 101–107].

3.2.1 Governing Equations

The equations governing multicomponent reactive flows express the conservation of species mass, momentum and energy and may be written in the form [7]

$$\partial_t \rho_k + \nabla\cdot(\rho_k \boldsymbol{v}) + \nabla\cdot\mathcal{F}_k = m_k \omega_k, \qquad k \in \mathfrak{S}, \tag{3.1}$$

$$\partial_t(\rho \boldsymbol{v}) + \nabla\cdot(\rho \boldsymbol{v}\otimes\boldsymbol{v} + p\boldsymbol{I}) + \nabla\cdot\boldsymbol{\Pi} = \rho \boldsymbol{b}, \tag{3.2}$$

$$\partial_t(\mathcal{E} + \tfrac{1}{2}\rho \boldsymbol{v}\cdot\boldsymbol{v}) + \nabla\cdot\big((\mathcal{E} + \tfrac{1}{2}\rho \boldsymbol{v}\cdot\boldsymbol{v} + p)\boldsymbol{v}\big) + \nabla\cdot(\boldsymbol{Q} + \boldsymbol{\Pi}\cdot\boldsymbol{v}) = \rho \boldsymbol{b}\cdot\boldsymbol{v}, \tag{3.3}$$

where ∂_t denotes the time derivative, ∇ the space derivative operator, ρ_k the mass density of the kth species assumed to be positive, $\boldsymbol{v}$ the mass average flow velocity, $\mathcal{F}_k$ the diffusion flux of the kth species, m_k the molar mass of the kth species, ω_k the molar production rate of the kth species, $\mathfrak{S} = \{1, \dots, n\}$ the set of species indices, $n \geqslant 1$ the number of species, $\rho = \sum_{k\in\mathfrak{S}} \rho_k$ the total mass density, $\otimes$ the tensor product symbol, p the pressure, $\boldsymbol{\Pi}$ the viscous tensor, $\boldsymbol{b}$ the force per unit mass acting on the species assumed to be species-independent, $\mathcal{E}$ the internal energy per unit volume and $\boldsymbol{Q}$ the heat flux. The species mass fluxes $\mathcal{F}_k$, $k \in \mathfrak{S}$, may further be written as $\mathcal{F}_k = \rho_k \mathbf{v}_k$, $k \in \mathfrak{S}$, where $\mathbf{v}_k$ is the species diffusion velocity of the kth species. The spatial dimension is denoted by d and the components of $\boldsymbol{v}$ and ∇ are $\boldsymbol{v} = (v_1, \dots, v_d)^t$ and $\nabla = (\partial_1, \dots, \partial_d)^t$ where v_i denotes the velocity in the ith spatial direction, ∂_i the derivation in the ith spatial direction and bold symbols are used for vector or tensor quantities in the physical space $\mathbb{R}^d$. These equations have to be completed by the relations expressing thermodynamic properties like p and $\mathcal{E}$, the chemical production rates ω_k, $k \in \mathfrak{S}$, and the transport fluxes $\boldsymbol{\Pi}$, $\mathbf{v}_k$, $k \in \mathfrak{S}$, and $\boldsymbol{Q}$.

The transport fluxes $\boldsymbol{\Pi}$, $\mathbf{v}_k$, $k \in \mathfrak{S}$, and $\boldsymbol{Q}$ due to macroscopic variable gradients may be written [5–7]

$$\boldsymbol{\Pi} = -\kappa\, \nabla\cdot\boldsymbol{v}\boldsymbol{I} - \eta\big(\nabla\boldsymbol{v} + \nabla\boldsymbol{v}^t - \tfrac{2}{3}(\nabla\cdot\boldsymbol{v})\boldsymbol{I}\big), \tag{3.4}$$

$$\mathbf{v}_k = -\sum_{l\in\mathfrak{S}} D_{kl}\widehat{\boldsymbol{d}}_l - \theta_k \nabla \log T, \qquad k \in \mathfrak{S}, \tag{3.5}$$

$$\boldsymbol{Q} = -\widehat{\lambda}\nabla T - p\sum_{k\in\mathfrak{S}} \theta_k \widehat{\boldsymbol{d}}_i + \sum_{k\in\mathfrak{S}} h_k \rho_k \mathbf{v}_k, \tag{3.6}$$

where κ denotes the volume viscosity, η the shear viscosity, $\boldsymbol{I}$ the d-dimensional identity tensor, D_{kl}, $k, l \in \mathfrak{S}$, the multicomponent diffusion coefficients, $\widehat{\boldsymbol{d}}_k$, $k \in \mathfrak{S}$, the unconstrained species diffusion driving forces, ^{t} the transposition operator, θ_k, $k \in \mathfrak{S}$, the thermal diffusion coefficients, h_k, $k \in \mathfrak{S}$, the species enthalpies per unit mass and $\widehat{\lambda}$ the partial thermal conductivity. The diffusion driving forces are given by

$$\widehat{\boldsymbol{d}}_k = \frac{\nabla p_k}{p}, \quad k \in \mathfrak{S},$$

where p_k is the partial pressure of the kth species, keeping in mind that the force acting on the species are species independent. Other alternative expressions also exist for multicomponent transport fluxes [5–7, 108–114]. The fluxes (3.5) and (3.6) appear to have a complex structure with a multicomponent diffusion matrix D that is generally full and irreducible and with the Soret and Dufour cross effects [5, 6, 108–114].

One of the strength of the kinetic theory is to provide the transport coefficients κ, η, $D = (D_{kl})_{k,l\in\mathfrak{S}}$, $\widehat{\lambda}$, and $\theta = (\theta_1, \dots, \theta_n)^t$, in terms of molecular properties. Such coefficients are first expressed in terms of solutions of systems of linearized Boltzmann equations. The linearized Boltzmann equation systems are then solved by using Galerkin type methods in such a way that the transport coefficients are obtained from solutions of transport linear systems [5, 6, 108–113]. These linear systems inherit important mathematical properties from the underlying kinetic framework [7, 108–113]. The transport linear systems may notably be solved by using efficient iterative techniques [108–113]. The resulting transport coefficients satisfy the symmetry properties, i.e., the Onsager reciprocal relations, the mass conservation constraints, as well as the required positivity properties and are smooth functions of the state variables.

The thermodynamic properties are written in terms of the state variables $\rho_1, \dots, \rho_n, T$ where T is the absolute temperature. Other state variables may be used as well and lead to slightly different mathematical formalisms [7]. The internal energy per unit volume $\mathcal{E}$, the pressure p, and the entropy per unit volume $\mathcal{S}$ are given by

$$\mathcal{E} = \sum_{k\in\mathfrak{S}} \rho_k e_k, \qquad p = \sum_{k\in\mathfrak{S}} RT\frac{\rho_k}{m_k}, \qquad \mathcal{S} = \sum_{k\in\mathfrak{S}} \rho_k s_k,$$

where e_k is the internal energy per unit mass of the kth species, R the gas constant, and s_k is the entropy per unit mass of the kth species. The internal energy e_k of the kth species is given by

$$e_k(T) = e_k^{\mathrm{st}} + \int_{T^{\mathrm{st}}}^{T} c_{\mathrm{v}k}(\tau)\, d\tau, \qquad k \in \mathfrak{S},$$

where e_k^{st} is the standard formation energy of the kth species at the standard temperature T^{st} and $c_{\mathrm{v}k}$ the constant volume specific heat of the kth species. The formation energy at zero temperature $e_k^0 = e_k(0)$ is then defined by $e_k^0 = e_k^{\mathrm{st}} - \int_0^{T^{\mathrm{st}}} c_{\mathrm{v}k}(\tau)\, d\tau$. The entropy per unit mass of the kth species reads

$$s_k(T, \rho_k) = s_k^{\mathrm{st}} + \int_{T^{\mathrm{st}}}^{T} \frac{c_{\mathrm{v}k}(T')}{T'}\, dT' - \frac{R}{m_k} \log\Big(\frac{\rho_k}{\rho_k^{\mathrm{st}}}\Big), \qquad k \in \mathfrak{S},$$

where s_k^{st} is the formation entropy of the kth species at the standard temperature T^{st} and standard pressure $p^{\mathrm{st}} = p^{\mathrm{atm}}$ and where $\rho_k^{\mathrm{st}} = p^{\mathrm{st}} m_k / RT^{\mathrm{st}}$ is the standard density of the kth species. One can also introduce the enthalpy per unit volume $\mathcal{H} = \sum_{k\in\mathfrak{S}} \rho_k h_k(T)$ with $h_k(T) = e_k(T) + RT/m_k$, $k \in \mathfrak{S}$, the Gibbs function per unit volume $\mathcal{G} = \sum_{k\in\mathfrak{S}} \rho_k g_k(T, \rho_k)$, with $g_k(T, \rho_k) = h_k(T) - T s_k(T, \rho_k)$, $k \in \mathfrak{S}$ as well as the reduced chemical potentials $\mu_k(T, \rho_k) = m_k g_k/RT$, $k \in \mathfrak{S}$. Introducing the potentials at unit concentrations $\mu_k^{\mathrm{u}}(T) = \mu_k(T, m_k)$, $k \in \mathfrak{S}$, the reduced chemical potentials may be written $\mu_k(T, \rho_k) = \mu_k^{\mathrm{u}}(T) + \log(\rho_k/m_k)$, $k \in \mathfrak{S}$. The species mass fractions y_k, partial pressures p_k, and mole fractions x_k, are then defined by $\mathsf{y}_k = \rho_k/\rho$, $p_k = \rho_k RT/m_k$, and $\mathsf{x}_k = p_k/p$, for $k \in \mathfrak{S}$. The species mole fractions may also be evaluated with $\mathsf{x}_k = m\mathsf{y}_k/m_k$ where m is the mean molar weight given by $\sum_{k\in\mathfrak{S}} \mathsf{y}_k/m = \sum_{k\in\mathfrak{S}} \mathsf{y}_k/m_k$. The mass vector is also defined by $\mathfrak{m} = \big(m_1, \dots, m_n\big)^t$ and the density vector by $\varrho = \big(\rho_1, \dots, \rho_n\big)^t$.

The species of the mixture are assumed to be constituted by atoms, and we denote by $\mathfrak{a}_{kl}$ the number of lth atom in the kth species, $\mathfrak{A} = \{1, \dots, n^{\mathrm{a}}\}$ the set of atom indices, and $n^{\mathrm{a}} \geqslant 1$ the number of atoms—or elements—in the mixture. The atom vectors $\mathfrak{a}_l$ are defined by $\mathfrak{a}_l = \big(\mathfrak{a}_{1l}, \dots, \mathfrak{a}_{nl}\big)^t$ for $l \in \mathfrak{A}$ and assumed to be linearly independent. We denote by $\mathcal{A}$ the vector space spanned by the atom vectors $\mathcal{A} = \mathrm{Span}\{\,\mathfrak{a}_l,\ l \in \mathfrak{A}\,\}$ and by $M = \mathrm{diag}(m_1, \dots, m_n)$ the mass matrix. Denoting by $\widetilde{m}_l$ the mass of the lth atom, we have the relations $m_k = \sum_{l\in\mathfrak{A}} \widetilde{m}_l \mathfrak{a}_{kl}$ in such a way that $\mathfrak{m} = \sum_{l\in\mathfrak{A}} \widetilde{m}_l \mathfrak{a}_l$ and $\mathbb{I} \in M^{-1}\mathcal{A}$ with $\mathbb{I} = (1, \dots, 1)^t \in \mathbb{R}^n$. We will denote by $\langle a, b\rangle$ the Euclidean scalar product between two vectors a and b.

We also consider—elementary—chemical reactions between the species that may be written formally

$$\sum_{k\in\mathfrak{S}} \nu_{ki}^{\mathrm{f}}\ \mathfrak{M}_k \rightleftarrows \sum_{k\in\mathfrak{S}} \nu_{ki}^{\mathrm{b}}\ \mathfrak{M}_k, \qquad i \in \mathfrak{R},$$

where $\mathfrak{M}_k$ is the chemical symbol of the kth species, ν_{ki}^{f} and ν_{ki}^{b} the forward and backward stoichiometric coefficients of the kth species in the ith chemical reaction, $\mathfrak{R} = \{1, \dots, n^{\mathrm{r}}\}$ the set of reaction indices, $n^{\mathrm{r}} \geqslant 1$ the number of chemical reactions, and $\nu_{ki} = \nu_{ki}^{\mathrm{b}} - \nu_{ki}^{\mathrm{f}}$ the overall stoichiometric coefficients of the kth species in the ith reaction. We introduce for convenience the vectors $\omega = \big(\omega_1, \dots, \omega_n\big)^t$, $\nu_i = (\nu_{1i}, \dots, \nu_{ni})^t$, $\nu_i^{\mathrm{f}} = (\nu_{1i}^{\mathrm{f}}, \dots, \nu_{ni}^{\mathrm{f}})^t$, $\nu_i^{\mathrm{b}} = (\nu_{1i}^{\mathrm{b}}, \dots, \nu_{ni}^{\mathrm{b}})^t$, $\mu = (\mu_1, \dots, \mu_n)^t$, $\mu^{\mathrm{u}} = (\mu_1^{\mathrm{u}}, \dots, \mu_n^{\mathrm{u}})^t$, and denote by $\mathcal{R}$ the vector space spanned by the reaction vectors $\mathcal{R} = \mathrm{Span}\{\,\nu_i,\ i \in \mathfrak{R}\,\}$. The molar production rates that we consider are the Maxwellian production rates obtained from the kinetic theory when the chemical characteristic times are larger than the mean free times of the molecules and the characteristic times of internal energy relaxation [7]. These rates ω_k, $k \in \mathfrak{S}$, are in the form

$$\omega_k = \sum_{i\in\mathfrak{R}} (\nu_{ki}^{\mathrm{b}} - \nu_{ki}^{\mathrm{f}})\tau_i, \qquad k \in \mathfrak{S}, \tag{3.7}$$

so that $\omega = \sum_{i\in\mathfrak{R}} \tau_i \nu_i$ where τ_i is the rate of progress of the ith reaction

$$\tau_i = \mathcal{K}_i^{\mathrm{s}}\big(\exp\langle\mu, \nu_i^{\mathrm{f}}\rangle - \exp\langle\mu, \nu_i^{\mathrm{b}}\rangle\big). \tag{3.8}$$

Such an expression for τ_i was first obtained by Marcelin [115, 116], later by Keizer [3], and then deduced from the kinetic theory [7, 117]. Using the relations $\mu_k(T, \rho_k) = \mu_k^{\mathrm{u}}(T) + \log(\rho_k/m_k)$, $k \in \mathfrak{S}$, the rates of progress are found to be compatible with the law of mass action with

$$\tau_i = \mathcal{K}_i^{\mathrm{f}} \prod_{l\in\mathfrak{S}} \left(\frac{\rho_l}{m_l}\right)^{\nu_{li}^{\mathrm{f}}} - \mathcal{K}_i^{\mathrm{b}} \prod_{l\in\mathfrak{S}} \left(\frac{\rho_l}{m_l}\right)^{\nu_{li}^{\mathrm{b}}}, \tag{3.9}$$

where $\mathcal{K}_i^{\mathrm{f}} = \mathcal{K}_i^{\mathrm{s}} \exp\langle \mu^{\mathrm{u}}, \nu_i^{\mathrm{f}}\rangle$ and $\mathcal{K}_i^{\mathrm{b}} = \mathcal{K}_i^{\mathrm{s}} \exp\langle \mu^{\mathrm{u}}, \nu_i^{\mathrm{b}}\rangle$ are the forward and backward rate constants of the ith reaction. The reaction constants $\mathcal{K}_i^{\mathrm{f}}$, $\mathcal{K}_i^{\mathrm{b}}$ and $\mathcal{K}_i^{\mathrm{s}}$ are functions of temperature and are Maxwellian averaged values of molecular chemical transition probabilities [7]. We also obtain the reciprocity relations between the reaction constants

$$\mathcal{K}_i^{\mathrm{e}}(T) = \frac{\mathcal{K}_i^{\mathrm{f}}(T)}{\mathcal{K}_i^{\mathrm{b}}(T)}, \qquad \log \mathcal{K}_i^{\mathrm{e}}(T) = -\langle \nu_i, \mu^{\mathrm{u}}\rangle, \qquad i \in \mathfrak{R}, \tag{3.10}$$

where $\mathcal{K}_i^{\mathrm{e}}(T)$ is the equilibrium constant of the ith reaction [7]. These reciprocity relations (3.10) may be seen as Onsager relations for chemistry and yield symmetric representations of the rates of progress. In practice, the forward reaction constants $\mathcal{K}_i^{\mathrm{f}}$, $i \in \mathfrak{R}$, are often approximated with Arrhenius law.

3.2.2 *Mathematical Framework*

The assumptions on the thermodynamic properties and on the transport coefficients may be extracted from the kinetic theory of dilute gases. These assumptions are recast in the following form (H_1–H_5) where the regularity class γ denotes an integer such that $\gamma \geq 3$. Throughout these notes, for any matrix A, we denote by $N(A)$ its nullspace, $R(A)$ its range, and $|A|$ its Frobenius norm given by $|A|^2 = \sum_{1\leq i,j\leq d} a_{ij}^2$ when $A = (a_{ij})_{i,j\in\mathfrak{S}}$.

(H_1) The molar masses m_k, $k \in \mathfrak{S}$, and the perfect gas constant R are positive constants. The formation energies e_k^{st}, $k \in \mathfrak{S}$, and entropies s_k^{st}, $k \in \mathfrak{S}$, are real constants. The specific heats $c_{\mathrm{v}k}$, $k \in \mathfrak{S}$, are $C^{\gamma-1}$ functions of temperature $T \in [0, \infty)$. There exist positive constants $\underline{c}_{\mathrm{v}}$ and $\overline{c}_{\mathrm{v}}$ such that $0 < \underline{c}_{\mathrm{v}} \leqslant c_{\mathrm{v}k}(T) \leqslant \overline{c}_{\mathrm{v}}$ for $T \geq 0$ and $k \in \mathfrak{S}$.

(H_2) The stoichiometric coefficients ν_{ki}^{f} and ν_{ki}^{b}, $k \in \mathfrak{S}$, $i \in \mathfrak{R}$, the atomic coefficients $\mathfrak{a}_{kl}$, $k \in \mathfrak{S}$, $l \in \mathfrak{A}$, are nonnegative integers. The atom vectors $\mathfrak{a}_l$, $l \in \mathfrak{A}$, are linearly independent and satisfy the atom conservation constraints $\langle \nu_i, \mathfrak{a}_l\rangle = 0$, $i \in \mathfrak{R}$, $l \in \mathfrak{A}$, where $\nu_i = \nu_i^{\mathrm{b}} - \nu_i^{\mathrm{f}}$, $i \in \mathfrak{R}$, are the reaction vectors. The atom masses $\widetilde{m}_l$, $l \in \mathfrak{A}$, are positive constants and the mass vector $\mathfrak{m}$ is given by $\mathfrak{m} = \sum_{l\in\mathfrak{A}} \widetilde{m}_l\, \mathfrak{a}_l$.

(H$_3$) The reaction constants $\mathcal{K}_i^{\mathrm{s}}$ are C^γ positive functions of $T > 0$ for $i \in \mathfrak{R}$.

(H$_4$) The multicomponent diffusion matrix $D = (D_{kl})_{k,l\in\mathfrak{S}}$, the thermal diffusion vector $\theta = (\theta_1, \dots, \theta_n)^t$, the volume viscosity κ, the shear viscosity η, and the partial thermal conductivity $\widehat{\lambda}$ are C^γ functions of $(\rho_1, \dots, \rho_n, T)$ for $T > 0$ and $\rho_i > 0$, $i \in \mathfrak{S}$. The coefficients satisfy the mass conservation constraints $N(D) = \mathbb{R}\mathsf{y}$, $R(D) = \mathsf{y}^\perp$, and $\theta \in \mathsf{y}^\perp$ where $\mathsf{y} = (\mathsf{y}_1, \dots, \mathsf{y}_n)^t$.

(H$_5$) The diffusion matrix D is symmetric positive semidefinite and its nullspace is $N(D) = \mathbb{R}\mathsf{y}$. Denoting by χ the unique solution of the linear system $\theta = D\chi$, under the constraint $\langle \mathbb{I}, \chi\rangle = 0$, the thermal conductivity given by $\lambda = \widehat{\lambda} - (p/T)\langle\theta, \chi\rangle$ is positive. The shear viscosity η is positive and the volume viscosity κ is nonnegative.

Assumptions (H$_1$ − H$_5$) are assumed to hold in the remaining part of the section whenever we consider multicomponent flows. All coefficients $\widehat{\lambda}$, κ η, and θ have smooth extensions to the domain $\rho_i \ge 0$, $i \in \mathfrak{S}$, and $\rho > 0$. This is also the case for the non diagonal coefficients D_{ij} for $i \neq j$ as well as the rescaled coefficients $\rho_i D_{ii}$ that have a finite positive limit when $\rho_i \to 0$ [109, 110]. The limiting behavior of transport coefficients for zero mass fractions lay out of the scope of the present paper and we refer to [7, 108–113]. We generally have $\mathcal{R} \subset \mathcal{A}^\perp$ but chemical reaction mechanisms are usually sufficiently rich so that reaction vectors ν_i, $i \in \mathfrak{R}$, are spanning the maximum space and $\mathcal{R} = \mathcal{A}^\perp$.

Using the volumetric Gibbs relation for multicomponent mixtures

$$\mathbb{D}\mathcal{S} = \frac{1}{T}\mathbb{D}\mathcal{E} - \sum_{k\in\mathfrak{S}} \frac{g_k}{T}\mathbb{D}\rho_k, \tag{3.11}$$

where $\mathbb{D}$ denotes the total derivative, as well as the conservation equations, the following balance equation for the entropy $\mathcal{S}$ is obtained

$$\begin{aligned}
\partial_t \mathcal{S} + \nabla\cdot(\boldsymbol{v}\mathcal{S}) + \nabla\cdot\Big(\frac{\boldsymbol{Q}}{T} - \sum_{k\in\mathfrak{S}}\frac{g_k}{T}\mathcal{F}_k\Big) &= \frac{1}{T}\Big(\kappa + \eta\big(\tfrac{2}{d} - \tfrac{2}{3}\big)\Big)(\nabla\cdot\boldsymbol{v})^2 \\
&+ \frac{\eta}{2T}\big|\nabla\boldsymbol{v} + \nabla\boldsymbol{v}^t - \tfrac{2}{d}(\nabla\cdot\boldsymbol{v})\,\boldsymbol{I}\big|^2 \\
&+ \frac{\lambda}{T^2}\nabla T\cdot\nabla T + \frac{p}{T}\sum_{k,l\in\mathfrak{S}} D_{kl}\big(\widehat{\boldsymbol{d}}_k + \chi_k\nabla\log T\big)\cdot\big(\widehat{\boldsymbol{d}}_l + \chi_l\nabla\log T\big) \\
&+ \sum_{i\in\mathfrak{R}} R\mathcal{K}_i^{\mathrm{s}}\big(\langle\mu, \nu_i^{\mathrm{f}}\rangle - \langle\mu, \nu_i^{\mathrm{b}}\rangle\big)\,\big(\exp\langle\mu, \nu_i^{\mathrm{f}}\rangle - \exp\langle\mu, \nu_i^{\mathrm{b}}\rangle\big).
\end{aligned} \tag{3.12}$$

The viscous tensor has been rewritten for convenience in the form

$$\boldsymbol{\Pi} = -\Big(\kappa + \eta\big(\tfrac{2}{d} - \tfrac{2}{3}\big)\Big)\nabla\cdot\boldsymbol{v}\,\boldsymbol{I} - \eta\Big(\nabla\boldsymbol{v} + \nabla\boldsymbol{v}^t - \tfrac{2}{d}(\nabla\cdot\boldsymbol{v})\,\boldsymbol{I}\Big),$$

keeping in mind that $1 \leq d \leq 3$. Entropy production in the right hand side of (3.12) thus appears to be a sum of nonnegative terms. The last term in (3.12) represents the entropy production due to chemical reactions $-R\langle\mu,\omega\rangle$. This expression of $\langle\mu,\omega\rangle$ allows to obtain the following equivalence results concerning chemical equilibrium.

Proposition 3.1 *Assume that the Properties (H_1–H_3) hold. Then for any thermodynamic state $(\rho_1,\ldots,\rho_n,T)\in(0,\infty)^{n+1}$ the following statements are equivalent.*

(i) *The entropy production due to chemistry vanishes* $-R\langle\mu,\omega\rangle = 0$.
(ii) *The reaction rates of progress vanish* $\tau_j = 0, \quad j\in\mathfrak{R}$.
(iii) *The species production rates vanish* $\omega_k = 0, \quad k\in\mathfrak{S}$.
(iv) *The reduced potential vector μ belongs to $\mathcal{R}^\perp$.*

A state $(\rho_1,\ldots,\rho_n,T)^t\in(0,\infty)^{n+1}$ that satisfies these equivalent properties is termed an equilibrium state. We are only interested here in positive equilibrium states such that $\rho_i > 0$, $i\in\mathfrak{S}$, that are in the interior of the composition space. Spurious points with zero mass fractions where the source terms ω_k, $k\in\mathfrak{S}$, also vanish—termed boundary equilibrium points—are of a different nature [7]. Properly structured chemical kinetic mechanisms automatically exclude such spurious points unless some element is missing in the mixture [7]. The existence of equilibrium points may be established by minimizing appropriate thermodynamic functions over atom conservation simplexes and we have in particular the following proposition [7, 31].

Proposition 3.2 *For $T^\star > 0$ and $\overline{\varrho} = (\overline{\rho}_1,\ldots,\overline{\rho}_n)^t\in(0,\infty)^n$ there exists a unique equilibrium point $\varrho^\star = (\rho_1^\star,\ldots,\rho_n^\star)^t\in(0,\infty)^n$ such that $\varrho^\star - \overline{\varrho}\in M\mathcal{R}$.*

Various stability inequalities may then be obtained near equilibrium as in the following proposition. Such inequalities may be extended over atom conservation manifolds provided no boundary equilibrium point exist [7].

Proposition 3.3 *Let $(\rho_1^\star,\ldots,\rho_n^\star,T^\star)\in(0,\infty)^{n+1}$ be an equilibrium state. Then, there exists a positive constant δ and a neighborhood of $(\rho_1^\star,\ldots,\rho_n^\star,T^\star)$ such that*

$$\delta\,\big|\omega\big|^2 \leq -\big\langle\mu-\mu^\star,\omega\big\rangle.$$

The production rates ω may further be written in a quasilinear form as a direct result of the expression of the rates of progress in terms of the reduced chemical potentials.

Lemma 3.1 *The chemical source term $\omega\in\mathbb{R}^n$ may be written*

$$\omega = -\Lambda\mu, \tag{3.13}$$

where the matrix $\Lambda\in\mathbb{R}^{n,n}$ is given by

$$\Lambda = \sum_{i\in\mathfrak{R}}\Lambda_i\,\nu_i\otimes\nu_i\,. \tag{3.14}$$

The coefficients $\Lambda_i = \mathcal{K}_i^{\mathrm{s}} \int_0^1 \exp\big(\langle \nu_i^{\mathrm{f}}, \mu\rangle + \xi\langle \nu_i^{\mathrm{b}} - \nu_i^{\mathrm{f}}, \mu\rangle\big)\, d\xi$ *are positive, the matrix* Λ *is symmetric positive semi-definite with nullspace* $N(\Lambda) = \mathcal{R}^{\perp}$ *and range* $R(\Lambda) = \mathcal{R}$ *in* $\mathbb{R}^n$ *so that* $N(\Lambda) = \mathcal{A}$ *when* $\mathcal{A} = \mathcal{R}^{\perp}$.

The equations governing multicomponent flows (3.1)–(3.3) are rewritten for mathematical convenience in the compact form

$$\partial_t \mathsf{u} + \sum_{i\in\mathfrak{D}} \partial_i \mathsf{F}_i + \sum_{i\in\mathfrak{D}} \partial_i \mathsf{F}_i^{\mathrm{d}} = \Omega, \tag{3.15}$$

where u is the conservative variable, ∂_i the spatial derivative operator in the ith spatial direction, $\mathfrak{D} = \{1, \ldots, d\}$ the indexing set of spatial directions, $d \in \{1, 2, 3\}$ the spatial dimension, F_i the convective flux in the ith direction, $\mathsf{F}_i^{\mathrm{d}}$ the dissipative flux in the ith direction, and Ω the source term. Letting $\mathsf{n} = n + d + 1$, the conservative variable $\mathsf{u} \in \mathbb{R}^{\mathsf{n}}$ and natural variable $\mathsf{z} \in \mathbb{R}^{\mathsf{n}}$ are given by

$$\mathsf{u} = \big(\rho_1, \ldots, \rho_n, \rho\boldsymbol{v}, \mathcal{E} + \tfrac{1}{2}\rho\boldsymbol{v}\cdot\boldsymbol{v}\big)^t, \qquad \mathsf{z} = \big(\rho_1, \ldots, \rho_n, \boldsymbol{v}, T\big)^t. \tag{3.16}$$

The velocity components of vectors in $\mathbb{R}^{\mathsf{n}} = \mathbb{R}^n \times \mathbb{R}^d \times \mathbb{R}$ are written as vectors of $\mathbb{R}^d$ and bold symbols are used for vector or tensor quantities in the physical space $\mathbb{R}^d$. The map $\mathsf{z} \to \mathsf{u}$ is a C^{γ} diffeomorphism from the open set $O_{\mathsf{z}} = (0, \infty)^n \times \mathbb{R}^d \times (0, \infty)$ onto a convex open set O_{u} of $\mathbb{R}^{\mathsf{n}}$ [7, 31].

The convective fluxes F_i, diffusive fluxes $\mathsf{F}_i^{\mathrm{d}}$ and source term Ω are given by

$$\mathsf{F}_i = \big(\rho_1 v_i, \ldots, \rho_n v_i, \rho v_i \boldsymbol{v} + p\boldsymbol{e}_i, (\mathcal{E} + \tfrac{1}{2}\rho|\boldsymbol{v}|^2 + p)v_i\big)^t,$$

$$\mathsf{F}_i^{\mathrm{d}} = \big(\mathcal{F}_{1i}, \ldots, \mathcal{F}_{ni}, \boldsymbol{\Pi}_i, Q_i + \boldsymbol{\Pi}_i\cdot\boldsymbol{v}\big)^t,$$

$$\Omega = \big(m_1\omega_1, \ldots, m_n\omega_n, \mathbf{0}, 0\big)^t,$$

where $\boldsymbol{e}_i$ denotes the ith basis vector in the physical space $\mathbb{R}^d$, v_i the velocity in the ith direction, $\mathcal{F}_{ki}$ the diffusion flux of the kth species in the ith direction, Q_i the heat flux in the ith direction, $\boldsymbol{\Pi} = (\Pi_{ij})_{i,j\in\mathfrak{D}}$ the viscous tensor, and $\boldsymbol{\Pi}_i$ the vector $\boldsymbol{\Pi}_i = (\Pi_{1i}, \ldots, \Pi_{di})^t$, so that $\boldsymbol{v} = (v_1, \ldots, v_d)^t$, $\mathcal{F}_k = \rho_k \mathbf{v}_k = (\mathcal{F}_{k1}, \ldots, \mathcal{F}_{kd})^t$, and $\boldsymbol{Q} = (Q_1, \ldots, Q_d)^t$. Since the diffusive fluxes, the heat flux and the viscous tensor are linear expressions in terms of the fluid variable gradients, they may be expressed in the form

$$\mathsf{F}_i^{\mathrm{d}} = -\sum_{j\in\mathfrak{D}} \mathsf{B}_{ij}(\mathsf{u})\partial_j \mathsf{u}, \qquad i \in \mathfrak{D},$$

where B_{ij}, $i, j \in \mathfrak{D}$, are uniquely defined matrices. The dissipation matrix B_{ij} relates the dissipative flux in the ith direction $\mathsf{F}_i^{\mathrm{d}}$ to the gradient of the conservative variable u in the jth spatial direction $\partial_j \mathsf{u}$. Denoting by $\mathsf{A}_i = \partial_{\mathsf{u}} \mathsf{A}_i$, $i \in \mathfrak{D}$, the convective flux Jacobian matrices we finally obtain the quasilinear system

$$\partial_t \mathsf{u} + \sum_{i\in\mathfrak{D}} \mathsf{A}_i(\mathsf{u})\partial_i \mathsf{u} = \sum_{i,j\in\mathfrak{D}} \partial_i \big(\mathsf{B}_{ij}(\mathsf{u})\partial_j \mathsf{u}\big) + \Omega(\mathsf{u}), \tag{3.17}$$

and the system coefficients $\mathsf{A}_i(\mathsf{u})$, $i \in \mathfrak{D}$, $\mathsf{B}_{ij}(\mathsf{u})$, $i, j \in \mathfrak{D}$, and $\Omega(\mathsf{u})$, are smooth functions of u on the open convex set O_{u}. The hyperbolic-parabolic mathematical structure of this quasilinear system is discussed in the next section.

3.2.3 Hyperbolic-parabolic Structure

We discuss symmetrization of the quasilinear system (3.17) by using entropic and normal variables. To this aim, a mathematical entropy σ is introduced that is compatible with convective, diffusive as well as source terms [118, 124]. In the following definition, properties ($\mathsf{E}_1-\mathsf{E}_2$) concerning the convective terms have been adapted from [16, 17], properties ($\mathsf{E}_3 - \mathsf{E}_4$) of dissipative terms have been adapted from [18, 19, 22, 27], properties ($\mathsf{E}_5 - \mathsf{E}_7$) have been adapted from [48] and we denote by Σ^{d-1} the sphere in d dimension. All properties ($\mathsf{E}_1 - \mathsf{E}_7$) may be linked to the kinetic theory derivation, the properties of linearized Boltzmann collision operators as well as to the reciprocity relations between transition probabilities.

Definition 3.1 Consider a C^γ function $\mathsf{u} \to \sigma(\mathsf{u})$ defined over the open convex domain O_{u}. The function σ is said to be an entropy function for the system (3.17) if the following properties hold.

(E_1) The Hessian matrix $\partial^2_{\mathsf{u}}\sigma = \partial_{\mathsf{u}}(\partial_{\mathsf{u}}\sigma)^t$ is positive definite over O_{u}.
(E_2) There exist C^γ functions $\mathsf{u} \to \mathsf{q}_i$, $i \in \mathfrak{D}$, such that $\partial_{\mathsf{u}}\sigma\, \mathsf{A}_i = \partial_{\mathsf{u}}\mathsf{q}_i$ over O_{u}.
(E_3) We have $\big(\mathsf{B}_{ij}\,(\partial^2_{\mathsf{u}}\sigma)^{-1}\big)^t = \mathsf{B}_{ji}\,(\partial^2_{\mathsf{u}}\sigma)^{-1}$ for $i, j \in \mathfrak{D}$ over O_{u}.
(E_4) The matrix $\sum_{i,j\in\mathfrak{D}} \mathsf{B}_{ij}\,(\partial^2_{\mathsf{u}}\sigma)^{-1}\,\xi_i\xi_j$ where $\boldsymbol{\xi} \in \Sigma^{d-1}$ is positive semi-definite over $\mathsf{u} \in O_{\mathsf{u}}$.
(E_5) There exists a fixed vector space $\mathcal{E} \subset \mathbb{R}^{\mathsf{n}}$ such that $\Omega \in \mathcal{E}^\perp$ for $\mathsf{u} \in O_{\mathsf{u}}$ and $\Omega = 0$ if and only if $(\partial_{\mathsf{u}}\sigma)^t \in \mathcal{E}$ and if and only if $\partial_{\mathsf{u}}\sigma\ \Omega = 0$.
(E_6) If $\Omega = 0$, then the matrix $\partial_{\mathsf{u}}\Omega\,(\partial^2_{\mathsf{u}}\sigma)^{-1}$ is symmetric and its nullspace is given by $N\big(\partial_{\mathsf{u}}\Omega\,(\partial^2_{\mathsf{u}}\sigma)^{-1}\big) = \mathcal{E}$.
(E_7) We have $\partial_{\mathsf{u}}\sigma\ \Omega \le 0$ over O_{u}.

Existence of an entropy is closely associated with symmetrization properties [16–19, 22, 27, 48, 52]. We do not encounter here the difficulty associated with nonideal fluids where only *local* symmetrizations are feasible and where $O_{\cdot}$ may not be convex [124].

Definition 3.2 Consider a $C^{\gamma-1}$ diffeomorphism $\mathsf{u} \to \mathsf{v}$ from O_{u} onto an open domain O_{v} and the system in the v variable

$$\widetilde{\mathsf{A}}_0(\mathsf{v})\partial_t \mathsf{v} + \sum_{i\in\mathfrak{D}} \widetilde{\mathsf{A}}_i(\mathsf{v})\partial_i \mathsf{v} - \sum_{i,j\in\mathfrak{D}} \partial_i\big(\widetilde{\mathsf{B}}_{ij}(\mathsf{v})\partial_j \mathsf{v}\big) - \widetilde{\Omega}(\mathsf{v}) = 0, \tag{3.18}$$

where $\widetilde{\mathsf{A}}_0 = \partial_{\mathsf{v}}\mathsf{u}$, $\widetilde{\mathsf{A}}_i = \mathsf{A}_i\partial_{\mathsf{v}}\mathsf{u} = \partial_{\mathsf{v}}\mathsf{F}_i$, $\widetilde{\mathsf{B}}_{ij} = \mathsf{B}_{ij}\partial_{\mathsf{v}}\mathsf{u}$, and $\widetilde{\Omega} = \Omega$, have regularity $\gamma - 2$. The system is said of the symmetric form if properties $(\mathsf{S}_1 - \mathsf{S}_7)$ hold.

(S_1) The matrix $\widetilde{\mathsf{A}}_0$ is symmetric positive definite for $\mathsf{v} \in \mathcal{O}_{\mathsf{v}}$.
(S_1) The matrices $\widetilde{\mathsf{A}}_i$, $i \in \mathfrak{D}$, are symmetric over $\mathcal{O}_{\mathsf{v}}$.
(S_3) We have $\widetilde{\mathsf{B}}_{ij}^t = \widetilde{\mathsf{B}}_{ji}$ for $i, j \in \mathfrak{D}$ over $\mathcal{O}_{\mathsf{v}}$.
(S_4) The matrix $\widetilde{\mathsf{B}}(\mathsf{v}, \boldsymbol{\xi}) = \sum_{i,j\in\mathfrak{D}} \widetilde{\mathsf{B}}_{ij}(\mathsf{v})\xi_i\xi_j$ where $\boldsymbol{\xi} \in \Sigma^{d-1}$ is positive semi-definite over $\mathsf{v} \in \mathcal{O}_{\mathsf{v}}$.
(S_5) There exists a fixed vector space $\mathcal{E} \subset \mathbb{R}^{\mathsf{n}}$ such that $\widetilde{\Omega} \in \mathcal{E}^{\perp}$ over $\mathcal{O}_{\mathsf{v}}$ and $\widetilde{\Omega}(\mathsf{v}) = 0$ if and only if $\mathsf{v} \in \mathcal{E}$ and if and only if $\langle \mathsf{v}, \widetilde{\Omega}(\mathsf{v})\rangle = 0$.
(S_6) If $\widetilde{\Omega}(\mathsf{v}) = 0$, then $\partial_{\mathsf{v}}\widetilde{\Omega}(\mathsf{v})$ is symmetric and $N\big(\partial_{\mathsf{v}}\widetilde{\Omega}(\mathsf{v})\big) = \mathcal{E}$.
(S_7) We have $\langle \mathsf{v}, \widetilde{\Omega}(\mathsf{v})\rangle \le 0$ over $\mathcal{O}_{\mathsf{v}}$.

The manifold $\mathcal{E}$ is naturally termed the equilibrium manifold since $\widetilde{\Omega}(\mathsf{v}) = 0$ when $\mathsf{v} \in \mathcal{E}$. The equivalence between symmetrization $(\mathsf{S}_1 - \mathsf{S}_7)$ and entropy $(\mathsf{E}_1 - \mathsf{E}_7)$ for hyperbolic-parabolic systems is obtained with $\mathsf{v} = (\partial_{\mathsf{u}}\sigma)^t$ [124].

Theorem 3.1 *Assume that the system* (3.17) *admits a C^γ entropy function σ defined over an open convex domain O_{u}. Then the system can be symmetrized with the entropic variable $\mathsf{v} = (\partial_{\mathsf{u}}\sigma)^t$. Conversely, assume that the system can be symmetrized with the $C^{\gamma-1}$ diffeomorphism $\mathsf{u} \to \mathsf{v}$. Then there exists a C^γ entropy σ over the open convex set O_{u} such that $\mathsf{v} = (\partial_{\mathsf{u}}\sigma)^t$.*

Sketch of the Proof The equivalence of $(\mathsf{S}_1 - \mathsf{S}_2)$ and $(\mathsf{E}_1 - \mathsf{E}_2)$ is classical and is essentially obtained with Poincaré lemma. Then $(\mathsf{S}_3 - \mathsf{S}_7)$ and $(\mathsf{E}_3 - \mathsf{E}_7)$ become identical statements with $\mathsf{v} = (\partial_{\mathsf{u}}\sigma)^t$. □

In the special situation of multicomponent flows the mathematical entropy is taken to be

$$\sigma = -\mathcal{S}/R, \tag{3.19}$$

where the $1/R$ factor is introduced for convenience.

Theorem 3.2 *Assume that $(H_1 - H_5)$ hold. Then the function $\sigma = -\mathcal{S}/R$ is a mathematical entropy for the system* (3.17) *and the entropic variable reads*

$$\mathsf{v} = (\partial_{\mathsf{u}}\sigma)^t = \frac{1}{RT}\Big(g_1 - \tfrac{1}{2}|\boldsymbol{v}|^2, \ldots, g_n - \tfrac{1}{2}|\boldsymbol{v}|^2, \boldsymbol{v}, -1\Big)^t. \tag{3.20}$$

The map $\mathsf{u} \to \mathsf{v}$ is a $C^{\gamma-1}$ diffeomorphism from O_{u} onto an open set O_{v} and the system in term of the v variable is in the symmetrized form (3.18).

The explicit form of the matrices $\widetilde{\mathsf{A}}_0$, $\widetilde{\mathsf{A}}_i$, $\widetilde{\mathsf{B}}_{ij}$ and of the source term $\widetilde{\Omega}$ that may be written in linearized form $\widetilde{\Omega} = -\widetilde{\mathsf{L}}\,\mathsf{v}$ may be found in the literature [7, 38] and the equilibrium manifold is given by $\mathcal{E} = (M\mathcal{R})^{\perp} \times \mathbb{R}^d \times \mathbb{R}$. Since we are considering ideal fluids, the symmetrizing change of variable $\mathsf{u} \to \mathsf{v}$ is one to one and is a global change of variable [7, 31]. For mixtures of ideal gases, there is also a uniqueness theorem for mathematical entropies that are independent of transport coefficients and the corresponding entropic variables are then proportional [118].

Theorem 3.3 *Let $\widetilde{\sigma}$ be a C^{γ} function defined on the open set O_{u} satisfying ($E_2 - E_3$) and such that $\partial^2_{\mathsf{u}}\widetilde{\sigma}$ is invertible. Assuming that $\widetilde{\sigma}$ is independent of the mass and heat diffusion parameters, then $\widetilde{\sigma}$ is in the form*

$$\widetilde{\sigma} = \alpha_{\mathcal{S}}\mathcal{S} + \sum_{i\in\mathfrak{S}} \alpha_i \rho_i + \alpha_{\boldsymbol{v}} \rho \boldsymbol{v} + \alpha_{\mathcal{E}}(\mathcal{E} + \tfrac{1}{2}\rho|\boldsymbol{v}|^2), \tag{3.21}$$

where $\alpha_{\mathcal{S}}$ α_i, $i \in \mathfrak{S}$, $\alpha_{\boldsymbol{v}}$ and $\alpha_{\mathcal{E}}$ are constants.

We now need to split the variables between hyperbolic and parabolic components. To this aim, we have to put the system in normal form, that is, in the form of a symmetric hyperbolic–parabolic composite system [18, 19, 22, 31].

Definition 3.3 Consider a symmetrized system as in Definition 3.2 and let $\mathsf{v} \to \mathsf{w}$ be a $C^{\gamma-1}$ diffeomorphism from the open set O_{v} onto an open set O_{w}. Letting $\mathsf{v} = \mathsf{v}(\mathsf{w})$ in the symmetrized system (3.18) and multiplying on the left side by $(\partial_{\mathsf{w}}\mathsf{v})^t$ we obtain a new system in the variable w

$$\overline{\mathsf{A}}_0(\mathsf{w})\partial_t \mathsf{w} + \sum_{i\in\mathfrak{D}} \overline{\mathsf{A}}_i(\mathsf{w})\partial_i \mathsf{w} - \sum_{i,j\in\mathfrak{D}} \partial_i\big(\overline{\mathsf{B}}_{ij}(\mathsf{w})\partial_j \mathsf{w}\big) - \overline{\Omega}(\mathsf{w}) = \overline{\mathsf{h}}(\mathsf{w}, \partial_{\mathsf{x}}\mathsf{w}), \tag{3.22}$$

where $\overline{\mathsf{A}}_0 = (\partial_{\mathsf{w}}\mathsf{v})^t\, \widetilde{\mathsf{A}}_0\, (\partial_{\mathsf{w}}\mathsf{v})$, $\overline{\mathsf{B}}_{ij} = (\partial_{\mathsf{w}}\mathsf{v})^t\, \widetilde{\mathsf{B}}_{ij}\, (\partial_{\mathsf{w}}\mathsf{v})$, $\overline{\mathsf{A}}_i = (\partial_{\mathsf{w}}\mathsf{v})^t\, \widetilde{\mathsf{A}}_i\, (\partial_{\mathsf{w}}\mathsf{v})$, $\overline{\Omega} = (\partial_{\mathsf{w}}\mathsf{v})^t\widetilde{\Omega}$, have at least regularity $\gamma - 2$ and where $\overline{\mathsf{h}}$ is quadratic in the gradients $\overline{\mathsf{h}} = -\sum_{i,j\in\mathfrak{D}} \partial_i(\partial_{\mathsf{w}}\mathsf{v})^t\, \widetilde{\mathsf{B}}_{ij}\, (\partial_{\mathsf{w}}\mathsf{v})\partial_j \mathsf{w}$. This system satisfies in particular properties ($\overline{\mathsf{S}}_1 - \overline{\mathsf{S}}_4$), that is, properties ($\mathsf{S}_1 - \mathsf{S}_4$) rewritten in terms of overbar matrices. This system (3.22) is said to be of the normal form if there exists a partition of $\{1, \dots, \mathsf{n}\}$ into $\mathrm{I} = \{1, \dots, \mathsf{n}_{\mathrm{I}}\}$ and $\mathrm{II} = \{\mathsf{n}_{\mathrm{I}} + 1, \dots, \mathsf{n}_{\mathrm{I}} + \mathsf{n}_{\mathrm{II}}\}$ with $\mathsf{n} = \mathsf{n}_{\mathrm{I}} + \mathsf{n}_{\mathrm{II}}$ such that the following properties hold.

(N_1) The matrices $\overline{\mathsf{A}}_0$ and $\overline{\mathsf{B}}_{ij}$ have the block structure

$$\overline{\mathsf{A}}_0 = \begin{bmatrix} \overline{\mathsf{A}}_0^{\mathrm{I,I}} & 0_{\mathsf{n}_{\mathrm{I}},\mathsf{n}_{\mathrm{II}}} \\ 0_{\mathsf{n}_{\mathrm{II}},\mathsf{n}_{\mathrm{I}}} & \overline{\mathsf{A}}_0^{\mathrm{II,II}} \end{bmatrix}, \qquad \overline{\mathsf{B}}_{ij} = \begin{bmatrix} 0_{\mathsf{n}_{\mathrm{I}},\mathsf{n}_{\mathrm{I}}} & 0_{\mathsf{n}_{\mathrm{I}},\mathsf{n}_{\mathrm{II}}} \\ 0_{\mathsf{n}_{\mathrm{II}},\mathsf{n}_{\mathrm{I}}} & \overline{\mathsf{B}}_{ij}^{\mathrm{II,II}} \end{bmatrix}.$$

(N_2) The matrix $\overline{\mathsf{B}}^{\mathrm{II,II}}(\mathsf{w}, \boldsymbol{\xi}) = \sum_{i,j\in\mathfrak{D}} \overline{\mathsf{B}}_{ij}^{\mathrm{II,II}}(\mathsf{w})\xi_i\xi_j$ is positive definite for $\mathsf{w} \in O_{\mathsf{w}}$ and $\boldsymbol{\xi} \in \Sigma^{d-1}$.

(N_3) We have $\overline{\mathsf{h}}(\mathsf{w}, \partial_{\mathsf{x}}\mathsf{w}) = \big(\overline{\mathsf{h}}_{\mathrm{I}}(\mathsf{w}, \partial_{\mathsf{x}}\mathsf{w}_{\mathrm{II}}), \overline{\mathsf{h}}_{\mathrm{II}}(\mathsf{w}, \partial_{\mathsf{x}}\mathsf{w})\big)^t$.
We have used here the vector and matrix block structure induced by the partitioning of $\mathbb{R}^{\mathsf{n}}$ into $\mathbb{R}^{\mathsf{n}} = \mathbb{R}^{\mathsf{n}_{\mathrm{I}}} \times \mathbb{R}^{\mathsf{n}_{\mathrm{II}}}$ so that we have $\mathsf{w} = (\mathsf{w}_{\mathrm{I}}, \mathsf{w}_{\mathrm{II}})^t$ for instance and denoted by $0_{i,j}$ the zero matrix with i lines and j columns.

A sufficient condition for system (3.18) to be recast into a normal form is that the nullspace naturally associated with dissipation matrices $\widetilde{\mathsf{B}}$ is a fixed subspace of $\mathbb{R}^{\mathsf{n}}$. This is Condition (N) introduced by Kawashima and Shizuta [22] which has been strengthened in [31].

(N) The nullspace $N(\widetilde{\mathsf{B}})$ of the matrix $\widetilde{\mathsf{B}}(\mathsf{v}, \boldsymbol{\xi}) = \sum_{i,j\in\mathfrak{D}} \widetilde{\mathsf{B}}_{ij}(\mathsf{v})\xi_i\xi_j$ does not depend on $\mathsf{v} \in \mathcal{O}_{\mathsf{v}}$ and $\boldsymbol{\xi} \in \Sigma^{d-1}$ and $\widetilde{\mathsf{B}}_{ij}(\mathsf{v})N(\widetilde{\mathsf{B}}) = 0$, for $i, j \in \mathfrak{D}$.
Letting $\mathsf{n}_{\mathrm{I}} = \dim\big(N(\widetilde{\mathsf{B}})\big)$ and $\mathsf{n}_{\mathrm{II}} = \mathsf{n} - \mathsf{n}_{\mathrm{I}}$ we denote by P an arbitrary constant nonsingular matrix of dimension n such that its first n_{I} columns span the nullspace $N(\widetilde{\mathsf{B}})$. In order to characterize more easily normal forms for symmetric systems of conservation laws satisfying N we may introduce the auxiliary variables [7, 31] $\mathsf{u}' = \mathsf{P}^t\mathsf{u}$ and $\mathsf{v}' = \mathsf{P}^{-1}\mathsf{v}$. The dissipation matrices corresponding to these auxiliary variables have nonzero coefficients only in the lower right block of size $\mathsf{n}_{\mathrm{II}} = \mathsf{n} - \mathsf{n}_{\mathrm{I}}$. Normal symmetric forms are then equivalently—and more easily—obtained from the v' symmetric equation [7, 31].

Theorem 3.4 *Consider a system of conservation laws* (3.18) *that is symmetric in the sense of Definition 3.2 and assume that the nullspace invariance property* (N) *is satisfied. Denoting by* $\mathsf{u}' = \mathsf{P}^t\mathsf{u}$ *and* $\mathsf{v}' = \mathsf{P}^{-1}\mathsf{v}$*, the auxiliary variable, any normal form of the system* (3.18) *is given by a change of variable in the form* $\mathsf{w} = \big(\mathfrak{F}_{\mathrm{I}}(\mathsf{u}'_{\mathrm{I}}), \mathfrak{F}_{\mathrm{II}}(\mathsf{v}'_{\mathrm{II}})\big)^t$ *where* $\mathfrak{F}_{\mathrm{I}}$ *and* $\mathfrak{F}_{\mathrm{II}}$ *are two diffeomorphisms of* $\mathbb{R}^{\mathsf{n}_{\mathrm{I}}}$ *and* $\mathbb{R}^{\mathsf{n}_{\mathrm{II}}}$*, respectively, and the quadratic residual* $\overline{\mathsf{h}}$ *is in the form*

$$\overline{\mathsf{h}} = \Big(0,\ \overline{\mathsf{h}}_{\mathrm{II}}(\mathsf{w}, \partial_{\mathsf{x}}\mathsf{w}_{\mathrm{II}})\Big)^t = \Big(0,\ \sum_{i,j\in\mathfrak{D}} \overline{\mathrm{M}}_{ij}^{\mathrm{II,II,II}}(\mathsf{w})\ \partial_i \mathsf{w}_{\mathrm{II}}\ \partial_j \mathsf{w}_{\mathrm{II}}\Big)^t, \tag{3.23}$$

where $\overline{\mathrm{M}}_{ij}^{\mathrm{II,II,II}}(\mathsf{w})$ *are third order tensors depending on* w *with regularity at least* $\gamma - 3$*. Finally, when* $\mathfrak{F}_{\mathrm{II}}$ *is linear, the quadratic residual* $\overline{\mathsf{h}}$ *is zero.*

The main interest of normal forms is that the resulting subsystem of partial differential equations governing the variable w_{I} is symmetric hyperbolic [18] whereas the subsystem governing w_{II} is symmetric strongly parabolic [18, 22]. For such symmetric second order systems governing w_{II}, Petrovsky parabolicity is then equivalent to strong parabolicity [124]. In the special situation of multicomponent flows the nullspace invariance property holds and the normal variables are easily obtained [31].

Lemma 3.2 *The nullspace of the matrix* $\widetilde{\mathsf{B}}(\mathsf{v}, \boldsymbol{\xi}) = \sum_{i,j\in\mathfrak{D}} \widetilde{\mathsf{B}}_{ij}(\mathsf{v})\xi_i\xi_j$ *is independent of* $\mathsf{v} \in \mathcal{O}_{\mathsf{v}}$ *and* $\boldsymbol{\xi} \in \Sigma^{d-1}$ *and given by* $N(\widetilde{\mathsf{B}}) = \mathbb{R}(\mathbb{I}, \mathbf{0}, 0)^t$ *and we have* $\widetilde{\mathsf{B}}_{ij}(\mathsf{v})N(\widetilde{\mathsf{B}}) = 0$, $i, j \in \mathfrak{D}$, *for* $\mathsf{v} \in \mathcal{O}_{\mathsf{v}}$.

Since $N(\widetilde{\mathsf{B}})$ is spanned by $\mathbb{I}' = (\mathbb{I}, \mathbf{0}, 0)^t$ we introduce the matrix P having columns $\mathsf{P} = [\mathbb{I}', \mathsf{e}_2, \dots \mathsf{e}_\mathsf{n}]$ where e_i, $1 \le i \le \mathsf{n}$, denote the basis vectors in $\mathbb{R}^\mathsf{n}$, as well as the auxiliary variables

$$\mathsf{u}' = \mathsf{P}^t\mathsf{u} = \mathsf{u}' = \Big(\rho, \rho_2, \dots, \rho_n, \rho\boldsymbol{v}, \mathcal{E} + \tfrac{1}{2}\rho|\boldsymbol{v}|^2\Big)^t,$$

$$\mathsf{v}' = \mathsf{P}^{-1}\mathsf{v} = \frac{1}{RT}\Big(g_1 - \tfrac{1}{2}|\boldsymbol{v}|^2, g_2 - g_1, \dots g_n - g_1, \boldsymbol{v}, -1\Big)^t.$$

We may then use identity maps for $\mathfrak{F}_{\mathrm{I}}$ and $\mathfrak{F}_{\mathrm{II}}$ and this yield the following normal form.

Theorem 3.5 *Assume that* $(H_1 - H_5)$ *hold and consider the normal variable*

$$\mathsf{w} = \Big(\rho, \frac{g_2 - g_1}{RT}, \dots, \frac{g_n - g_1}{RT}, \frac{\boldsymbol{v}}{RT}, \frac{-1}{RT}\Big)^t. \tag{3.24}$$

Then the map $\mathsf{v} \longmapsto \mathsf{w}$ *is a the diffeomorphism from* $\mathcal{O}_{\mathsf{v}}$ *onto the open set* $\mathcal{O}_{\mathsf{w}} = (0, \infty) \times \mathbb{R}^{n-1} \times \mathbb{R}^d \times (-\infty, 0)$ *and the system of partial differential equations may be written in the normal form* (3.22).

The coefficients $\overline{\mathsf{A}}_0$, $\overline{\mathsf{A}}_i$, $\overline{\mathsf{B}}_{ij}$ as well as the quasilinear form of the source term $\overline{\Omega} = -\overline{\mathsf{L}}\mathsf{w}$ may be found in the literature [7, 54, 124]. This normal form is used in particular in the following sections and the corresponding quadratic residual vanishes $\overline{\mathsf{h}} = 0$ for this particular normal form. The source term $\overline{\Omega}$ may be written in the quasilinear form

$$\overline{\Omega}(\mathsf{w}) = -\overline{\mathsf{L}}(\mathsf{w})\,\mathsf{w},$$

after some algebra by using Lemma 3.1. The matrix $\overline{\mathsf{L}}(\mathsf{w})$ is a symmetric positive semi-definite matrix of size n with a fixed nullspace $N(\overline{\mathsf{L}}) = \mathcal{E}$ of dimension $\mathsf{n}_\mathrm{e} = \dim(\mathcal{E})$ with $1 \le \mathsf{n}_\mathrm{e} < \mathsf{n}$ and the map $\mathsf{w} \mapsto \overline{\mathsf{L}}(\mathsf{w})$ is C^γ over $\mathcal{O}_{\mathsf{w}}$. More details on the matrix $\overline{\mathsf{L}}$ and the equilibrium manifold may be found in the literature [54].

3.2.4 Local Existence

We discuss local existence of strong solutions for quasilinear systems in normal form [7, 18]. To this aim we consider an equilibrium state $\rho_1^\star, \dots, \rho_n^\star, T^\star$ obtained from Proposition 3.2, the associated variable $\mathsf{z}^\star = (\rho_1^\star, \dots, \rho_n^\star, \boldsymbol{v}^\star, T^\star)^t$ with $\boldsymbol{v}^\star = 0$, and denote by $\mathsf{w}^\star \in \mathcal{O}_{\mathsf{w}}$ the corresponding normal variable $\mathsf{w}(\mathsf{z}^\star)$. We are interested in local in time solutions such that $\mathsf{w} - \mathsf{w}^\star$ remains in the Sobolev space H^l where $l > [d/2] + 1$ and the regularity class γ is assumed to be such that $\gamma - 2 \geq l + 2$.

In order to establish a local existence result for strong solutions we need to establish a priori estimates for linearized equations. We include a small parameter $\epsilon_d \in (0, 1]$ in the following estimates for future use concerning the relaxation of internal energy in Sect. 3.3 and this incidentally yields new results concerning multicomponent fluids with small transport coefficients. We thus consider a system of linearized equations in the form

$$\overline{\mathsf{A}}_0(\mathsf{w})\partial_t\widetilde{\mathsf{w}} + \sum_{i\in\mathfrak{D}} \overline{\mathsf{A}}_i(\mathsf{w})\partial_i\widetilde{\mathsf{w}} - \epsilon_d \sum_{i,j\in\mathfrak{D}} \overline{\mathsf{B}}_{ij}(\mathsf{w})\partial_i\partial_j\widetilde{\mathsf{w}} + \overline{\mathsf{L}}(\mathsf{w})\widetilde{\mathsf{w}} = \mathsf{f} + \epsilon_d\mathsf{g}. \tag{3.25}$$

The variables w and $\widetilde{\mathsf{w}}$ are assumed to have $\mathsf{n} \geq 1$ components and $\mathbb{R}^{\mathsf{n}}$ is decomposed into $\mathbb{R}^{\mathsf{n}} = \mathbb{R}^{\mathsf{n}_{\mathrm{I}}} \times \mathbb{R}^{\mathsf{n}_{\mathrm{II}}}$ with $\mathsf{n} = \mathsf{n}_{\mathrm{I}} + \mathsf{n}_{\mathrm{II}}$. The bloc structure induced by the partitioning of $\mathbb{R}^{\mathsf{n}}$ is used in the following so that $\widetilde{\mathsf{w}}$ is decomposed into $\widetilde{\mathsf{w}} = (\widetilde{\mathsf{w}}_{\mathrm{I}}, \widetilde{\mathsf{w}}_{\mathrm{II}})^t$ with $\widetilde{\mathsf{w}}_{\mathrm{I}} \in \mathbb{R}^{\mathsf{n}_{\mathrm{I}}}$ and $\widetilde{\mathsf{w}}_{\mathrm{II}} \in \mathbb{R}^{\mathsf{n}_{\mathrm{II}}}$. The integer l_0 denotes the minimum integer for an embedding of the Sobolev space $H^{l_0}(\mathbb{R}^d)$ into the space of bounded continuous functions $l_0 = [d/2] + 1$ and l is an integer such that $l \geq l_0 + 1$.

The required assumptions essentially state that the linearized equations are obtained from a normal form. The matrix $\overline{\mathsf{A}}_0$ is symmetric, positive definite, with a block diagonal structure $\overline{\mathsf{A}}_0 = \mathrm{diag}(\overline{\mathsf{A}}_0^{\mathrm{I,I}}, \overline{\mathsf{A}}_0^{\mathrm{II,II}})$. The matrices $\overline{\mathsf{A}}_i(\mathsf{w}) \in \mathbb{R}^{\mathsf{n,n}}$, $i \in \mathfrak{D}$, are symmetric, the dissipation matrices are such that $(\overline{\mathsf{B}}_{ij}^{\mathrm{d}})^t = \overline{\mathsf{B}}_{ji}^{\mathrm{d}}$, $i, j \in \mathfrak{D}$, have nonzero components only into the right lower $\overline{\mathsf{B}}_{ij}^{\mathrm{d\,II,II}}$ blocks, and $\overline{\mathsf{B}}^{\mathrm{d\,II,II}} = \sum_{i,j\in\mathfrak{D}} \overline{\mathsf{B}}_{ij}^{\mathrm{d\,II,II}}\xi_i\xi_j$ is positive definite for $\boldsymbol{\xi} \in \Sigma^{d-1}$ where Σ^{d-1} is the sphere in d dimension. The matrix $\overline{\mathsf{L}} = \overline{\mathsf{L}}(\mathsf{w})$ is positive semidefinite and has a fixed nullspace $\overline{\mathcal{E}}$. The orthogonal projector onto $\overline{\mathcal{E}}^{\perp}$, the range of $\overline{\mathsf{L}}$, is denoted by π. The system coefficients $\overline{\mathsf{A}}_0$, $\overline{\mathsf{A}}_i$, $\overline{\mathsf{B}}_{ij}^{\mathrm{d}}$, $i, j \in \mathfrak{D}$, and $\overline{\mathsf{L}}$ are smooth functions of $\mathsf{w} \in \mathbb{R}^{\mathsf{n}}$ defined over an open set O_{w} with at least C^{l+2} regularity assuming the regularity class γ is such that $\gamma - 2 \geq l + 2$. The right hand sides f and g as well as w are given functions of time and space defined over $[0, \bar{\tau}] \times \mathbb{R}^d$ for some given positive time $\bar{\tau} > 0$.

We denote by $|\bullet|_l$ the norm in the Sobolev space $H^l(\mathbb{R}^d)$ and otherwise $|\bullet|_A$ in the functional space A. If $\alpha = (\alpha_1, \ldots, \alpha_d) \in \mathbb{N}^d$ is a multiindex, we denote as usual by ∂^α the differential operator $\partial_1^{\alpha_1} \cdots \partial_d^{\alpha_d}$ and by $|\alpha|$ its order $|\alpha| = \alpha_1 + \cdots + \alpha_d$. The square of kth derivatives of a scalar function ϕ, like T, ρ, or v_i, $1 \leq i \leq d$, is defined by

$$|\partial^k\phi|^2 = \sum_{|\alpha|=k} \frac{|\alpha|!}{\alpha!}(\partial^\alpha\phi)^2 = \sum_{1\leq i_1,\ldots,i_k\leq d} (\partial_{i_1}\cdots\partial_{i_k}\phi)^2, \tag{3.26}$$

where $|\alpha|!/\alpha!$ are the multinomial coefficients and similarly, for a vector function like $\boldsymbol{v}$ we define $|\partial^k\boldsymbol{v}|^2 = \sum_{1\leq i\leq d} |\partial^k v_i|^2$. Finally, for any map $\phi : [0, \bar{\tau}]\times\mathbb{R}^d \to$

$\mathbb{R}^{\mathsf{n}}$ where $\bar{\tau} > 0$ is positive and for any $\tau \in [0, \bar{\tau}]$, we denote by $\phi(\tau)$ the partial map $\mathsf{x} \to \phi(\tau, \mathsf{x})$ defined over $\mathbb{R}^d$.

We assume that w is such that

$$\begin{cases} \mathsf{w}_{\mathrm{I}} - \mathsf{w}_{\mathrm{I}}^{\star} \in C^0\big([0, \bar{\tau}], H^l\big) \cap C^1\big([0, \bar{\tau}], H^{l-1}\big), \\ \mathsf{w}_{\mathrm{II}} - \mathsf{w}_{\mathrm{II}}^{\star} \in C^0\big([0, \bar{\tau}], H^l\big) \cap C^1\big([0, \bar{\tau}], H^{l-2}\big) \cap L^2\big((0, \bar{\tau}), H^{l+1}\big), \end{cases} \tag{3.27}$$

that $\partial_t \mathsf{w} \in L^2\big((0, \bar{\tau}), H^{l-1}\big)$ and the quantities M and M_{t} are defined by

$$M^2 = \sup_{0 \le \tau \le \bar{\tau}} |\mathsf{w}(\tau) - \mathsf{w}^{\star}|_l^2, \qquad M_{\mathrm{t}}^2 = \int_0^{\bar{\tau}} |\partial_t \mathsf{w}(\tau)|_{l-1}^2 \, d\tau. \tag{3.28}$$

It is assumed that O_0 is an open set such that $O_0 \subset \overline{O}_0 \subset O_{\mathsf{w}}$, that a_1 is such that $0 < a_1 < \mathrm{dist}(\overline{O}_0, \partial O_{\mathsf{w}})$, and the open set O_1 is defined by

$$O_1 = \{\, \mathsf{w} \in O_{\mathsf{w}};\ \mathrm{dist}(\mathsf{w}, \overline{O}_0) < a_1 \,\}. \tag{3.29}$$

It is further assumed that $\mathsf{w}_0(\mathsf{x}) = \mathsf{w}(0, \mathsf{x}) \in O_0$, and that $\mathsf{w}(t, \mathsf{x}) \in O_1$, for $(t, \mathsf{x}) \in [0, \bar{\tau}] \times \mathbb{R}^d$ so that the values of w are controlled with the open set O_1.

Let now l' be such that $1 \le l' \le l$ and assume that the right hand sides $\mathsf{f} \in \mathbb{R}^{\mathsf{n}}$ and $\mathsf{g} \in \mathbb{R}^{\mathsf{n}}$ are such that

$$\mathsf{f} \in C^0\big([0, \bar{\tau}], H^{l'-1}\big) \cap L^1\big((0, \bar{\tau}), H^{l'}\big), \tag{3.30}$$

$$\mathsf{g} \in C^0\big([0, \bar{\tau}], H^{l'-1}\big), \qquad \mathsf{g}_{\mathrm{I}} = 0. \tag{3.31}$$

Denoting by $\widetilde{\mathsf{w}}^{\star}$ a constant state in $\overline{\mathcal{E}}$ so that $\overline{\mathsf{L}}(\mathsf{w})\widetilde{\mathsf{w}}^{\star} = 0$ for any $\mathsf{w} \in O_{\mathsf{w}}$, a priori estimates of $\widetilde{\mathsf{w}} - \widetilde{\mathsf{w}}^{\star}$ are obtained in the following theorem.

Theorem 3.6 *Let $\widetilde{\mathsf{w}}$ be a solution of the linearized system* (3.25) *with*

$$\begin{aligned} &\widetilde{\mathsf{w}}_{\mathrm{I}} - \widetilde{\mathsf{w}}_{\mathrm{I}}^{\star} \in C^0\big([0, \bar{\tau}], H^{l'}\big) \cap C^1\big([0, \bar{\tau}], H^{l'-1}\big), \\ &\widetilde{\mathsf{w}}_{\mathrm{II}} - \widetilde{\mathsf{w}}_{\mathrm{II}}^{\star} \in C^0\big([0, \bar{\tau}], H^{l'}\big) \cap C^1\big([0, \bar{\tau}], H^{l'-2}\big) \cap L^2\big((0, \bar{\tau}), H^{l'+1}\big), \end{aligned} \tag{3.32}$$

where $1 \le l' \le l$ and denote by $\widetilde{\mathsf{w}}_0$ the initial state $\widetilde{\mathsf{w}}_0(\mathsf{x}) = \widetilde{\mathsf{w}}(0, \mathsf{x})$. Then there exists constants $\mathsf{c}_1(O_1) \ge 1$ and $\mathsf{c}_2(O_1, M) \ge 1$, with $\mathsf{c}_2(O_1, M)$ increasing with M, independent of $\epsilon_d \in (0, 1]$, such that for any $t \in [0, \bar{\tau}]$

$$\begin{aligned} &\sup_{0 \le \tau \le t} |\widetilde{\mathsf{w}}(\tau) - \widetilde{\mathsf{w}}^{\star}|_{l'}^2 + \epsilon_d \int_0^t |\widetilde{\mathsf{w}}_{\mathrm{II}}(\tau) - \widetilde{\mathsf{w}}_{\mathrm{II}}^{\star}|_{l'+1}^2 \, d\tau + \int_0^t |\pi\, \widetilde{\mathsf{w}}|_{l'}^2 \, d\tau \\ &\le \mathsf{c}_1^2 \exp\big(\mathsf{c}_2\big(t + M_{\mathrm{t}}\sqrt{t}\big)\big) \Big(|\widetilde{\mathsf{w}}_0 - \widetilde{\mathsf{w}}^{\star}|_{l'}^2 + \mathsf{c}_2 \Big\{ \int_0^t |\mathsf{f}|_{l'} \, d\tau \Big\}^2 + \epsilon_d \mathsf{c}_2 \int_0^t |\mathsf{g}_{\mathrm{II}}|_{l'-1}^2 \, d\tau \Big). \end{aligned} \tag{3.33}$$

Sketch of the Proof It is sufficient to consider smooth solutions of (3.25) since one may use mollifiers and convolution operators [18]. Letting $\delta\widetilde{\mathsf{w}} = \widetilde{\mathsf{w}} - \widetilde{\mathsf{w}}^{\star}$ and defining the auxiliary norms $E_k^2(\phi) = \sum_{0\leq|\alpha|\leq k} \frac{|\alpha|!}{\alpha!} \int_{\mathbb{R}^d} \big\langle \overline{\mathsf{A}}_0(\mathsf{w})\partial^\alpha\phi, \partial^\alpha\phi\big\rangle\, d\mathsf{x}$ the main step is to derive an inequality in the form

$$\partial_t E_{l'}^2(\delta\widetilde{\mathsf{w}}) + \delta_1\epsilon_d|\delta\widetilde{\mathsf{w}}_{\mathrm{II}}|_{l'+1}^2 \leq \mathsf{c}_2\big(1 + |\partial_t\mathsf{w}|_{l-1}\big)E_{l'}^2(\delta\widetilde{\mathsf{w}}) + \mathsf{c}_2|\mathsf{f}|_{l'}E_{l'}(\delta\widetilde{\mathsf{w}}) + \mathsf{c}_2\epsilon_d|\mathsf{g}_{\mathrm{II}}|_{l'-1}^2,$$

and then to use Gronwall Lemma. The zeroth order inequality is obtained after the scalar multiplication of (3.25) by $\widetilde{\mathsf{w}}$, using the symmetry properties, Sobolev inequalities, and Gårding inequality [18, 53, 54]. We may indeed write

$$\langle\delta\widetilde{\mathsf{w}}, \overline{\mathsf{A}}_0(\mathsf{w})\partial_t\delta\widetilde{\mathsf{w}}\rangle = \tfrac{1}{2}\partial_t\langle\delta\widetilde{\mathsf{w}}, \overline{\mathsf{A}}_0(\mathsf{w})\delta\widetilde{\mathsf{w}}\rangle - \tfrac{1}{2}\langle\delta\widetilde{\mathsf{w}}, \partial_t\overline{\mathsf{A}}_0(\mathsf{w})\delta\widetilde{\mathsf{w}}\rangle,$$

and estimate $\partial_t\overline{\mathsf{A}}_0(\mathsf{w}) = \partial_{\mathsf{w}}\overline{\mathsf{A}}_0\,\partial_t\mathsf{w}$ with $|\partial_t\overline{\mathsf{A}}_0|_{L^\infty} \leq \mathsf{c}_0|\partial_t\overline{\mathsf{A}}_0|_{l-1} \leq \mathsf{c}_1|\partial_t\mathsf{w}|_{l-1}$. Similarly, we may use

$$\langle\delta\widetilde{\mathsf{w}}, \overline{\mathsf{A}}_i(\mathsf{w})\partial_i\delta\widetilde{\mathsf{w}}\rangle = \tfrac{1}{2}\partial_i\langle\delta\widetilde{\mathsf{w}}, \overline{\mathsf{A}}_i(\mathsf{w})\delta\widetilde{\mathsf{w}}\rangle - \tfrac{1}{2}\langle\delta\widetilde{\mathsf{w}}, \partial_i\overline{\mathsf{A}}_i(\mathsf{w})\delta\widetilde{\mathsf{w}}\rangle,$$

with $|\overline{\mathsf{A}}_i|_{L^\infty} \leq \mathsf{c}_1$ and $|\partial_i\overline{\mathsf{A}}_i|_{L^\infty} \leq \mathsf{c}_2$ and use Gårding inequality

$$\delta_1|\phi_{\mathrm{II}}|_1^2 \leq \sum_{i,j\in\mathfrak{D}}\int_{\mathbb{R}^d}\langle\overline{\mathsf{B}}_{ij}^{\mathrm{II},\mathrm{II}}(\mathsf{w})\partial_j\phi_{\mathrm{II}}, \partial_i\phi_{\mathrm{II}}\rangle\, d\mathsf{x} + \mathsf{c}_2|\phi_{\mathrm{II}}|_0^2 \qquad \phi_{\mathrm{II}} \in H^1(\mathbb{R}^d).$$

In order to obtain the l' estimate, we form the derived equation

$$\overline{\mathsf{A}}_0(\mathsf{w})\partial_t\partial^\alpha\widetilde{\mathsf{w}} + \sum_{i\in\mathfrak{D}}\overline{\mathsf{A}}_i(\mathsf{w})\partial_i\partial^\alpha\widetilde{\mathsf{w}} - \epsilon_d\sum_{i,j\in\mathfrak{D}}\overline{\mathsf{B}}_{ij}(\mathsf{w})\partial_i\partial_i\partial^\alpha\widetilde{\mathsf{w}} + \overline{\mathsf{L}}(\mathsf{w})\partial^\alpha\widetilde{\mathsf{w}} = \mathsf{f}^\alpha + \epsilon_d\mathsf{g}^\alpha, \tag{3.34}$$

where

$$\mathsf{f}^\alpha = \overline{\mathsf{A}}_0\partial^\alpha\big(\overline{\mathsf{A}}_0^{-1}\mathsf{f}\big) - \sum_{i\in\mathfrak{D}}\overline{\mathsf{A}}_0\big[\partial^\alpha, \overline{\mathsf{A}}_0^{-1}\overline{\mathsf{A}}_i\big]\partial_i\widetilde{\mathsf{w}} - \overline{\mathsf{A}}_0\big[\partial^\alpha, \overline{\mathsf{A}}_0^{-1}\overline{\mathsf{L}}\big]\widetilde{\mathsf{w}},$$

$$\mathsf{g}^\alpha = \overline{\mathsf{A}}_0\partial^\alpha\big(\overline{\mathsf{A}}_0^{-1}\mathsf{g}\big) + \sum_{i,j\in\mathfrak{D}}\overline{\mathsf{A}}_0\big[\partial^\alpha, \overline{\mathsf{A}}_0^{-1}\overline{\mathsf{B}}_{ij}\big]\partial_i\partial_j\widetilde{\mathsf{w}}.$$

One then multiply by $\partial^\alpha\delta\widetilde{\mathsf{w}}$ and by $|\alpha|!/\alpha!$, integrate over $\mathbb{R}^d$, sum over $1 \leq |\alpha| \leq l'$, and add zeroth order estimate. The terms arising from the left hand side are treated as for the zeroth order estimates whereas those arising from the right hand side are majorized by using commutator's and we refer to [7, 18, 32, 53]. □

As is classical, by using the linearized estimates, a local existence theorem may be obtained [7, 18, 31, 32, 53].

Theorem 3.7 *Let $d \geq 1$ and $l \geq l_0 + 1$, be integers with $l_0 = [d/2] + 1$, and let $b > 0$ be given. Let $\mathcal{O}_0$ such that $\overline{\mathcal{O}}_0 \subset \mathcal{O}_{\mathsf{w}}$, d_1 such that $0 < d_1 < d(\overline{\mathcal{O}}_0, \partial\mathcal{O}_{\mathsf{w}})$, and define $\mathcal{O}_1 = \{\, \mathsf{w} \in \mathcal{O}_{\mathsf{w}};\ d(\mathsf{w}, \overline{\mathcal{O}}_0) < d_1 \,\}$. There exists $\bar{\tau} > 0$ depending on $\mathcal{O}_1$ and b, independent of ϵ_d, such that for any w_0 with*

$$|\mathsf{w}_0 - \mathsf{w}^\star|_l^2 < b^2, \tag{3.35}$$

and $\mathsf{w}_0 \in \overline{\mathcal{O}}_0$, there exists a unique local solution w to the system

$$\overline{\mathsf{A}}_0(\mathsf{w})\partial_t \mathsf{w} + \sum_{i\in\mathfrak{D}} \overline{\mathsf{A}}_i(\mathsf{w})\partial_i \mathsf{w} - \epsilon_d \sum_{i,j\in\mathfrak{D}} \partial_i\big(\overline{\mathsf{B}}_{ij}(\mathsf{w})\partial_j \mathsf{w}\big) + \overline{\mathsf{L}}(\mathsf{w})\,\mathsf{w} = \epsilon_d \overline{\mathsf{h}}(\mathsf{w}, \partial_{\mathsf{x}}\mathsf{w}_{\mathrm{II}}), \tag{3.36}$$

with $\mathsf{w}(0, \mathsf{x}) = \mathsf{w}_0(\mathsf{x})$, such that $\mathsf{w}(t, \mathsf{x}) \in \mathcal{O}_1$ for $t \in [0, \bar{\tau}]$, $\mathsf{x} \in \mathbb{R}^d$, and

$$\begin{aligned}
&\mathsf{w}_{\mathrm{I}} - \mathsf{w}_{\mathrm{I}}^\star \in C^0\big([0, \bar{\tau}], H^l\big) \cap C^1\big([0, \bar{\tau}], H^{l-1}\big),\\
&\mathsf{w}_{\mathrm{II}} - \mathsf{w}_{\mathrm{II}}^\star \in C^0\big([0, \bar{\tau}], H^l\big) \cap C^1\big([0, \bar{\tau}], H^{l-2}\big) \cap L^2\big((0, \bar{\tau}), H^{l+1}\big).
\end{aligned}$$

In addition, there exists $\mathfrak{c}_{\mathrm{loc}} > 0$ which only depend on $\mathcal{O}_1$ and b, such that

$$\begin{aligned}
\sup_{0\leq\tau\leq\bar{\tau}} |\mathsf{w}(\tau) - \mathsf{w}^\star|_l^2 &+ \epsilon_d \int_0^{\bar{\tau}} |\mathsf{w}_{\mathrm{II}}(\tau) - \mathsf{w}_{\mathrm{II}}^\star|_{l+1}^2\, d\tau\\
&+ \int_0^{\bar{\tau}} |\pi\, \mathsf{w}(\tau)|_l^2\, d\tau \leq \mathfrak{c}_{\mathrm{loc}}^2 |\mathsf{w}_0 - \mathsf{w}^\star|_l^2.
\end{aligned} \tag{3.37}$$

Sketch of the Proof Linearizing the system of equation, the solution is then a fixed point of the map $\mathsf{w} \mapsto \widetilde{\mathsf{w}}$. Using the linearized estimates, a set $\mathsf{X}_{\bar{\tau}}^l\big(\mathcal{O}_1, \overline{M}, \overline{M}_1\big)$ is found to be invariant under the map $\mathsf{w} \mapsto \widetilde{\mathsf{w}}$ for suitable $\overline{M}$ and $\overline{M}_1$ and $\bar{\tau}$ small enough. The invariant set $\mathsf{X}_{\bar{\tau}}^l\big(\mathcal{O}_1, \overline{M}, \overline{M}_1\big)$ is defined by $\mathsf{w}_{\mathrm{I}} - \mathsf{w}_{\mathrm{I}}^\star \in C^0\big([0, \bar{\tau}], H^l\big) \cap C^1\big([0, \bar{\tau}], H^{l-1}\big)$, $\mathsf{w}_{\mathrm{II}} - \mathsf{w}_{\mathrm{II}}^\star \in C^0\big([0, \bar{\tau}], H^l\big) \cap C^1\big([0, \bar{\tau}], H^{l-2}\big) \cap L^2\big((0, \bar{\tau}), H^{l+1}\big)$, $\partial_t \mathsf{w} \in L^2\big([0, \bar{\tau}], H^{l-1}\big)$, $\mathsf{w}(t, \mathsf{x}) \in \mathcal{O}_1$, and

$$\sup_{0\leq\tau\leq\bar{\tau}} |\mathsf{w}(\tau) - \mathsf{w}^\star|_l^2 + \epsilon_d \int_0^{\bar{\tau}} |\mathsf{w}_{\mathrm{II}}(\tau) - \mathsf{w}_{\mathrm{II}}^\star|_{l+1}^2\, d\tau \leq \overline{M}^2,$$

$$\int_0^{\bar{\tau}} |\partial_t \mathsf{w}(\tau)|_{l-1}^2\, d\tau \leq \overline{M}_1^2.$$

The set $\mathsf{X}_{\bar{\tau}}^l\big(\mathcal{O}_1, \overline{M}, \overline{M}_1\big)$ is shown to be invariant by $\mathsf{w} \mapsto \widetilde{\mathsf{w}}$ for $\overline{M} = 2\mathfrak{c}_1 b$, $\overline{M}_1 = \mathfrak{c}_2^2 2\mathfrak{c}_1 b$ and $\bar{\tau}$ small enough using the a priori estimates for linearized equations. The successive approximations $\{\mathsf{w}^k\}_{k\geq 0}$ defined by $\mathsf{w}^0 = \mathsf{w}^\star$ and

$\mathsf{w}^{k+1} = \widetilde{\mathsf{w}}^k$ are then well defined. The iterates are convergent for a lower norm as obtained with the linearized estimates applied to $\mathsf{w}^{k+1} - \mathsf{w}^k$. The limit function $\mathsf{w} = \lim_{k\to\infty} \mathsf{w}^k$ is then shown to have all the required properties using standard arguments of functional analysis. □

From the structure of the linearized system of equations we also obtain an estimate in the form $\int_0^{\bar{\tau}} |\partial_t \mathsf{w}(\tau)|_{l-1}^2 \, d\tau \le C|\mathsf{w}_0 - \mathsf{w}^\star|_l^2$ for the time derivative. It is also possible to establish the following stability theorem concerning local solutions [18, 53].

Theorem 3.8 *Let $d \ge 1$ and $l \ge l_0 + 1$ be integers and let $b > 0$ be given and consider the perturbed system of equations*

$$\overline{A}_0(\mathsf{w})\partial_t \mathsf{w} + \sum_{i\in\mathfrak{D}} \overline{A}_i(\mathsf{w})\partial_i \mathsf{w} - \epsilon_d \sum_{i,j\in\mathfrak{D}} \partial_i\big(\overline{B}_{ij}(\mathsf{w})\partial_j \mathsf{w}\big) - \epsilon_d \overline{h}(\mathsf{w}, \partial_x \mathsf{w})$$

$$+ \overline{L}(\mathsf{w})\mathsf{w} = f + \epsilon_d g, \tag{3.38}$$

where $\overline{h} = -\sum_{i,j\in\mathfrak{D}} \partial_i (\partial_{\mathsf{w}} \mathsf{v})^t \, (\partial_{\mathsf{v}} \mathsf{w})^t \overline{B}_{ij} \, \partial_j \mathsf{w}$ and where for some positive $\bar{\tau}_{\mathrm{m}} > 0$

$$f \in C^0\big([0, \bar{\tau}_{\mathrm{m}}], H^{l-1}\big) \cap L^1\big([0, \bar{\tau}_{\mathrm{m}}], H^l\big), \tag{3.39}$$

$$g \in C^0\big([0, \bar{\tau}_{\mathrm{m}}], H^{l-1}\big), \qquad g_{\mathrm{I}} = 0. \tag{3.40}$$

Let $\mathcal{O}_0$ be given such that $\overline{\mathcal{O}}_0 \subset \mathcal{O}_{\mathsf{w}}$, d_1 such that $0 < d_1 < d(\overline{\mathcal{O}}_0, \partial\mathcal{O}_{\mathsf{w}})$, and define

$$\mathcal{O}_1 = \{\, \mathsf{w} \in \mathcal{O}_{\mathsf{w}};\ d(\mathsf{w}, \overline{\mathcal{O}}_0) < d_1 \,\}.$$

There exists $\bar{\tau}$ with $0 < \bar{\tau} \le \bar{\tau}_{\mathrm{m}}$ and $\chi > 0$ depending on $\mathcal{O}_1$ and b, and independent of $\epsilon_d \in (0, 1]$, such that for any w_0 with $\mathsf{w}_0 \in \overline{\mathcal{O}}_0$ and any f and g satisfying (3.39)–(3.40) *with*

$$|\mathsf{w}_0 - \mathsf{w}^\star|_l^2 < b^2, \quad \Big\{\int_0^t |f|_l \, d\tau\Big\}^2 + \int_0^t |f|_{l-1}^2 \, d\tau < \chi b^2, \quad \int_0^t |g|_{l-1}^2 \, d\tau < \chi b^2, \tag{3.41}$$

there exists a unique local solution w to the perturbed system (3.38) *with initial condition $\mathsf{w}(0, x) = \mathsf{w}_0(x)$, $x \in \mathbb{R}^d$ such that $\mathsf{w}(t, x) \in \mathcal{O}_1$, $t \in [0, \bar{\tau}]$, $x \in \mathbb{R}^d$, and*

$$\mathsf{w}_{\mathrm{I}} - \mathsf{w}_{\mathrm{I}}^\star \in C^0\big([0, \bar{\tau}], H^l\big) \cap C^1\big([0, \bar{\tau}], H^{l-1}\big),$$
$$\mathsf{w}_{\mathrm{II}} - \mathsf{w}_{\mathrm{II}}^\star \in C^0\big([0, \bar{\tau}], H^l\big) \cap C^1\big([0, \bar{\tau}], H^{l-2}\big) \cap L^2\big((0, \bar{\tau}), H^{l+1}\big).$$

In addition, there exists $c_{\mathrm{loc}} > 0$ only depending on $\mathcal{O}_1$ and b, such that

$$\sup_{0\le\tau\le\bar\tau} |\mathsf{w}(\tau)-\mathsf{w}^\star|_l^2 + \epsilon_d \int_0^{\bar\tau} |\mathsf{w}_{\rm II}(\tau)-\mathsf{w}_{\rm II}^\star|_{l+1}^2\, d\tau + \int_0^{\bar\tau} |\pi\,\mathsf{w}(\tau)|_l^2\, d\tau \le$$

$$c_{\rm loc}^2\Big(|\mathsf{w}_0-\mathsf{w}^\star|_l^2 + \Big\{\int_0^{\bar\tau}|\mathsf{f}|_l\, d\tau\Big\}^2 + \epsilon_d\int_0^{\bar\tau}|\mathsf{g}|_{l-1}^2\, d\tau\Big), \tag{3.42}$$

$$\int_0^t |\partial_t \mathsf{w}(\tau)|_{l-1}^2\, d\tau$$

$$\le c_{\rm loc}^2\Big(|\mathsf{w}_0-\mathsf{w}^\star|_l^2 + \Big\{\int_0^{\bar\tau}|\mathsf{f}|_l\, d\tau\Big\}^2 + \int_0^{\bar\tau}|\mathsf{f}|_{l-1}^2\, d\tau + \epsilon_d\int_0^{\bar\tau}|\mathsf{g}|_{l-1}^2\, d\tau\Big). \tag{3.43}$$

Moreover, if w *and* w' *correspond to two different initial conditions and different perturbations, letting* $\delta\mathsf{w}=\mathsf{w}-\mathsf{w}'$, $\delta\mathsf{f}=\mathsf{f}-\mathsf{f}'$, $\delta\mathsf{g}=\mathsf{g}-\mathsf{g}'$, *then*

$$\sup_{0\le\tau\le\bar\tau} |\delta\mathsf{w}(\tau)|_{l-1}^2 + \epsilon_d \int_0^{\bar\tau} |\delta\mathsf{w}_{\rm II}(\tau)|_l^2\, d\tau + \int_0^{\bar\tau} |\pi\,\delta\mathsf{w}(\tau)|_{l-1}^2\, d\tau$$

$$\le c_{\rm loc}^2\Big(|\delta\mathsf{w}_0|_{l-1}^2 + \Big\{\int_0^{\bar\tau}|\delta\mathsf{f}|_{l-1}\, d\tau\Big\}^2 + \epsilon_d\int_0^{\bar\tau}|\delta\mathsf{g}|_{l-2}^2\, d\tau\Big). \tag{3.44}$$

This stability result will be useful for investigating the apparition of the volume viscosity coefficient in Sect. 3.3. Local strong solutions may also be obtained by using an existence theorem from Volpert and Hujaev [26, 30].

3.2.5 Asymptotic Stability

We consider an equilibrium state $\mathsf{w}^\star \in \overline{\mathcal{E}}$ such that $\overline{\Omega}(\mathsf{w}^\star) = 0$ and investigate global existence of solutions and asymptotic stability of the constant state $\mathsf{w}^\star$. The results are then applied to the situation of multicomponent fluids and new results are obtained by taking into account the parameter $\epsilon_d \in (0,1]$. Investigating smooth global solutions around constant equilibrium states requires the linearized normal forms to be strictly dissipative [18, 19].

If we linearize system (3.22) around the constant stationary state $\mathsf{w}^\star$, we obtain a linear system in the variable $\delta\mathsf{w} = \mathsf{w}-\mathsf{w}^\star$ in the form

$$\overline{\mathsf{A}}_0(\mathsf{w}^\star)\partial_t\delta\mathsf{w} + \sum_{i\in\mathfrak{D}}\overline{\mathsf{A}}_i(\mathsf{w}^\star)\partial_i\delta\mathsf{w} - \epsilon_d\sum_{i,j\in\mathfrak{D}}\overline{\mathsf{B}}_{ij}(\mathsf{w}^\star)\partial_i\partial_j\delta\mathsf{w} + \overline{\mathsf{L}}(\mathsf{w}^\star)\delta\mathsf{w} = 0. \tag{3.45}$$

By Fourier transform, the spectral problem associated with (3.45) reads

$$\alpha\overline{\mathsf{A}}_0(\mathsf{w}^\star)\phi + \Big(\mathrm{i}\zeta\overline{\mathsf{A}}(\mathsf{w}^\star,\boldsymbol{\xi}) + \zeta^2\overline{\mathsf{B}}(\mathsf{w}^\star,\boldsymbol{\xi}) + \overline{\mathsf{L}}(\mathsf{w}^\star)\Big)\phi = 0, \tag{3.46}$$

where $\zeta \in \mathbb{R}$, $\mathrm{i}^2 = -1$, $\boldsymbol{\xi} \in \Sigma^{d-1}$, $\phi \in \mathbb{C}^{\mathsf{n}}$ and

$$\overline{\mathsf{A}}(\mathsf{w}^\star,\boldsymbol{\xi}) = \sum_{i\in\mathfrak{D}} \overline{\mathsf{A}}_i(\mathsf{w}^\star)\xi_i, \qquad \overline{\mathsf{B}}(\mathsf{w}^\star,\boldsymbol{\xi}) = \sum_{i,j\in\mathfrak{D}} \overline{\mathsf{B}}_{ij}(\mathsf{w}^\star)\xi_i\xi_j.$$

The strict dissipativity of the linearized system is now the object of the following proposition where we denote by $\mathcal{S}(\zeta, w)$ the set of complex numbers α such that there exists $\phi \in \mathbb{C}^{\mathsf{n}}$, $\phi \neq 0$, satisfying (3.46).

Theorem 3.9 *Assume that the matrix $\overline{\mathsf{A}}_0(\mathsf{w}^\star)$ is symmetric positive definite, that the matrices $\overline{\mathsf{A}}_i(\mathsf{w}^\star)$, $i \in \mathfrak{D}$, are symmetric, that the reciprocity relations $\overline{\mathsf{B}}_{ij}{}^t(\mathsf{w}^\star) = \overline{\mathsf{B}}_{ji}(\mathsf{w}^\star)$, $i, j \in \mathfrak{D}$ hold, that the matrix $\overline{\mathsf{B}}(\mathsf{w}^\star,\boldsymbol{\xi}) = \sum_{i,j\in\mathfrak{D}} \overline{\mathsf{B}}_{ij}(\mathsf{w}^\star)\xi_i\xi_j$ is positive semi-definite for $\boldsymbol{\xi} \in \Sigma^{d-1}$, that $\overline{\mathsf{L}}(\mathsf{w}^\star)$ is symmetric positive semi-definite, and denote $\overline{\mathsf{A}}(\mathsf{w}^\star,\boldsymbol{\xi}) = \sum_{i\in\mathfrak{D}} \overline{\mathsf{A}}_i(\mathsf{w}^\star)\xi_i$. The system* (3.45) *is said to be strictly dissipative at $\mathsf{w}^\star$ when any of the following equivalent properties holds.*

(K_1) *There exists a compensating matrix C^∞ map $K : \Sigma^{d-1} \to \mathbb{R}^{d,d}$ such that for any $\boldsymbol{\xi} \in \Sigma^{d-1}$ the product $K(\boldsymbol{\xi})\overline{\mathsf{A}}_0(\mathsf{w}^\star)$ is skew-symmetric, $K(-\boldsymbol{\xi}) = -K(\boldsymbol{\xi})$, and $K(\boldsymbol{\xi})\overline{\mathsf{A}}_i(\mathsf{w}^\star,\boldsymbol{\xi}) + \overline{\mathsf{B}}(\mathsf{w}^\star,\boldsymbol{\xi}) + \overline{\mathsf{L}}(\mathsf{w}^\star)$ is positive definite.*

(K_2) *For any $\zeta \in \mathbb{R}$, $\zeta \neq 0$, and any $\boldsymbol{\xi} \in \Sigma^{d-1}$, the eigenvalues $\alpha \in \mathcal{S}(\zeta,\boldsymbol{\xi})$ have a negative real part $\Re(\alpha) < 0$.*

(K_3) *Let $\phi \in \mathbb{R}^{\mathsf{n}}\backslash\{0\}$ such that $\overline{\mathsf{B}}(\mathsf{w}^\star,\boldsymbol{\xi})\phi = 0$ and $\overline{\mathsf{L}}(\mathsf{w}^\star)\phi = 0$ for some $\boldsymbol{\xi} \in \Sigma^{d-1}$. Then we have $\zeta\overline{\mathsf{A}}_0(\mathsf{w}^\star)\phi + \overline{\mathsf{A}}(\mathsf{w}^\star,\boldsymbol{\xi})\phi \neq 0$ for any $\zeta \in \mathbb{R}$.*

(K_4) *There exists $\delta > 0$ such that for any $\zeta \in \mathbb{R}$, $\boldsymbol{\xi} \in \Sigma^{d-1}$, the eigenvalue $\alpha \in \mathcal{S}(\zeta,\boldsymbol{\xi})$ have their real part majorized by $\Re(\alpha) \leq -\delta\frac{|\zeta|^2}{1+|\zeta|^2}$.*

Using the contrapositive of (K_3), a physical interpretation of the 'Kawashima condition' ($\mathsf{K}_1-\mathsf{K}_4$) is that the waves $\phi\ \exp(\zeta t + \mathsf{x}\cdot\boldsymbol{\xi})$ associated with the hyperbolic operator $\overline{\mathsf{A}}_0(\mathsf{w}^\star)\partial_t + \sum_{i\in\mathfrak{D}} \overline{\mathsf{A}}_i(\mathsf{w}^\star)\partial_i$ lead to dissipation since there are not in the nullspace of $\overline{\mathsf{B}}$ and $\overline{\mathsf{L}}$. Only the symmetric part of the product $K(\boldsymbol{\xi})\,\overline{\mathsf{A}}(\mathsf{w}^\star,\boldsymbol{\xi})$ plays a role in (K_1). It is also possible to relate the Kawashima condition to Kalmann properties [37]. It is not known in general if the matrix $K(\boldsymbol{\xi})$ may be written

$$K(\boldsymbol{\xi}) = \sum_{j\in\mathfrak{D}} K_j\xi_j, \tag{3.47}$$

where the matrices K_j, $j \in \mathfrak{D}$ are independent of ξ, although it is generally possible to obtain compensating matrices in this form in many practical applications. In the following, for the sake of simplicity, we will use in various places such constant compensating matrices K_j, $\mathfrak{D}$.

Global solutions around equilibrium states may be obtained by using the local strict dissipative properties for multicomponent flows. Assuming that $\mathsf{w}^\star$ is a chemical equilibrium state as in Proposition 3.2, then the linearized normal form at $\mathsf{w}^\star$ is strictly dissipative [7, 18, 22, 31, 32]. The strict dissipativity property then implies global existence and asymptotic stability of equilibrium states [18, 31, 32]. In the following proposition we extend previous results concerning multicomponent flows by taking into account a small diffusion parameter $\epsilon_d \in (0, 1]$. The initial state and the solutions need then to be rescaled by the small diffusion parameter $\epsilon_d \in (0, 1]$. More specifically, the initial solution is assumed to be in the form

$$\mathsf{w}_0 = \mathsf{w}^\star + \epsilon_d(\widetilde{\mathsf{w}}_0 - \mathsf{w}^\star), \tag{3.48}$$

where $\widetilde{\mathsf{w}}_0 - \mathsf{w}^\star \in H^l$ so that $\mathsf{w}_0 - \mathsf{w}^\star = \epsilon_d(\widetilde{\mathsf{w}}_0 - \mathsf{w}^\star)$.

In order to establish a global existence result, it is necessary to establish a few technical lemmas. Defining the relative entropy $\overline{\sigma}$ as

$$\overline{\sigma}(\mathsf{u}) = \sigma(\mathsf{u}) - \sigma(\mathsf{u}^\star) - \partial_{\mathsf{u}}\sigma(\mathsf{u}^\star)(\mathsf{u} - \mathsf{u}^\star), \qquad \mathsf{u} \in \mathcal{O}_{\mathsf{u}}.$$

Then $\overline{\sigma}$ satisfies the governing equation

$$\partial_t\overline{\sigma} + \sum_{i\in\mathfrak{D}} \partial_i\big(\mathsf{q}_i - \mathsf{q}_i^\star - (\partial_{\mathsf{u}}\sigma)^\star(\mathsf{F}_i - \mathsf{F}_i{}^\star)\big) - \sum_{i,j\in\mathfrak{D}} \partial_i\big((\partial_{\mathsf{u}}\sigma - \partial_{\mathsf{u}}\sigma^\star)\mathsf{B}_{ij}\partial_j\mathsf{u}\big)$$
$$+ \sum_{i,j\in\mathfrak{D}} \langle \overline{\mathsf{B}}_{ij}{}^{\mathrm{II,II}} \partial_j\mathsf{w}_{\mathrm{II}}, \partial_i\mathsf{w}_{\mathrm{II}}\rangle - \partial_{\mathsf{u}}\sigma\,\Omega = 0.$$

From the convexity of σ and the structure of the source term, the following lemma is easily established [54].

Lemma 3.3 *There exists* $\mathcal{O}_{\overline{\sigma}} = \{\, \mathsf{w} \in \mathcal{O}_{\mathsf{w}},\ |\mathsf{w} - \mathsf{w}^\star| < r \,\} \subset \mathcal{O}_{\mathsf{w}}$ *and* $0 < \delta_{\overline{\sigma}} < 1$ *such that*

$$\delta_{\overline{\sigma}}|\mathsf{w} - \mathsf{w}^\star|^2 \le \overline{\sigma}\big(u(\mathsf{w})\big) \le (1/\delta_{\overline{\sigma}})|\mathsf{w} - \mathsf{w}^\star|^2, \qquad \mathsf{w} \in \mathcal{O}_{\overline{\sigma}},$$

$$\delta_{\overline{\sigma}}|\pi\,\mathsf{w}|^2 \le -\partial_u\sigma\big(u(\mathsf{w})\big)\,\Omega\big(u(\mathsf{w})\big), \qquad \mathsf{w} \in \mathcal{O}_{\overline{\sigma}},$$

Since $l \ge l_0$ we deduce that there exists $b_{\overline{\sigma}}$ such that $N_l(t) \le b_{\overline{\sigma}}$ implies $\mathsf{w}(\tau, \mathsf{x}) \in \mathcal{O}_{\overline{\sigma}}$ for $0 \le \tau \le t$ and $\mathsf{x} \in \mathbb{R}^d$. It is now necessary to establish a priori estimates with constants independent of the time interval $[0, \overline{\tau}]$. It is assumed that for some time interval $\overline{\tau}$

$$\mathsf{w}_{\mathrm{I}} - \mathsf{w}_{\mathrm{I}}^\star \in C^0\big([0, \overline{\tau}], H^l\big) \cap C^1\big([0, \overline{\tau}], H^{l-1}\big), \qquad \partial_{\mathsf{x}}\mathsf{w}_{\mathrm{I}} \in L^2\big((0, \overline{\tau}), H^{l-1}\big),$$

$$\mathsf{w}_{\mathrm{II}} - \mathsf{w}_{\mathrm{II}}^\star \in C^0\big([0, \overline{\tau}], H^l\big) \cap C^1\big([0, \overline{\tau}], H^{l-2}\big), \qquad \partial_{\mathsf{x}}\mathsf{w}_{\mathrm{II}} \in L^2\big((0, \overline{\tau}), H^l\big).$$

We introduce the quantity $N_l(t)$ for $0 \le t \le \bar{t}$

$$N_l^2(t) = \sup_{0\le\tau\le t} |\mathsf{w}(\tau) - \mathsf{w}^\star|_l^2 + \epsilon_d \int_0^t \left(|\partial_{\mathsf{x}} \mathsf{w}_{\mathrm{I}}(\tau)|_{l-1}^2 + |\partial_{\mathsf{x}} \mathsf{w}_{\mathrm{II}}(\tau)|_l^2\right) d\tau.$$

Theorem 3.10 *Assume that* w *is a solution of* (3.36) *over* $[0, \bar{\tau}] \times \mathbb{R}^d$. *There exists contains* $b_N < b_{\overline{\sigma}}$ *and* $\mathsf{c}_N \ge 1$, *independent of* $\bar{\tau}$ *and* ϵ_d, *such that*

$$N_l(\bar{\tau}) \le \epsilon_d b_N \quad \Longrightarrow \quad N_l(\bar{\tau}) \le \mathsf{c}_N |\mathsf{w}_0 - \mathsf{w}^\star|_l$$

Proof The main steps of the proof consist in proving that near equilibrium

$$N_l^2(t) \le C' |\mathsf{w}_0 - \mathsf{w}^\star|_l^2 + \frac{1}{\epsilon_d} N_l(t)^3.$$

At zeroth order, we may integrate the relative entropy balance and use $\overline{\mathsf{B}}_{ij} = \overline{\mathsf{B}}_{ij}^{\,\star} + (\overline{\mathsf{B}}_{ij} - \overline{\mathsf{B}}_{ij}^{\,\star})$, Parseval identity, and the entropic inequalities of Lemma 3.3, to get that

$$|\delta\mathsf{w}|_0^2 + \epsilon_d \int_0^t |\partial_{\mathsf{x}} \mathsf{w}_{\mathrm{II}}|_0^2 \, d\tau + \int_0^t |\pi \mathsf{w}|_0^2 \, d\tau \le \mathsf{c}_1 |\delta \mathsf{w}_0|_0^2 + \mathsf{c}_1 N_l^3(t).$$

Using the derived Eq. (3.34) with $\mathsf{f} = 0$ and $\mathsf{g} = \overline{\mathsf{h}}$ with commutator estimates, we next obtain an lth order entropic type estimate in the form

$$|\delta\mathsf{w}|_l^2 + \epsilon_d \int_0^t |\partial_{\mathsf{x}} \mathsf{w}_{\mathrm{II}}|_l^2 \, d\tau + \int_0^t |\pi \mathsf{w}|_l^2 \, d\tau \le \mathsf{c}_1 |\delta \mathsf{w}_0|_l^2 + \frac{1}{\epsilon_d} \mathsf{c}_2 N_l(t)^3.$$

In order to estimate the time integral of the hyperbolic variable we use the compensating matrices K_j, $l \in \mathfrak{D}$, with the linearized equation

$$\overline{\mathsf{A}}_0^{\,\star} \partial_t \mathsf{w} + \sum_{i\in\mathfrak{D}} \overline{\mathsf{A}}_i^{\,\star} \partial_i \mathsf{w} - \epsilon_d \sum_{i,j\in\mathfrak{D}} \overline{\mathsf{B}}_{ij}^{\,\star} \partial_i \partial_j \mathsf{w} + \overline{\mathsf{L}}^\star \mathsf{w} = \mathsf{h}^{\mathrm{lin}}, \tag{3.49}$$

where $\overline{\mathsf{A}}_0^{\,\star} = \overline{\mathsf{A}}_0(\mathsf{w}^\star)$, $\overline{\mathsf{A}}_i^{\,\star} = \overline{\mathsf{A}}_i(\mathsf{w}^\star)$, $\overline{\mathsf{B}}_{ij}^{\,\star} = \overline{\mathsf{B}}_{ij}(\mathsf{w}^\star)$, $\overline{\mathsf{L}}^\star = \overline{\mathsf{L}}(\mathsf{w}^\star)$ and

$$\begin{aligned}\mathsf{h}^{\mathrm{lin}} = &- \sum_{i\in\mathfrak{D}} (\overline{\mathsf{A}}_0^{\,\star} \overline{\mathsf{A}}_0^{\,-1} \overline{\mathsf{A}}_i - \overline{\mathsf{A}}_i^{\,\star}) \partial_i \mathsf{w} + \epsilon_d \sum_{i,j\in\mathfrak{D}} (\overline{\mathsf{A}}_0^{\,\star} \overline{\mathsf{A}}_0^{\,-1} \overline{\mathsf{B}}_{ij} - \overline{\mathsf{B}}_{ij}^{\,\star}) \partial_i \partial_j \mathsf{w} \\ &- (\overline{\mathsf{A}}_0^{\,\star} \overline{\mathsf{A}}_0^{\,-1} \overline{\mathsf{L}} - \overline{\mathsf{L}}^\star) \pi \mathsf{w} + \sum_{i,j\in\mathfrak{D}} \overline{\mathsf{A}}_0^{\,\star} \overline{\mathsf{A}}_0^{\,-1} \partial_i \overline{\mathsf{B}}_{ij} \partial_j \mathsf{w} + \epsilon_d \overline{\mathsf{A}}_0^{\,\star} \overline{\mathsf{A}}_0^{\,-1} \overline{\mathsf{h}}.\end{aligned}$$

We then apply ∂^α to Eq. (3.49), multiply on the left by $K_j|\alpha|!/\alpha!$ and $\partial_j\partial^\alpha \mathsf{w}$, sum over $0 \le |\alpha| \le l-1$ and $j \in \mathfrak{D}$, and integrate over $\mathsf{x} \in \mathbb{R}^d$ and $\tau \in [0,t]$. The time derivative terms are controlled with

$$2\langle K_j\overline{\mathsf{A}}_0{}^\star \partial_t\partial^\alpha \mathsf{w}, \partial_j\partial^\alpha \mathsf{w}\rangle = \partial_t\langle K_j\overline{\mathsf{A}}_0{}^\star \partial^\alpha \delta\mathsf{w}, \partial_j\partial^\alpha \mathsf{w}\rangle + \partial_j\langle K_j\overline{\mathsf{A}}_0{}^\star \partial_t\partial^\alpha \mathsf{w}, \partial^\alpha \delta\mathsf{w}\rangle,$$

whereas for the time integral of hyperbolic terms we may further use the properties of the compensating matrices

$$\delta_1 \int_0^t |\partial_{\mathsf{x}}\mathsf{w}|_{l-1}^2\,d\tau - \mathsf{c}_1\int_0^t |\partial_{\mathsf{x}}\mathsf{w}_{\mathrm{II}}|_{l-1}^2\,d\tau - \mathsf{c}_1\int_0^t |\pi\mathsf{w}|_l^2\,d\tau$$
$$\le \sum_{\substack{i,j\in\mathfrak{D}\\ 0\le|\alpha|\le l-1}} \int_0^t\int_{\mathbb{R}^d} \frac{|\alpha|!}{\alpha!}\langle K_j\overline{\mathsf{A}}_i{}^\star \partial_i\partial^\alpha \mathsf{w}, \partial_j\partial^\alpha \mathsf{w}\rangle\,d\mathsf{x}\,d\tau.$$

The terms arising from the dissipation matrices of the left hand side (3.49) and the cubic terms arising from $\mathsf{h}^{\mathrm{lin}}$ are easily controlled and combining these estimates with the lth entropic estimate completes the proof. □

Theorem 3.11 *Let $d \geqslant 1$, $l \geqslant [d/2]+2$, and consider the equations governing multicomponent reactive flows in normal form $\mathsf{w} = (\mathsf{w}_{\mathrm{I}}, \mathsf{w}_{\mathrm{II}})^t$ with*

$$\mathsf{w}_{\mathrm{I}} = \rho, \qquad \mathsf{w}_{\mathrm{II}} = \Big(\frac{g_2-g_1}{RT}, \dots, \frac{g_n-g_1}{RT}, \frac{\boldsymbol{v}}{RT}, \frac{-1}{RT}\Big)^t. \tag{3.50}$$

Assume that the initial condition is in the form $\mathsf{w}_0 = \mathsf{w}^\star + \epsilon_d(\widetilde{\mathsf{w}}_0 - \mathsf{w}^\star)$. There exists $\bar{b} > 0$ and $\bar{c} > 0$ such that, if $\|\mathsf{w}_0 - \mathsf{w}^\star\|_l < \epsilon_d\bar{b}$, then the Cauchy problem with $\mathsf{w}(0,x) = \mathsf{w}_0(x)$ has a global solution with

$$\mathsf{w}_{\mathrm{I}} - \mathsf{w}_{\mathrm{I}}^\star \in C^0\big([0,\infty); H^l\big) \cap C^1\big([0,\infty); H^{l-1}\big), \tag{3.51}$$

$$\mathsf{w}_{\mathrm{II}} - \mathsf{w}_{\mathrm{II}}^\star \in C^0\big([0,\infty); H^l\big) \cap C^1\big([0,\infty); H^{l-2}\big), \tag{3.52}$$

and we have the estimates

$$\|\mathsf{w}(t) - \mathsf{w}^\star\|_l^2 + \epsilon_d\int_0^t \big(\|\nabla \mathsf{w}_{\mathrm{I}}(\tau)\|_{l-1}^2 + \|\nabla \mathsf{w}_{\mathrm{II}}(\tau)\|_l^2\big)\,d\tau \leqslant \bar{c}^2\|\mathsf{w}_0 - \mathsf{w}^\star\|_l^2,$$

and $\sup_{\mathbb{R}^d}|\mathsf{w}(t) - \mathsf{w}^\star|$ goes to zero as $t \to \infty$.

Sketch of the Proof The existence proof mainly consists in using the local existence theorem repeatedly combined to the a priori estimates independent of the time interval using strict dissipativity [18, 27, 31, 32, 53].

More specifically, the exists a local solution for $b = b_{\overline{\sigma}}$ over the time interval $[0,\bar{\tau}]$ with $N_l(\bar{\tau}) \le \mathsf{c}_{\mathrm{loc}}|\mathsf{w}_0 - \mathsf{w}^\star|_l$ and $\mathsf{c}_{\mathrm{loc}} \ge 1$ depends on O_1 and $b = b_{\overline{\sigma}}$. We now define $\bar{b}$ by

$$\bar{b} = \min\Big(\frac{b_N}{\mathsf{c}_{\mathrm{loc}}}, \frac{b_N}{\mathsf{c}_N\sqrt{1+\mathsf{c}_{\mathrm{loc}}^2}}\Big),$$

and assuming that $|\mathsf{w}_0 - \mathsf{w}^\star|_l < \epsilon_d \bar{b}$ then

$$N_l(\bar{\tau}) \le \mathsf{c}_{\mathrm{loc}}|\mathsf{w}_0 - \mathsf{w}^\star|_l < \mathsf{c}_{\mathrm{loc}}\epsilon_d\bar{b} \le \epsilon_d b_N < \epsilon_d b_{\overline{\sigma}},$$

and from the estimates of Theorem 3.10 $N_l(\bar{\tau}) \le \mathsf{c}_N|\mathsf{w}_0 - \mathsf{w}^\star|_l < \epsilon_d\mathsf{c}_N\bar{b}$.

One can now start again from $\mathsf{w}(\bar{\tau})$ at $\bar{\tau}$ since $|\mathsf{w}(\bar{\tau}) - \mathsf{w}^\star|_l \le N_l(\bar{\tau}) < \epsilon_d b_{\overline{\sigma}}$ and this yields a solution on $[\bar{\tau}, 2\bar{\tau}]$. Defining

$$N_l^2(\bar{\tau}, 2\bar{\tau}) = \sup_{\bar{\tau}\le\tau\le 2\bar{\tau}} |\mathsf{w}(\tau) - \mathsf{w}^\star|_l^2 + \epsilon_d \int_{\bar{\tau}}^{2\bar{\tau}} \big(|\partial_{\mathsf{x}}\mathsf{w}_{\mathrm{I}}(\tau)|_{l-1}^2 + |\partial_{\mathsf{x}}\mathsf{w}_{\mathrm{II}}(\tau)|_l^2\big)\,d\tau,$$

we then have $N_l(\bar{\tau}, 2\bar{\tau}) \le \mathsf{c}_{\mathrm{loc}}|\mathsf{w}(\bar{\tau}) - \mathsf{w}^\star|_l \le \mathsf{c}_{\mathrm{loc}}N_l(\bar{\tau})$. This yields

$$N_l(2\bar{\tau}) \le (1+\mathsf{c}_{\mathrm{loc}}^2)^{1/2}N_l(\bar{\tau}) < (1+\mathsf{c}_{\mathrm{loc}}^2)^{1/2}\mathsf{c}_N\epsilon_d\bar{b} \le \epsilon_d b_N < \epsilon_d b_{\overline{\sigma}},$$

as well as $N_l(2\bar{\tau}) \le \mathsf{c}_N\epsilon_d\bar{b} < \epsilon_d b_N$ from Theorem 3.10. Since $N_l(2\bar{\tau}) < \epsilon_d b_{\overline{\sigma}}$ we have $\mathsf{w}(2\bar{\tau}) \in \mathcal{O}_{\overline{\sigma}}$ and one can start again from $\mathsf{w}(2\bar{\tau})$ with a local solution over $[2\bar{\tau}, 3\bar{\tau}]$ and an induction completes the proof. The proofs are rescaled versions of the classical proofs [7, 18, 27, 31, 32, 53] used in the standard situation $\epsilon_d = 1$. □

3.3 Relaxation of Internal Energy

The relaxation of internal energy is of fundamental importance for fluids with internal degrees of freedom and is related to the concept of volume viscosity as established by the kinetic theory of polyatomic gases [5, 6, 40–46]. The volume viscosity coefficient is of the same order as the shear viscosity coefficient for polyatomic gases as established by the kinetic theory [5–7, 44, 45] and experimental measurements and its impact in fluid mechanics has also been established [55, 56]. This is a strong motivation for investigating mathematically the fast relaxation of internal energy as well as the concept of volume viscosity in nonequilibrium fluid models.

We study in this section a mathematical model for the relaxation of translational and internal temperatures in a nonequilibrium gas with a single internal energy mode. In the more general situation where there are several internal energy processes for the relaxation of internal energy, or for state-to-state models, the relaxation regime leads to the same type of relaxation equation [44, 45]. Symmetrization of the corresponding system of partial differential equations leads to hyperbolic-parabolic systems with stiff sources [47–54]. We prove local in time existence as well as error estimates between the out of equilibrium solution and the one-temperature equilibrium fluid solution and justify the apparition of volume viscosity terms. In the situation of well prepared data, the proofs are adapted from those for nonstiff models [52, 53, 60]. In the situation of ill prepared data, multiscale expansions taking into account the existence of initial layers are obtained by using a previously established theorem valid for a family of hyperbolic-parabolic problems [60].

3.3.1 Governing Equations

In a nonequilibrium gas with internal degrees of freedom, the conservation of mass, momentum, total energy and internal energy may be written [44, 45]

$$\partial_t \rho + \boldsymbol{\nabla}\cdot(\rho \boldsymbol{v}) = 0, \tag{3.53}$$

$$\partial_t(\rho \boldsymbol{v}) + \boldsymbol{\nabla}\cdot(\rho \boldsymbol{v}\otimes \boldsymbol{v} + p\boldsymbol{I}) + \boldsymbol{\nabla}\cdot\boldsymbol{\Pi} = 0, \tag{3.54}$$

$$\partial_t(\mathcal{E}_{\mathrm{tr}} + \mathcal{E}_{\mathrm{in}} + \tfrac{1}{2}\rho|\boldsymbol{v}|^2) + \boldsymbol{\nabla}\cdot\big(\boldsymbol{v}(\mathcal{E}_{\mathrm{tr}} + \mathcal{E}_{\mathrm{in}} + \tfrac{1}{2}\rho|\boldsymbol{v}|^2 + p)\big) + \boldsymbol{\nabla}\cdot(\boldsymbol{Q}_{\mathrm{tr}} + \boldsymbol{Q}_{\mathrm{in}} + \boldsymbol{\Pi}\cdot\boldsymbol{v}) = 0, \tag{3.55}$$

$$\partial_t \mathcal{E}_{\mathrm{in}} + \boldsymbol{\nabla}\cdot(\boldsymbol{v}\mathcal{E}_{\mathrm{in}}) + \boldsymbol{\nabla}\cdot\boldsymbol{Q}_{\mathrm{in}} = \omega_{\mathrm{in}}, \tag{3.56}$$

where ∂_t denotes the time derivative operator, $\boldsymbol{\nabla}$ the space derivative operator, ρ the mass density, $\boldsymbol{v}$ the fluid velocity, $\otimes$ the tensor product symbol, p the pressure, $\boldsymbol{I}$ the unit tensor in the physical space $\mathbb{R}^d$, $\boldsymbol{\Pi}$ the viscous tensor, $\mathcal{E}_{\mathrm{tr}}$ the internal energy of translational origin per unit volume, $\mathcal{E}_{\mathrm{in}}$ the internal energy of internal origin per unit volume, $\boldsymbol{Q}_{\mathrm{tr}}$ the heat flux of translational origin, $\boldsymbol{Q}_{\mathrm{in}}$ the heat flux of internal origin, and ω_{in} the energy exchange rate. We denote by $d \geq 1$ the dimension of the physical space so that $\boldsymbol{v} = (v_1, \dots, v_d)^t$, $\boldsymbol{\nabla} = (\partial_1, \dots, \partial_d)^t$, $\boldsymbol{Q}_{\mathrm{tr}} = (Q_{\mathrm{tr},1}, \dots, Q_{\mathrm{tr},d})^t$ and $\boldsymbol{Q}_{\mathrm{in}} = (Q_{\mathrm{in},1}, \dots, Q_{\mathrm{in},d})^t$ with v_i denoting the velocity in the ith spatial direction, ∂_i the derivation in the ith spatial direction, with similar notation for the fluxes and bold symbols are used for vector or tensor quantities in the physical space $\mathbb{R}^d$. These Eqs. (3.53)–(3.56) may further be combined in order to form the evolution equation for the internal energy of translational origin

$$\partial_t \mathcal{E}_{\mathrm{tr}} + \boldsymbol{\nabla}\cdot(\boldsymbol{v}\mathcal{E}_{\mathrm{tr}}) + \boldsymbol{\nabla}\cdot\boldsymbol{Q}_{\mathrm{tr}} = -p\boldsymbol{\nabla}\cdot\boldsymbol{v} - \boldsymbol{\Pi}:\boldsymbol{\nabla}\boldsymbol{v} - \omega_{\mathrm{in}}. \tag{3.57}$$

The Eqs. (3.53)–(3.56) have to be completed by relations expressing the thermodynamic properties $\mathcal{E}_{\mathrm{in}}$, $\mathcal{E}_{\mathrm{tr}}$, and p, the rate of energy exchange ω_{in}, and the transport fluxes $\boldsymbol{\Pi}$, $\boldsymbol{Q}_{\mathrm{in}}$ and $\boldsymbol{Q}_{\mathrm{tr}}$.

In the framework of the kinetic theory of polyatomic gases out of thermodynamic equilibrium, the heat fluxes $\boldsymbol{Q}_{\mathrm{tr}}$ and $\boldsymbol{Q}_{\mathrm{in}}$ are in the form [44, 45]

$$\boldsymbol{Q}_{\mathrm{tr}} = - \lambda_{\mathrm{tr,tr}} \boldsymbol{\nabla} T_{\mathrm{tr}} - \lambda_{\mathrm{tr,in}} \boldsymbol{\nabla} T_{\mathrm{in}}, \tag{3.58}$$

$$\boldsymbol{Q}_{\mathrm{in}} = - \lambda_{\mathrm{in,tr}} \boldsymbol{\nabla} T_{\mathrm{tr}} - \lambda_{\mathrm{in,in}} \boldsymbol{\nabla} T_{\mathrm{in}}, \tag{3.59}$$

where $\lambda_{\mathrm{tr,tr}}$, $\lambda_{\mathrm{tr,in}}$, $\lambda_{\mathrm{in,tr}}$, and $\lambda_{\mathrm{in,in}}$ denote the thermal conductivities. Both temperature gradients are involved in both fluxes even though the cross thermal conductivities $\lambda_{\mathrm{tr,in}}$ and $\lambda_{\mathrm{in,tr}}$ are generally smaller than $\lambda_{\mathrm{tr,tr}}$ and $\lambda_{\mathrm{in,in}}$ [44, 45]. On the other hand, the viscous tensor is given by

$$\boldsymbol{\Pi} = -\eta\big(\boldsymbol{\nabla} \boldsymbol{v} + (\boldsymbol{\nabla} \boldsymbol{v})^t - \tfrac{2}{3}(\boldsymbol{\nabla}\cdot\boldsymbol{v})\boldsymbol{I}\big), \tag{3.60}$$

where η denotes the shear viscosity and the factor $2/3$ arise from the dimension of the velocity space of the underlying kinetic framework. It will be assumed in the following that the dimension of the physical space d is such that $d \leq 3$. We also denote by Π_{ij}, $1 \leq i, j \leq d$, the components of the viscous tensor $\boldsymbol{\Pi}$.

The thermal conductivities $\lambda_{\mathrm{tr,tr}}$, $\lambda_{\mathrm{tr,in}}$, $\lambda_{\mathrm{in,tr}}$, and $\lambda_{\mathrm{in,in}}$ and the shear viscosity η are defined in a kinetic framework in terms of bracket products between solutions of integral linearized Boltzmann equation [44, 45]. The mathematical structure of the transport coefficients presented in the following is extracted from these kinetic expressions. It is also noted with (3.60) that the viscous tensor $\boldsymbol{\Pi}$ does not present a volume viscosity term and our aim is to investigate the apparition of such a contribution in the one-temperature equilibrium limit model when the energy exchanges become fast.

The pressure p, the total internal energy per unit volume $\mathcal{E} = \mathcal{E}_{\mathrm{tr}} + \mathcal{E}_{\mathrm{in}}$, the internal energy of translational origin per unit volume $\mathcal{E}_{\mathrm{tr}}$, and the internal energy of internal origin per unit volume $\mathcal{E}_{\mathrm{in}}$ are in the form

$$p = \rho \mathfrak{r} T_{\mathrm{tr}}, \qquad \mathcal{E} = \rho e, \qquad \mathcal{E}_{\mathrm{tr}} = \rho e_{\mathrm{tr}}, \qquad \mathcal{E}_{\mathrm{in}} = \rho e_{\mathrm{in}}, \tag{3.61}$$

where $\mathfrak{r} = R/m$ denotes the gas constant per unit mass, e the total internal energy per unit mass, e_{tr} the internal energy of translational origin per unit mass, and e_{in} the internal energy of internal origin per unit mass. These energies $e(T_{\mathrm{tr}}, T_{\mathrm{in}})$, $e_{\mathrm{tr}}(T_{\mathrm{tr}})$, and $e_{\mathrm{in}}(T_{\mathrm{in}})$ are given by

$$e = e_{\mathrm{tr}} + e_{\mathrm{in}}, \qquad e_{\mathrm{tr}} = c_{\mathrm{v,tr}} T_{\mathrm{tr}}, \qquad e_{\mathrm{in}} = e_{\mathrm{in}}^{\mathrm{st}} + \int_{T^{\mathrm{st}}}^{T_{\mathrm{in}}} c_{\mathrm{in}}(\tau)\, d\tau, \tag{3.62}$$

where $c_{\mathrm{v,tr}} = \frac{3}{2}\mathfrak{r}$ denotes the translational heat at constant volume per unit mass, T_{tr} the translational temperature, c_{in} the internal heat per unit mass, T_{in} the internal temperature, T^{st} the standard temperature, and $e_{\mathrm{in}}^{\mathrm{st}}$ the internal formation energy at the standard temperature T^{st}. We will also use in the following the translational heat at constant pressure per unit mass $c_{p,\mathrm{tr}} = \frac{5}{2}\mathfrak{r}$ and the formation energy at zero temperature $e_{\mathrm{in}}^0 = e_{\mathrm{in}}(0) = e_{\mathrm{in}}^{\mathrm{st}} - \int_0^{T^{\mathrm{st}}} c_{\mathrm{in}}(\tau)\, d\tau$.

The total entropy per unit volume $\mathcal{S} = \mathcal{S}_{tr} + \mathcal{S}_{in}$, the translational entropy per unit volume $\mathcal{S}_{tr}$ and the internal entropy per unit volume $\mathcal{S}_{in}$ are given by

$$\mathcal{S} = \rho s, \qquad \mathcal{S}_{tr} = \rho s_{tr}, \qquad \mathcal{S}_{in} = \rho s_{in}, \tag{3.63}$$

where s denotes the total entropy per unit mass, s_{tr} the translational entropy per unit mass, and s_{in} the internal entropy per unit mass. These entropies per unit mass $s = s(\rho, T_{tr}, T_{in})$, $s_{tr}(\rho, T_{tr})$, and $s_{in}(T_{in})$ are in the form $s = s_{tr} + s_{in}$ and

$$s_{tr} = s_{tr}^{st} + c_{v,tr} \log\Big(\frac{T_{tr}}{T^{st}}\Big) - \mathfrak{r} \log\Big(\frac{\rho}{\rho^{st}}\Big), \qquad s_{in} = s_{in}^{st} + \int_{T^{st}}^{T_{in}} \frac{c_{in}(\tau)}{\tau}\, d\tau, \tag{3.64}$$

where s_{tr}^{st} denotes the translational formation entropy at the standard temperature T^{st} and pressure p^{st}, $\rho^{st} = mp^{st}/(RT^{st})$ the standard mass density, m the molar mass of the gas, and s_{in}^{st} the internal formation entropy at the standard temperature T^{st}. We also introduce the translational and internal Gibbs functions per unit mass $g_{tr} = e_{tr} + \mathfrak{r}T_{tr} - T_{tr}s_{tr}$ and $g_{in} = e_{in} - T_{in}s_{in}$. From the definition of $\mathcal{S}_{tr}$ and $\mathcal{S}_{in}$ it is further obtained that $T_{tr}\,\mathbb{D}\mathcal{S}_{tr} = \mathbb{D}\mathcal{E}_{tr} - g_{tr}\,\mathbb{D}\rho$ and $T_{in}\,\mathbb{D}\mathcal{S}_{in} = \mathbb{D}\mathcal{E}_{in} - g_{in}\,\mathbb{D}\rho$ where $\mathbb{D}$ denotes the total differential, so that the volumetric Gibbs relation out of thermodynamic equilibrium may be written [44, 45]

$$\mathbb{D}\mathcal{S} = \frac{1}{T_{tr}}\mathbb{D}\mathcal{E}_{tr} + \frac{1}{T_{in}}\mathbb{D}\mathcal{E}_{in} - \Big(\frac{g_{tr}}{T_{tr}} + \frac{g_{in}}{T_{in}}\Big)\mathbb{D}\rho. \tag{3.65}$$

The rate of energy exchange between the translational and internal degrees of freedom ω_{in} may finally be written in the relaxation form [44, 45]

$$\omega_{in} = \frac{\rho c_{in}}{\tau_{in}}(T_{tr} - T_{in}), \tag{3.66}$$

where τ_{in} denotes the energy exchange time. The fast relaxation limit is then obtained as the relaxation time τ_{in} goes to zero.

3.3.2 *Mathematical Framework*

The mathematical assumptions naturally associated with fluids out of thermodynamic equilibrium are presented. The rescaled energy exchange time $\overline{\tau}_{in}$ as well as the rescaled transport coefficients $\overline{\eta}$, $\overline{\lambda}_{tr,tr}$, $\overline{\lambda}_{tr,in}$, $\overline{\lambda}_{in,tr}$, and $\overline{\lambda}_{in,in}$ are introduced in order to investigate the fast relaxation limit and $\gamma \geq 3$ denotes the regularity class of thermodynamic functions [7, 44, 45, 123].

(H$_1$) The formation energy e_{in}^{st} and formation entropies s_{tr}^{st} and s_{in}^{st} are real constants. The mass per unit mole m, the gas constant R, and the gas constant per unit mass $\mathfrak{r} = R/m$ are positive. The internal species heat per unit mass

$c_{\rm in}$ is a $C^{\gamma-1}$ function of the internal temperature $T_{\rm in}$ over $[0,\infty)$ and there exist constants $\underline{c}$ and $\overline{c}$ such that $0 < \underline{c} \leqslant c_{\rm in}(T_{\rm in}) \leqslant \overline{c}$ for all $T_{\rm in} \geqslant 0$.

(H$_2$) The energy exchange rate $\tau_{\rm in}(p, T_{\rm tr}, T_{\rm in})$ is in the form

$$\tau_{\rm in} = \epsilon\overline{\tau}_{\rm in} = \epsilon\frac{p^{\rm st}\overline{\tau}^{\rm st}_{\rm in}}{p}, \tag{3.67}$$

where $\epsilon \in (0, 1]$ is a small parameter, $\overline{\tau}_{\rm in}(p, T_{\rm tr}, T_{\rm in}) = p^{\rm st}\overline{\tau}^{\rm st}_{\rm in}/p$ the rescaled energy exchange time and $\overline{\tau}^{\rm st}_{\rm in}(T_{\rm tr}, T_{\rm in})$ the rescaled energy exchange time at the standard pressure $p^{\rm st}$ which only depends on $T_{\rm tr}$ and $T_{\rm in}$. The rescaled time $\overline{\tau}^{\rm st}_{\rm in}$ is a positive C^γ function of both temperatures $T_{\rm tr}, T_{\rm in} \in (0,\infty)$.

The internal specific heat remains bounded away from zero since we consider a polyatomic gas governed by Boltzmann type statistics [44, 45]. The specific heat could also depend on both temperatures but such a dependence would easily be included. The fact that the quantity $p\tau_{\rm in}$ only depends on $(T_{\rm tr}, T_{\rm in})$ is a direct consequence from the kinetic theory and will imply that the volume viscosity only depends on $(T_{\rm tr}, T_{\rm in})$. The assumptions associated with the transport coefficients are the following.

(H$_1$) The coefficients η, $\lambda_{\rm tr,tr}$, $\lambda_{\rm tr,in}$, $\lambda_{\rm in,tr}$, and $\lambda_{\rm in,in}$ are in the form

$$\eta = \epsilon_d\overline{\eta}, \quad \lambda_{\rm tr,tr} = \epsilon_d\overline{\lambda}_{\rm tr,tr}, \quad \lambda_{\rm tr,in} = \epsilon_d\overline{\lambda}_{\rm tr,in},$$
$$\lambda_{\rm in,tr} = \epsilon_d\overline{\lambda}_{\rm in,tr}, \quad \lambda_{\rm in,in} = \epsilon_d\overline{\lambda}_{\rm in,in}, \tag{3.68}$$

where $\epsilon_d \in (0, 1]$ denotes a positive parameter, and $\overline{\eta}$, $\overline{\lambda}_{\rm tr,tr}$, $\overline{\lambda}_{\rm tr,in}$, $\overline{\lambda}_{\rm in,tr}$, and $\overline{\lambda}_{\rm in,in}$ the rescaled transport coefficients. These coefficients $\overline{\eta}$, $\overline{\lambda}_{\rm tr,tr}$, $\overline{\lambda}_{\rm tr,in}$, $\overline{\lambda}_{\rm in,tr}$, and $\overline{\lambda}_{\rm in,in}$ are C^γ functions of both temperatures $T_{\rm tr}, T_{\rm in} \in (0,\infty)$.

(H$_4$) For any $T_{\rm tr}, T_{\rm in} \in (0,\infty)$, the matrix

$$\begin{bmatrix} T_{\rm in}^2\overline{\lambda}_{\rm in,in} & T_{\rm tr}^2\overline{\lambda}_{\rm in,tr} \\ T_{\rm in}^2\overline{\lambda}_{\rm tr,in} & T_{\rm tr}^2\overline{\lambda}_{\rm tr,tr} \end{bmatrix}, \tag{3.69}$$

is symmetric positive definite. In the viscous tensor (3.60), the coefficient η is positive and the dimension d is such that $1 \le d \le 3$.

The symmetry properties of the matrix (3.69) deduced from the kinetic theory may also be interpreted as Onsager type relations associated with the variables $1/T_{\rm in}$ and $1/T_{\rm tr}$. Assumptions (H$_1$ – H$_4$) are assumed to hold in the remaining part of the section whenever we consider the two-temperature relaxation model. From the differential of entropy (3.65), after some algebra, the following balance of entropy is obtained [44, 45]. This equation shows that—physical—entropy production is nonnegative

$$\partial_t \mathcal{S} + \nabla\cdot(\boldsymbol{v}\mathcal{S}) + \nabla\cdot\Big(\frac{\boldsymbol{Q}_{\rm tr}}{T_{\rm tr}} + \frac{\boldsymbol{Q}_{\rm in}}{T_{\rm in}}\Big) =$$

$$\frac{\epsilon_d\bar{\lambda}_{\rm tr,tr}}{T_{\rm tr}^2}|\nabla T_{\rm tr}|^2 + \epsilon_d\Big(\frac{\bar{\lambda}_{\rm tr,in}}{T_{\rm tr}^2} + \frac{\bar{\lambda}_{\rm in,tr}}{T_{\rm in}^2}\Big)\nabla T_{\rm in}\cdot\nabla T_{\rm tr} + \frac{\epsilon_d\bar{\lambda}_{\rm in,in}}{T_{\rm in}^2}|\nabla T_{\rm in}|^2$$

$$+ \frac{\epsilon_d\bar{\eta}}{T_{\rm tr}}\Big(\frac{1}{d} - \frac{1}{3}\Big)(\nabla\cdot\boldsymbol{v})^2 + \frac{\epsilon_d\bar{\eta}}{2T_{\rm tr}}\big|\nabla\boldsymbol{v} + \nabla\boldsymbol{v} - \tfrac{2}{d}\nabla\cdot\boldsymbol{v}\,\boldsymbol{I}\big|^2 + \frac{1}{\epsilon}\frac{\rho c_{\rm in}}{\overline{\tau}_{\rm in}}\frac{(T_{\rm tr}-T_{\rm in})^2}{T_{\rm tr}T_{\rm in}}. \tag{3.70}$$

Letting $\mathsf{n} = d + 3$, the conservative variable $\mathsf{u} \in \mathbb{R}^{\mathsf{n}}$ associated with the Eqs. (3.53)–(3.55) and the corresponding natural variable $\mathsf{z} \in \mathbb{R}^{\mathsf{n}}$ are

$$\mathsf{u} = \big(\rho,\, \rho\boldsymbol{v},\, \mathcal{E}_{\rm tr} + \mathcal{E}_{\rm in} + \tfrac{1}{2}\rho\boldsymbol{v}\cdot\boldsymbol{v},\, \mathcal{E}_{\rm in}\big)^t, \qquad \mathsf{z} = \big(\rho,\, \boldsymbol{v},\, T_{\rm tr},\, T_{\rm in}\big)^t. \tag{3.71}$$

For convenience, the velocity components of vectors in $\mathbb{R}^{\mathsf{n}} = \mathbb{R}\times\mathbb{R}^d\times\mathbb{R}^2$ are written as vectors of $\mathbb{R}^d$. Assuming that (H_1) holds, the map $\mathsf{z} \longmapsto \mathsf{u}$ is a C^γ diffeomorphism from the open set $\mathcal{O}_{\mathsf{z}} = (0,\infty)\times\mathbb{R}^d\times(0,\infty)^2$ onto the open set $\mathcal{O}_{\mathsf{u}}$ of $\mathbb{R}^{\mathsf{n}}$ given by

$$\mathcal{O}_{\mathsf{u}} = \big\{\mathsf{u} = (\mathsf{u}_\rho, \mathsf{u}_{\boldsymbol{v}}, \mathsf{u}_{\rm tl}, \mathsf{u}_{\rm in})^t \in \mathbb{R}^{\mathsf{n}};\ \mathsf{u}_\rho > 0,\ \mathsf{u}_{\rm tl} > f(\mathsf{u}_\rho, \mathsf{u}_{\boldsymbol{v}}, \mathsf{u}_{\rm in}),\ \mathsf{u}_{\rm in} > \mathsf{u}_\rho e_{\rm in}^0\big\}, \tag{3.72}$$

where $f(\mathsf{u}_\rho, \mathsf{u}_{\boldsymbol{v}}, \mathsf{u}_{\rm in}) = \mathsf{u}_{\rm in} + \frac{1}{2}\mathsf{u}_{\boldsymbol{v}}\cdot\mathsf{u}_{\boldsymbol{v}}/\mathsf{u}_\rho$ and the open set $\mathcal{O}_{\mathsf{u}}$ is convex.

The equations modeling fluids out of thermodynamic equilibrium may then be written in the compact form

$$\partial_t\mathsf{u} + \sum_{i\in\mathfrak{D}}\partial_i\mathsf{F}_i + \epsilon_d\sum_{i\in\mathfrak{D}}\partial_i\mathsf{F}_i^{\rm d} - \frac{1}{\epsilon}\Omega = 0, \tag{3.73}$$

where F_i is the convective flux in the ith direction, ϵ_d the Knudsen number, $\mathsf{F}_i^{\rm d}$ the rescaled dissipative flux in the ith direction, ϵ the relaxation parameter, Ω the rescaled source term, and $\mathfrak{D} = \{1,\ldots,d\}$ the indexing set of spatial dimensions. From the governing Eqs. (3.53)–(3.55) the convective flux F_i, the dissipative flux $\epsilon_d\mathsf{F}_i^{\rm d}$, and the source term Ω are given by

$$\mathsf{F}_i = \big(\rho v_i,\ \rho\boldsymbol{v}v_i + p\boldsymbol{e}_i,\ (\mathcal{E}_{\rm tr} + \mathcal{E}_{\rm in} + \tfrac{1}{2}\rho|\boldsymbol{v}|^2 + p)v_i,\ \mathcal{E}_{\rm in}v_i\big)^t, \tag{3.74}$$

$$\epsilon_d\mathsf{F}_i^{\rm d} = \big(0,\ \boldsymbol{\Pi}_i,\ Q_{{\rm tr},i} + Q_{{\rm in},i} + \boldsymbol{\Pi}_i\cdot\boldsymbol{v},\ Q_{{\rm in},i}\big)^t, \tag{3.75}$$

$$\frac{1}{\epsilon}\Omega = \big(0, \boldsymbol{0}, 0, \omega_{\rm in}\big)^t, \tag{3.76}$$

where $\boldsymbol{e}_i$ denotes the basis vectors of $\mathbb{R}^d$ and $\boldsymbol{\Pi}_i = (\Pi_{1i},\ldots,\Pi_{di})^t$. From the expressions of the viscous tensor and of the heat fluxes we deduce that the dissipative fluxes $\mathsf{F}_i^{\mathrm{d}}$ may be written in the form $\mathsf{F}_i^{\mathrm{d}} = -\sum_{j\in\mathfrak{D}} \mathsf{B}_{ij}(\mathsf{u})\partial_j\mathsf{u}$ where the uniquely defined dissipation matrix B_{ij} relates the rescaled flux $\mathsf{F}_i^{\mathrm{d}}$ in the ith direction to the gradient of the conservative variable $\partial_j\mathsf{u}$ in the jth direction. Further introducing the Jacobian matrices of the convective fluxes $\mathsf{A}_i = \partial_{\mathsf{u}}\mathsf{F}_i$ the governing equations are finally rewritten in the form of a quasilinear second order system

$$\partial_t\mathsf{u} + \sum_{i\in\mathfrak{D}} \mathsf{A}_i(\mathsf{u})\partial_i\mathsf{u} - \epsilon_d \sum_{i,j\in\mathfrak{D}} \partial_i\big(\mathsf{B}_{ij}(\mathsf{u})\partial_j\mathsf{u}\big) - \frac{1}{\epsilon}\Omega(\mathsf{u}) = 0, \tag{3.77}$$

where $\mathsf{u} \in \mathcal{O}_{\mathsf{u}}$, $\mathcal{O}_{\mathsf{u}}$ is an open convex set of $\mathbb{R}^{\mathsf{n}}$, $\mathsf{n} \geq 1$, and ϵ_d and ϵ are positive parameters.

3.3.3 Equations at Equilibrium

The limiting equation governing one-temperature fluids obtained when $\epsilon \to 0$ are presented in this section. The local thermal equilibrium temperature T is defined as the unique scalar T such that

$$e_{\mathrm{tr}}(T) + e_{\mathrm{in}}(T) = e_{\mathrm{tr}}(T_{\mathrm{tr}}) + e_{\mathrm{in}}(T_{\mathrm{in}}), \tag{3.78}$$

keeping in mind that $e_{\mathrm{tr}}(T) + e_{\mathrm{in}}(T)$ is an increasing function of T since $c_{\mathrm{v,tr}}$ and c_{in} are both positive. The temperature T is thus a C^{γ} function of $(T_{\mathrm{tr}}, T_{\mathrm{in}})$ and is the temperature that would be obtained locally at thermal equilibrium $T_{\mathrm{tr}} = T_{\mathrm{in}}$ assuming that the internal energy $e_{\mathrm{tr}} + e_{\mathrm{in}}$ is kept fixed. The function $e_{\mathrm{in}}(T_{\mathrm{in}})$ is generally nonlinear but we may write that $e_{\mathrm{in}}(T) - e_{\mathrm{in}}(T_{\mathrm{in}}) = (T - T_{\mathrm{in}})\widetilde{c}$. where $\widetilde{c}_{\mathrm{in}} = \int_0^1 c_{\mathrm{in}}\big(T_{\mathrm{in}} + \theta(T - T_{\mathrm{in}})\big)\, d\theta$. The relation $e_{\mathrm{tr}}(T_{\mathrm{tr}}) - e_{\mathrm{tr}}(T) = e_{\mathrm{in}}(T) - e_{\mathrm{in}}(T_{\mathrm{in}})$ may thus be recast in the form $(T_{\mathrm{tr}} - T)c_{\mathrm{v,tr}} = (T - T_{\mathrm{in}})\widetilde{c}_{\mathrm{in}}$ or $(T_{\mathrm{tr}} - T)\widetilde{c}_{\mathrm{v}} = (T_{\mathrm{tr}} - T_{\mathrm{in}})\widetilde{c}_{\mathrm{in}}$ where $\widetilde{c}_{\mathrm{v}}(T_{\mathrm{tr}}, T_{\mathrm{in}}) = c_{\mathrm{v,tr}} + \widetilde{c}_{\mathrm{in}}(T_{\mathrm{tr}}, T_{\mathrm{in}})$.

After some calculus, the following equation is obtained for the temperature difference $T_{\mathrm{tr}} - T_{\mathrm{in}}$

$$\partial_t(T_{\mathrm{tr}} - T_{\mathrm{in}}) + \boldsymbol{v}\cdot\nabla(T_{\mathrm{tr}} - T_{\mathrm{in}}) = -\frac{p\nabla\cdot\boldsymbol{v}}{\rho c_{\mathrm{v,tr}}}$$
$$- \frac{1}{\rho c_{\mathrm{v,tr}}}\Big(\boldsymbol{\Pi}:\nabla\boldsymbol{v} + \nabla\cdot\boldsymbol{Q}_{\mathrm{tr}} - \frac{c_{\mathrm{v,tr}}}{c_{\mathrm{in}}}\nabla\cdot\boldsymbol{Q}_{\mathrm{in}}\Big) - \frac{c_{\mathrm{v}}}{c_{\mathrm{v,tr}}}\frac{T_{\mathrm{tr}} - T_{\mathrm{in}}}{\tau_{\mathrm{in}}},$$

where we have defined $c_{\mathrm{v}}(T_{\mathrm{in}}) = c_{\mathrm{v,tr}} + c_{\mathrm{in}}(T_{\mathrm{in}})$. This is a relaxation equation which yields the approximation $T_{\mathrm{tr}} - T_{\mathrm{in}} \simeq -\tau_{\mathrm{in}}\, p\nabla\cdot\boldsymbol{v}/\rho c_{\mathrm{v}} = -\epsilon\overline{\tau}_{\mathrm{in}}\, p\nabla\cdot\boldsymbol{v}/\rho c_{\mathrm{v}}$ valid at leading order in ϵ, after discarding relaxation initial layers. Defining the nonequilibrium volume viscosity by [44, 45]

$$\kappa = \kappa(T_{\rm tr}, T_{\rm in}) = \frac{\mathfrak{r}\,\widetilde{c}_{\rm in}\, p\tau_{\rm in}}{c_{\rm v}\widetilde{c}_{\rm v}} = \frac{\mathfrak{r}\,\widetilde{c}_{\rm in}\, p^{\rm st}\tau_{\rm in}^{\rm st}}{c_{\rm v}\widetilde{c}_{\rm v}} = \epsilon\bar{\kappa},$$

we thus obtain at leading order that $\rho\mathfrak{r}(T_{\rm tr} - T) \simeq -\kappa\boldsymbol{\nabla}\cdot\boldsymbol{v} = -\epsilon\bar{\kappa}\boldsymbol{\nabla}\cdot\boldsymbol{v}$. In the fast relaxation limit with $\epsilon \to 0$ and $\tau_{\rm in} = \epsilon\overline{\tau}_{\rm in}$ we thus have $T_{\rm tr} - T = O(\epsilon)$ as well as $T_{\rm tr} - T_{\rm in} = O(\epsilon)$ and both temperatures $T_{\rm tr}$ and $T_{\rm in}$ converge towards the thermal equilibrium temperature T. After some algebra we thus obtain formally that

$$\rho\mathfrak{r}(T_{\rm tr} - T) = -\kappa_{\rm e}(T)\boldsymbol{\nabla}\cdot\boldsymbol{v} + O\big(\epsilon(\epsilon + \epsilon_d)\big), \tag{3.79}$$

where $\kappa_{\rm e}(T) = \kappa(T, T)$. In the momentum equation, the pressure term $\rho\mathfrak{r}T_{\rm tr}$ may then be written $\rho\mathfrak{r}T - \kappa_{\rm e}\boldsymbol{\nabla}\cdot\boldsymbol{v}$ and the pressure tensor $\rho\mathfrak{r}T_{\rm tr}\boldsymbol{I} + \boldsymbol{\Pi}$ is asymptotically in the form

$$\rho\mathfrak{r}T_{\rm tr}\boldsymbol{I} + \boldsymbol{\Pi} = \rho\mathfrak{r}T\boldsymbol{I} - \kappa_{\rm e}(\boldsymbol{\nabla}\cdot\boldsymbol{v})\boldsymbol{I} - \eta_{\rm e}\big(\boldsymbol{\nabla}\boldsymbol{v} + (\boldsymbol{\nabla}\boldsymbol{v})^t - \tfrac{2}{3}(\boldsymbol{\nabla}\cdot\boldsymbol{v})\boldsymbol{I}\big) + O\big(\epsilon(\epsilon + \epsilon_d)\big),$$

where $\eta_{\rm e}(T) = \eta(T, T)$. This is in agreement with classic one-temperature models where the pressure p is evaluated at the thermal equilibrium temperature T and the viscous tensor includes a volume viscosity term $-\kappa_{\rm e}(\boldsymbol{\nabla}\cdot\boldsymbol{v})\boldsymbol{I}$.

The limiting equations modeling one-temperature fluids are [7]

$$\partial_t\rho_{\rm e} + \boldsymbol{\nabla}\cdot(\rho_{\rm e}\boldsymbol{v}_{\rm e}) = 0, \tag{3.80}$$

$$\partial_t(\rho_{\rm e}\boldsymbol{v}_{\rm e}) + \boldsymbol{\nabla}\cdot(\rho_{\rm e}\boldsymbol{v}_{\rm e}\otimes\boldsymbol{v}_{\rm e} + p_{\rm e}\boldsymbol{I}) + \boldsymbol{\nabla}\cdot\boldsymbol{\Pi}_{\rm e} = 0, \tag{3.81}$$

$$\partial_t(\mathcal{E}_{\rm e} + \tfrac{1}{2}\rho|\boldsymbol{v}_{\rm e}|^2) + \boldsymbol{\nabla}\cdot\big(\boldsymbol{v}_{\rm e}(\mathcal{E}_{\rm e} + \tfrac{1}{2}\rho|\boldsymbol{v}_{\rm e}|^2 + p_{\rm e})\big) + \boldsymbol{\nabla}\cdot(\boldsymbol{Q}_{\rm e} + \boldsymbol{\Pi}_{\rm e}\cdot\boldsymbol{v}_{\rm e}) = 0, \tag{3.82}$$

where the subscript e denotes thermodynamic equilibrium, $\rho_{\rm e}$ the mass density, $\boldsymbol{v}_{\rm e}$ the fluid velocity, $p_{\rm e}$ the pressure, $\boldsymbol{\Pi}_{\rm e}$ the viscous tensor involving the volume viscosity, $\mathcal{E}_{\rm e}$ the internal energy per unit volume, and $\boldsymbol{Q}_{\rm e}$ the heat flux.

The equilibrium viscous tensor is in the form

$$\boldsymbol{\Pi}_{\rm e} = -\kappa_{\rm e}(T_{\rm e})\,(\boldsymbol{\nabla}\cdot\boldsymbol{v}_{\rm e})\boldsymbol{I} - \eta_{\rm e}(T_{\rm e})\,\big(\boldsymbol{\nabla}\boldsymbol{v}_{\rm e} + (\boldsymbol{\nabla}\boldsymbol{v}_{\rm e})^t - \tfrac{2}{3}(\boldsymbol{\nabla}\cdot\boldsymbol{v}_{\rm e})\boldsymbol{I}\big), \tag{3.83}$$

where $\eta_{\rm e}(T_{\rm e}) = \eta(T_{\rm e}, T_{\rm e})$ and $\kappa_{\rm e}(T_{\rm e}) = \kappa(T_{\rm e}, T_{\rm e})$, and the heat flux is given by

$$\boldsymbol{Q}_{\rm e} = -\lambda_{\rm e}(T_{\rm e})\boldsymbol{\nabla}T_{\rm e}, \tag{3.84}$$

with $\lambda_{\rm e}(T_{\rm e}) = \lambda_{\rm tr,tr}(T_{\rm e}, T_{\rm e}) + \lambda_{\rm tr,in}(T_{\rm e}, T_{\rm e}) + \lambda_{\rm in,tr}(T_{\rm e}, T_{\rm e}) + \lambda_{\rm in,in}(T_{\rm e}, T_{\rm e})$. The equilibrium fluid thus corresponds to a compressible one-temperature fluid.

The pressure $p_{\rm e}$ and the internal energy per unit volume $\mathcal{E}_{\rm e}$ are in the form $p_{\rm e} = \rho_{\rm e}\mathfrak{r}T_{\rm e}$ and $\mathcal{E}_{\rm e} = \rho_{\rm e}e_{\rm e}$ where $e_{\rm e}(T_{\rm e})$ denotes the total internal energy per unit mass given by

$$e_e(T_e) = e_e^{st} + \int_{T^{st}}^{T_e} c_{ve}(\tau)\, d\tau, \tag{3.85}$$

where $c_{ve}(T_e) = c_{v,tr} + c_{in}(T_e)$ denotes the heat at constant volume per unit mass, T_e the equilibrium temperature, e_e^{st} formation energy at the standard temperature and we have $e_e(T_e) = e_{tr}(T_e) + e_{in}(T_e)$. We will also use in the following the heat at constant pressure per unit mass $c_{pe} = c_{ve} + \mathfrak{r}$ and the formation energy at zero temperature $e_e^0 = e_e(0)$. The entropy per unit volume $\mathcal{S}_e$ is given by $\mathcal{S}_e = \rho_e s_e$ where s_e denotes the entropy per unit mass. The entropy per unit mass $s_e(\rho_e, T_e)$ is in the form

$$s_e(T_e, \rho_e) = s_{e,st} + \int_{T_{st}}^{T_e} \frac{c_{ve}(\tau)}{\tau}\, d\tau - \mathfrak{r} \log\Big(\frac{\rho_e}{\rho_{st}}\Big), \tag{3.86}$$

where $s_{e,st}$ denotes the formation entropy at the standard temperature T_{st} and pressure p_{st}, ρ_{st} the mass density at the standard state, and we have $s_e(\rho_e, T_e) = s_{tr}(\rho_e, T_e) + s_{in}(T_e)$. Similarly, the Gibbs function per unit mass reads $g_e = e_e + \mathfrak{r}T_e - T_e s_e$ so that $g_e(\rho_e, T_e) = g_{tr}(\rho_e, T_e) + g_{in}(T_e)$. Gibbs relation at thermodynamic equilibrium is then obtained in the form $T_e \mathbb{D}\mathcal{S}_e = \mathbb{D}\mathcal{E}_e - g_e \mathbb{D}\rho_e$. From the differential of entropy at equilibrium after some algebra, the following balance equation is obtained

$$\partial_t \mathcal{S}_e + \nabla\cdot(\boldsymbol{v}_e \mathcal{S}_e) + \nabla\cdot\Big(\frac{\boldsymbol{Q}_e}{T_e}\Big) = -\frac{\boldsymbol{Q}_e\cdot\nabla T_e}{T_e^2} - \frac{\boldsymbol{\Pi}_e : \nabla \boldsymbol{v}_e}{T_e}, \tag{3.87}$$

so that entropy production is nonnegative and only due to gradients.

Letting $\mathsf{n}_e = d + 2$, the conservative variable $\mathsf{u}_e \in \mathbb{R}^{\mathsf{n}_e}$ associated with Eqs. (3.80)–(3.82) is

$$\mathsf{u}_e = \big(\rho_e, \rho_e \boldsymbol{v}_e, \mathcal{E}_e + \tfrac{1}{2}\rho_e \boldsymbol{v}_e\cdot\boldsymbol{v}_e\big)^t, \tag{3.88}$$

and the corresponding natural variable $\mathsf{z}_e \in \mathbb{R}^{\mathsf{n}_e}$ reads $\mathsf{z}_e = \big(\rho_e, \boldsymbol{v}_e, T_e\big)^t$. The map $\mathsf{z}_e \to \mathsf{u}_e$ is a C^γ diffeomorphism from O_{z_e} onto O_{u_e}. Introducing the convective and dissipative fluxes of the equilibrium fluid model (3.80)–(3.82)

$$\mathsf{F}_i{}^{e} = \big(\rho_e v_{ei}, \rho_e \boldsymbol{v}_e v_{ei} + p_e \boldsymbol{e}_i, (\mathcal{E}_e + \tfrac{1}{2}\rho_e |\boldsymbol{v}_e|^2 + p_e) v_{ei}\big)^t, \tag{3.89}$$

$$\epsilon_d \mathsf{F}_i^{e\,diss} = \big(0, \boldsymbol{\Pi}_{ei}, Q_{ei} + \boldsymbol{\Pi}_{ei}\cdot\boldsymbol{v}_e\big)^t, \tag{3.90}$$

using straightforward notation, the system at equilibrium may be rewritten in qualilinear form

$$\partial_t \mathsf{u}_e + \sum_{i\in\mathfrak{D}} \mathsf{A}_i{}^{e}(\mathsf{u}_e)\partial_i \mathsf{u}_e - \epsilon_d \sum_{i,j\in\mathfrak{D}} \partial_i\big(\mathsf{B}_{ij}{}^{e}(\mathsf{u}_e)\partial_j \mathsf{u}_e\big) = 0, \tag{3.91}$$

where $\mathsf{A}_i{}^{\mathrm{e}}$, $i \in \mathfrak{D}$, denote the Jacobian matrices $\mathsf{A}_i{}^{\mathrm{e}} = \partial_{\mathsf{u}_{\mathrm{e}}} \mathsf{F}_i{}^{\mathrm{e}}$ and $\mathsf{B}_{ij}{}^{\mathrm{e}}$, $i, j \in \mathfrak{D}$, the dissipation matrices at equilibrium with $\mathsf{F}_i^{\mathrm{e\,diss}} = -\sum_{j\in\mathfrak{D}} \mathsf{B}_{ij}{}^{\mathrm{e}} \partial_j \mathsf{u}_{\mathrm{e}}$ [7, 18, 22]. A careful analysis reveals that the limit $\epsilon \to 0$ is a Chapman-Enskog procedure so that a double Chapman-Enskog procedure is indeed used with first $\epsilon_d \to 0$ and next $\epsilon \to 0$ [52, 53].

3.3.4 *Hyperbolic-parabolic Structure with Stiff Sources*

In order to symmetrize the equations governing two-temperature fluids, the mathematical entropy out of equilibrium is taken to be $\sigma = -\mathcal{S}/\mathfrak{r}$ where the $1/\mathfrak{r}$ factor is introduced for convenience.

Theorem 3.12 *Assume that* $(H_1 - H_4)$ *hold. Then the function* $\sigma = -\mathcal{S}/\mathfrak{r}$ *is a mathematical entropy for the system* (3.53)–(3.55) *and the corresponding entropic variable reads*

$$\mathsf{v} = (\partial_{\mathsf{u}}\sigma)^t = \frac{1}{\mathfrak{r}}\Big(\frac{g_{\mathrm{tr}}}{T_{\mathrm{tr}}} + \frac{g_{\mathrm{in}}}{T_{\mathrm{in}}} - \frac{\frac{1}{2}|\boldsymbol{v}|^2}{T_{\mathrm{tr}}}, \frac{\boldsymbol{v}}{T_{\mathrm{tr}}}, -\frac{1}{T_{\mathrm{tr}}}, \frac{1}{T_{\mathrm{tr}}} - \frac{1}{T_{\mathrm{in}}}\Big)^t. \tag{3.92}$$

The map $\mathsf{u} \to \mathsf{v}$ *is a* $C^{\gamma-1}$ *diffeomorphism from* O_{u} *given in* (3.72) *onto the open set* O_{v} *given by* $O_{\mathsf{v}} = \{\mathsf{v} \in \mathbb{R}^n;\ \mathsf{v}_{d+2} < 0,\ \mathsf{v}_{d+2} + \mathsf{v}_{d+3} < 0\}$. *The system written in terms of the entropic variable* v *is of the symmetric form with a source term in quasilinear form*

$$\widetilde{A}_0(\mathsf{v})\partial_t \mathsf{v} + \sum_{i\in\mathfrak{D}} \widetilde{A}_i(\mathsf{v})\partial_i \mathsf{v} - \epsilon_d \sum_{i,j\in\mathfrak{D}} \partial_i\big(\widetilde{B}_{ij}(\mathsf{v})\partial_j \mathsf{v}\big) + \frac{1}{\epsilon}\widetilde{L}(\mathsf{v})\mathsf{v} = 0, \tag{3.93}$$

where $\widetilde{A}_0 = \partial_{\mathsf{v}}\mathsf{u}$, $\widetilde{A}_i = A_i\partial_{\mathsf{v}}\mathsf{u}$, $\widetilde{B}_{ij} = B_{ij}\partial_{\mathsf{v}}\mathsf{u}$, $\widetilde{\Omega} = \Omega = -\widetilde{L}(\mathsf{v})\mathsf{v}$ *and the coefficients have regularity* $C^{\gamma-1}$. *The source term is naturally in relaxation form for the two-temperature model. The equilibrium manifold is given by* $\mathcal{E} = \mathbb{R}\times\mathbb{R}^d\times\mathbb{R}\times\{0\}$ *and the fast manifold by* $\mathcal{E}^\perp = \mathbb{R}\mathsf{e}_{d+3}$, *the projector by* $\widetilde{\pi} = \mathsf{e}_{d+3}\otimes\mathsf{e}_{d+3}$ *where* $\mathsf{e}_1, \dots, \mathsf{e}_n$ *denote the basis vectors of* $\mathbb{R}^n$, *and the matrix* $\widetilde{L}$ *associated with the source term* $\widetilde{\Omega}$ *by* $\widetilde{L}(\mathsf{v}) = \mathfrak{r}\frac{\rho c_{\mathrm{in}}}{\tau_{\mathrm{in}}} T_{\mathrm{in}} T_{\mathrm{tr}}\, \mathsf{e}_{d+3}\otimes\mathsf{e}_{d+3}$.

The system coefficients have been evaluated in the literature and are omitted [52, 53]. We investigate the nullspace invariance property for the symmetrized system (3.93) modeling fluids out of thermodynamic equilibrium and then evaluate a convenient normal form.

Lemma 3.4 *For any* $\mathsf{v} \in O_{\mathsf{v}}$ *and* $\boldsymbol{\xi} \in \Sigma^{d-1}$ *the nullspace of the matrix* $\widetilde{B}(\mathsf{v}, \boldsymbol{\xi}) = \sum_{i,j\in\mathfrak{D}} \widetilde{B}_{ij}(\mathsf{v})\xi_i\xi_j$ *is given by* $N(\widetilde{B}) = \mathbb{R}(1, \mathbf{0}, 0, 0)^t$ *and we have* $\widetilde{B}_{ij}(\mathsf{v})N(\widetilde{B}) = 0$ *for* $i, j \in \mathfrak{D}$ *and* $\mathsf{v} \in O_{\mathsf{v}}$.

Since $N(\widetilde{\mathsf{B}})$ is spanned by $(1, \mathbf{0}, 0, 0)^t$, we may use the identity matrix I_{n} as the auxiliary matrix P in Theorem 3.4 and the corresponding auxiliary variables u' and v' then coincide with u and v, respectively. From Theorem 3.4 and the relations (3.71) and (3.92) all normal variables are thus in the form $\big(\mathfrak{F}_{\mathrm{I}}(\rho), \mathfrak{F}_{\mathrm{II}}(\boldsymbol{v}, T_{\mathrm{in}}, T_{\mathrm{tr}})\big)^t$ where $\mathfrak{F}_{\mathrm{I}}$ and $\mathfrak{F}_{\mathrm{II}}$ are diffeomorphisms in $\mathbb{R}$ and $\mathbb{R}^{d+2}$, respectively. The natural variable z is in particular a normal variable but for convenience we select the normal variable

$$\mathsf{w} = \Big(\rho, \boldsymbol{v}, -\frac{1}{T}, \frac{1}{T_{\mathrm{tr}}} - \frac{1}{T_{\mathrm{in}}}\Big)^t. \tag{3.94}$$

The last component of w goes to zero with the relaxation time and the other components $\big(\rho, \boldsymbol{v}, -\frac{1}{T}\big)^t$ will converge towards the corresponding normal variable at thermodynamic equilibrium $\mathsf{w}_{\mathrm{e}} = \big(\rho_{\mathrm{e}}, \boldsymbol{v}_{\mathrm{e}}, -\frac{1}{T_{\mathrm{e}}}\big)^t$. The corresponding normal form also guarantees the convenient commutation relation $\overline{\mathsf{A}}_0\pi = \pi\overline{\mathsf{A}}_0$ between the matrix $\overline{\mathsf{A}}_0$ and the orthogonal projector π onto $\overline{\mathfrak{E}}^{\perp}$.

Theorem 3.13 *Assume that $(H_1 - H_4)$ hold. Then the map $\mathsf{v} \to \mathsf{w}$ is a $C^{\gamma-1}$ diffeomorphism from O_{v} onto the open set O_{w} given by*

$$O_{\mathsf{w}} = (0, \infty)\times\mathbb{R}^d\times\mathbb{R}\times(-\infty, 0). \tag{3.95}$$

The system written in the w variable is in the normal form with a source term in quasilinear form

$$\overline{\mathsf{A}}_0(\mathsf{w})\partial_t \mathsf{w} + \sum_{i\in\mathfrak{D}} \overline{\mathsf{A}}_i(\mathsf{w})\partial_i \mathsf{w} - \epsilon_d \sum_{i,j\in\mathfrak{D}} \partial_i\big(\overline{\mathsf{B}}_{ij}(\mathsf{w})\partial_j \mathsf{w}\big) + \frac{1}{\epsilon}\overline{\mathsf{L}}(\mathsf{w})\mathsf{w} = \epsilon_d \overline{\mathsf{h}}(\mathsf{w}, \partial_x \mathsf{w}_{\mathrm{II}}), \tag{3.96}$$

where $\overline{\mathsf{A}}_0 = (\partial_{\mathsf{w}}\mathsf{v})^t\,\widetilde{\mathsf{A}}_0\,\partial_{\mathsf{w}}\mathsf{v}$, $\overline{\mathsf{A}}_i = (\partial_{\mathsf{w}}\mathsf{v})^t\,\widetilde{\mathsf{A}}_i\,\partial_{\mathsf{w}}\mathsf{v}$, $\overline{\mathsf{B}}_{ij}=(\partial_{\mathsf{w}}\mathsf{v})^t\,\widetilde{\mathsf{B}}_{ij}\,\partial_{\mathsf{w}}\mathsf{v}$, and $\overline{\Omega}=(\partial_{\mathsf{w}}\mathsf{v})^t\,\Omega = -\overline{\mathsf{L}}(\mathsf{w})\mathsf{w}$, have regularity $\gamma - 1$ and $\overline{\mathsf{h}} = -\sum_{i,j\in\mathfrak{D}} \partial_i(\partial_{\mathsf{w}}\mathsf{v})^t\,\widetilde{\mathsf{B}}_{ij}\,\partial_{\mathsf{w}}\mathsf{v}\,\partial_j \mathsf{w}$. The matrix $\overline{\mathsf{L}}$ is symmetric positive semidefinite with a fixed nullspace $\overline{\mathfrak{E}}$ given by $\overline{\mathfrak{E}} = \mathbb{R}\times\mathbb{R}^d\times\mathbb{R}\times\{0\}$ and the orthogonal projector onto $\overline{\mathfrak{E}}^{\perp}$ is $\pi = \mathsf{e}_{d+3}\otimes\mathsf{e}_{d+3}$.

The quadratic source term is also in the form

$$\overline{\mathsf{h}}(\mathsf{w}, \partial_{\mathsf{x}}\mathsf{w}_{\mathrm{II}}) = \sum_{i,j\in\mathfrak{D}} \overline{\mathrm{M}}_{ij}(\mathsf{w})\partial_i\mathsf{w}\partial_j\mathsf{w} = \Big(0, \sum_{i,j\in\mathfrak{D}} \overline{\mathrm{M}}_{ij}^{\mathrm{II,II,II}}(\mathsf{w})\partial_i\mathsf{w}_{\mathrm{II}}\partial_j\mathsf{w}_{\mathrm{II}},\Big)^t \tag{3.97}$$

where $\overline{\mathrm{M}}_{ij}$ are third order tensors depending on w with at least regularity $\gamma - 3$.

The system coefficients have been evaluated in the literature and are omitted [52, 53] . The commutation property $\overline{\mathsf{A}}_0\pi = \pi\overline{\mathsf{A}}_0$ also holds and will be convenient in order to estimate the 'fast component' $\pi\mathsf{w}/\epsilon$ of the normal variable [53]. We further need to obtain a normal form for the system of partial differential equations modeling fluids at thermodynamic equilibrium (3.80)–(3.82). We use the mathematical entropy $\sigma_{\rm e} = -\mathcal{S}_{\rm e}/\mathfrak{r}$ where the $1/\mathfrak{r}$ factor is introduced for convenience.

Theorem 3.14 *Assume that* $(H_1 - H_4)$ *hold. Then the function* $\sigma_{\rm e} = -\mathcal{S}_{\rm e}/\mathfrak{r}$ *is a mathematical entropy for the system* (3.80)–(3.82) *and the corresponding entropic variable is*

$$\mathsf{v}_{\rm e} = (\partial_{\mathsf{u}_{\rm e}}\sigma_{\rm e})^t = \frac{1}{\mathfrak{r}}\Big(\frac{g_{\rm e}}{T_{\rm e}} - \frac{\frac{1}{2}|\boldsymbol{v}_{\rm e}|^2}{T_{\rm e}}, \frac{\boldsymbol{v}_{\rm e}}{T_{\rm e}}, -\frac{1}{T_{\rm e}}\Big)^t. \tag{3.98}$$

The map $\mathsf{u}_{\rm e} \to \mathsf{v}_{\rm e}$ *is a* $C^{\gamma-1}$ *diffeomorphism from* $O_{\mathsf{u}_{\rm e}}$ *onto the open set* $O_{\mathsf{v}_{\rm e}} = \mathbb{R}\times\mathbb{R}^d\times(-\infty, 0)$. *The system written in terms of the entropic variable* $\mathsf{v}_{\rm e}$ *is of the symmetric form*

$$\widetilde{A}_0^{\rm e}(\mathsf{v}_{\rm e})\partial_t \mathsf{v}_{\rm e} + \sum_{i\in\mathfrak{D}} \widetilde{A}_i^{\rm e}(\mathsf{v}_{\rm e})\partial_i \mathsf{v}_{\rm e} - \epsilon_d \sum_{i,j\in\mathfrak{D}} \partial_i\big(\widetilde{B}_{ij}^{\rm e}(\mathsf{v}_{\rm e})\partial_j \mathsf{v}_{\rm e}\big) = 0, \tag{3.99}$$

where $\widetilde{A}_0^{\rm e} = \partial_{\mathsf{v}_{\rm e}}\mathsf{u}_{\rm e}$, $\widetilde{A}_i^{\rm e} = A_i{}^{\rm e}\partial_{\mathsf{v}_{\rm e}}\mathsf{u}_{\rm e}$, $\widetilde{B}_{ij}^{\rm e} = B_{ij}{}^{\rm e}\partial_{\mathsf{v}_{\rm e}}\mathsf{u}_{\rm e}$ *have regularity* $\gamma - 1$. *The system's coefficients have been evaluated in the literature and are omitted [52, 53].*

The nullspace invariance property for the symmetrized system (3.99) modeling fluids at thermodynamic equilibrium is easily established with $N(\widetilde{\mathsf{B}}^{\rm e}) = \mathbb{R}(1, \mathbf{0}, 0,)^t$. Since $N(\widetilde{\mathsf{B}}^{\rm e})$ is spanned by $(1, \mathbf{0}, 0,)^t$, we may use the identity matrix $I_{n_{\rm e}}$ as the auxiliary matrix $\mathsf{P}_{\rm e}$ in Theorem 3.4 and select the convenient normal variable

$$\mathsf{w}_{\rm e} = \Big(\rho_{\rm e}, \boldsymbol{v}_{\rm e}, -\frac{1}{T_{\rm e}}\Big)^t. \tag{3.100}$$

Theorem 3.15 *Assume that* $(H_1 - H_4)$ *hold. Then the map* $\mathsf{v}_{\rm e} \to \mathsf{w}_{\rm e}$ *is a* $C^{\gamma-1}$ *diffeomorphism from* $O_{\mathsf{v}_{\rm e}}$ *onto the open set* $O_{\mathsf{w}_{\rm e}} = (0, \infty)\times\mathbb{R}^d\times(-\infty, 0)$. *The system written in the* $\mathsf{w}_{\rm e}$ *variable is of the normal form*

$$\overline{A}_0{}^{\rm e}(\mathsf{w}_{\rm e})\partial_t \mathsf{w}_{\rm e} + \sum_{i\in\mathfrak{D}} \overline{A}_i{}^{\rm e}(\mathsf{w}_{\rm e})\partial_i \mathsf{w}_{\rm e} - \epsilon_d \sum_{i,j\in\mathfrak{D}} \partial_i\big(\overline{B}_{ij}{}^{\rm e}(\mathsf{w}_{\rm e})\partial_j \mathsf{w}_{\rm e}\big) = \epsilon_d \overline{h}_{\rm e}(\mathsf{w}_{\rm e}, \partial_{\boldsymbol{x}}\mathsf{w}_{\rm IIe}), \tag{3.101}$$

where $\overline{A}_0{}^{\rm e} = (\partial_{\mathsf{w}_{\rm e}}\mathsf{v}_{\rm e})^t\, \widetilde{A}_0^{\rm e}\, \partial_{\mathsf{w}_{\rm e}}\mathsf{v}_{\rm e}$, $\overline{A}_i{}^{\rm e}(\partial_{\mathsf{w}_{\rm e}}\mathsf{v}_{\rm e})^t\, \widetilde{A}_i^{\rm e}\, \partial_{\mathsf{w}_{\rm e}}\mathsf{v}_{\rm e}$, $\overline{B}_{ij}{}^{\rm e} = (\partial_{\mathsf{w}_{\rm e}}\mathsf{v}_{\rm e})^t\, \widetilde{B}_{ij}^{\rm e}\, \partial_{\mathsf{w}_{\rm e}}\mathsf{v}_{\rm e}$, *have regularity* $\gamma - 1$ *and* $\overline{h}_{\rm e} = -\sum_{i,j\in\mathfrak{D}} \partial_i(\partial_{\mathsf{w}_{\rm e}}\mathsf{v}_{\rm e}^t)\, \widetilde{B}_{ij}^{\rm e}\, \partial_{\mathsf{w}_{\rm e}}\mathsf{v}_{\rm e}\, \partial_j \mathsf{w}_{\rm e}$.

The system coefficients have been evaluated in the literature and are omitted [52, 53] .

3.3.5 *Local Existence*

In order to establish the local existence of solutions, we have to obtain a priori estimates for linearized equations in the form

$$\overline{\mathsf{A}}_0(\mathsf{w})\partial_t\widetilde{\mathsf{w}}+\sum_{i\in\mathfrak{D}}\overline{\mathsf{A}}_i(\mathsf{w})\partial_i\widetilde{\mathsf{w}}-\epsilon_d\sum_{i,j\in\mathfrak{D}}\overline{\mathsf{B}}_{ij}(\mathsf{w})\partial_i\partial_j\widetilde{\mathsf{w}}+\frac{1}{\epsilon}\overline{\mathsf{L}}(\mathsf{w})\widetilde{\mathsf{w}}=\mathsf{f}+\epsilon_d\mathsf{g}. \tag{3.102}$$

The main difference with the hyperbolic-parabolic linearized systems (3.25) investigated in Sect. 3.2 is the presence of the stiff source terms. The a priori estimates are established for general systems and then applied to the special situation of two-temperature fluids.

It is assumed that $\mathsf{w}=(\mathsf{w}_{\mathrm{I}},\mathsf{w}_{\mathrm{II}})^t\in\mathcal{O}_{\mathsf{w}}$ where $\mathcal{O}_{\mathsf{w}}$ is an open set of $\mathbb{R}^{\mathsf{n}}$, w_{I} are the hyperbolic components, w_{II} the parabolic components, and $\epsilon_d,\epsilon\in(0,1]$ are two positive parameters. The dimensions of the hyperbolic and parabolic components are denoted by n_{I} and n_{II} respectively and $\mathsf{n}=\mathsf{n}_{\mathrm{I}}+\mathsf{n}_{\mathrm{II}}$. The matrices $\overline{\mathsf{A}}_0$, $\overline{\mathsf{A}}_i$, $\overline{\mathsf{B}}_{ij}$, and $\overline{\mathsf{L}}$ are assumed to have at least regularity $l+2$ with $l\geq l_0+1$ and $l_0=[d/2]+1$. The matrix $\overline{\mathsf{A}}_0$ is symmetric positive definite, the matrices $\overline{\mathsf{A}}_0$ are symmetric, $\overline{\mathsf{B}}_{ij}$ satisfy $\overline{\mathsf{B}}_{ij}^{\,t}=\overline{\mathsf{B}}_{ji}$, and $\overline{\mathsf{L}}$ is positive semi-definite with a fixed nullspace $\overline{\mathcal{E}}$. The matrices $\overline{\mathsf{A}}_0$ and $\overline{\mathsf{B}}_{ij}$, $i,j\in\mathfrak{D}$, have the bloc structure

$$\overline{\mathsf{A}}_0=\begin{bmatrix}\overline{\mathsf{A}}_0^{\,\mathrm{I,I}} & 0_{n_{\mathrm{I}},n_{\mathrm{II}}}\\ 0_{n_{\mathrm{II}},n_{\mathrm{I}}} & \overline{\mathsf{A}}_0^{\,\mathrm{II,II}}\end{bmatrix},\qquad \overline{\mathsf{B}}_{ij}=\begin{bmatrix}0_{n_{\mathrm{I}},n_{\mathrm{I}}} & 0_{n_{\mathrm{I}},n_{\mathrm{II}}}\\ 0_{n_{\mathrm{II}},n_{\mathrm{I}}} & \overline{\mathsf{B}}_{ij}^{\,\mathrm{II,II}}\end{bmatrix},$$

and $\overline{\mathsf{B}}^{\mathrm{II,II}}(\mathsf{w},\boldsymbol{\xi})=\sum_{i,j\in\mathfrak{D}}\overline{\mathsf{B}}_{ij}^{\,\mathrm{II,II}}(\mathsf{w})\xi_i\xi_j$ is positive definite for $\mathsf{w}\in\mathcal{O}_{\mathsf{w}}$ and $\boldsymbol{\xi}\in\Sigma^{d-1}$. Denoting by π the orthogonal projector onto the orthogonal of the equilibrium manifold $\overline{\mathcal{E}}^{\perp}$, we assume that $\overline{\mathsf{A}}_0$ satisfies the compatibility condition

$$\pi\,\overline{\mathsf{A}}_0(\mathsf{w})=\overline{\mathsf{A}}_0(\mathsf{w})\,\pi,\qquad \mathsf{w}\in\mathcal{O}_{\mathsf{w}}. \tag{3.103}$$

We are only interested in well prepared initial data in this section, that is, we assume that the initial condition w_0 is close to the equilibrium manifold $\overline{\mathcal{E}}$ in such a way that $\pi\mathsf{w}_0$ is small and the situation of ill prepared initial data will be investigated separately.

We denote by $\mathsf{u}^\star$ and $\mathsf{w}^\star$ corresponding constant equilibrium states in the u, and w variables respectively, so that $\mathsf{w}^\star\in\mathcal{O}_{\mathsf{w}}\cap\overline{\mathcal{E}}$ and $\pi\,\mathsf{w}^\star=0$. We denote by $|\bullet|_l$ the norm in the Sobolev space $H^l(\mathbb{R}^d)$ and otherwise $|\bullet|_A$ in the functional space A. If $\alpha=(\alpha_1,\ldots,\alpha_d)\in\mathbb{N}^d$ is a multiindex, we denote as usual by ∂^α the differential operator $\partial_1^{\alpha_1}\cdots\partial_d^{\alpha_d}$ and by $|\alpha|$ its order $|\alpha|=\alpha_1+\cdots+\alpha_d$. The square of kth derivatives of a scalar function ϕ, like T, ρ, or v_i, $1\leq i\leq d$, is defined as in (3.26) and for any map $\phi:[0,\bar{\tau}]\times\mathbb{R}^d\to\mathbb{R}^{\mathsf{n}}$ where $\bar{\tau}>0$ is positive and for any $\tau\in[0,\bar{\tau}]$, we denote by $\phi(\tau)$ the partial map $\mathsf{x}\to\phi(\tau,\mathsf{x})$ defined over $\mathbb{R}^d$.

For a given $\bar{\tau} > 0$ and $l \geq l_0 + 1$ where $l_0 = [d/2] + 1$, we assume that w is such that

$$\begin{cases} \mathsf{w}_{\mathrm{I}} - \mathsf{w}_{\mathrm{I}}^\star \in C^0\big([0, \bar{\tau}], H^l\big) \cap C^1\big([0, \bar{\tau}], H^{l-1}\big), \\ \mathsf{w}_{\mathrm{II}} - \mathsf{w}_{\mathrm{II}}^\star \in C^0\big([0, \bar{\tau}], H^l\big) \cap C^1\big([0, \bar{\tau}], H^{l-2}\big) \cap L^2\big((0, \bar{\tau}), H^{l+1}\big), \end{cases} \tag{3.104}$$

and such that $\partial_t \mathsf{w} \in L^2\big((0, \bar{\tau}), H^{l-1}\big)$. We define the quantities M and M_{t} by

$$M^2 = \sup_{0 \leq \tau \leq \bar{\tau}} |\mathsf{w}(\tau) - \mathsf{w}^\star|_l^2, \qquad M_{\mathrm{t}} = \int_0^{\bar{\tau}} |\partial_t \mathsf{w}(\tau)|_{l-1}^2 \, d\tau. \tag{3.105}$$

We consider $\mathcal{O}_0$ such that $\overline{\mathcal{O}}_0 \subset \mathcal{O}_{\mathsf{w}}$, d_1 such that $0 < d_1 < d(\overline{\mathcal{O}}_0, \partial \mathcal{O}_{\mathsf{w}})$, and define

$$\mathcal{O}_1 = \{\, \mathsf{w} \in \mathcal{O}_{\mathsf{w}};\ d(\mathsf{w}, \overline{\mathcal{O}}_0) < d_1 \,\}. \tag{3.106}$$

It is also assumed that w_0 and w are such that

$$\mathsf{w}_0(\mathsf{x}) = \mathsf{w}(0, \mathsf{x}) \in \mathcal{O}_0, \qquad \mathsf{w}(t, \mathsf{x}) \in \mathcal{O}_1, \qquad t \in [0, \bar{\tau}], \quad \mathsf{x} \in \mathbb{R}^d. \tag{3.107}$$

The following estimates take into account the presence of stiff sources and differ from previous estimates by the inclusion of new terms associated with the fast variable $\pi \mathsf{w}/\epsilon$ as well as for the time derivatives which cannot anymore be estimated directly from the governing equations.

Theorem 3.16 *Let $l \geq l_0 + 1$, with $l_0 = [d/2] + 1$, and assume that a solution $\widetilde{\mathsf{w}}$ of the linearized system* (3.102)*, is such that*

$$\begin{aligned} &\widetilde{\mathsf{w}}_{\mathrm{I}} - \widetilde{\mathsf{w}}_{\mathrm{I}}^\star \in C^0\big([0, \bar{\tau}], H^l\big) \cap C^1\big([0, \bar{\tau}], H^{l-1}\big), \\ &\widetilde{\mathsf{w}}_{\mathrm{II}} - \widetilde{\mathsf{w}}_{\mathrm{II}}^\star \in C^0\big([0, \bar{\tau}], H^l\big) \cap C^1\big([0, \bar{\tau}], H^{l-2}\big) \cap L^2\big((0, \bar{\tau}), H^{l+1}\big), \end{aligned} \tag{3.108}$$

where $\widetilde{\mathsf{w}}^\star = (\widetilde{\mathsf{w}}_{\mathrm{I}}^\star, \widetilde{\mathsf{w}}_{\mathrm{II}}^\star)^t$ is a constant state $\widetilde{\mathsf{w}}^\star \in \overline{\mathcal{E}}$. Further assume that

$$\mathsf{f} \in C^0\big([0, \bar{\tau}], H^{l-1}\big) \cap L^1\big([0, \bar{\tau}], H^l\big), \tag{3.109}$$

$$\mathsf{g} \in C^0\big([0, \bar{\tau}], H^{l-1}\big), \qquad \mathsf{g}_{\mathrm{I}} = 0, \tag{3.110}$$

and denote by $\widetilde{\mathsf{w}}_0$ the initial state $\widetilde{\mathsf{w}}_0(\mathsf{x}) = \widetilde{\mathsf{w}}(0, \mathsf{x})$. Then there exists constants $\mathsf{c}_1(\mathcal{O}_1) \geq 1$ and $\mathsf{c}_2(\mathcal{O}_1, M) \geq 1$, with $\mathsf{c}_2(\mathcal{O}_1, M)$ increasing with M, and independent of ϵ_d and ϵ, such that for any $t \in [0, \bar{\tau}]$

$$\sup_{0 \leq \tau \leq t} \Big\{ |\widetilde{\mathsf{w}}(\tau) - \widetilde{\mathsf{w}}^\star|_l^2 + \frac{1}{\epsilon} |\pi \widetilde{\mathsf{w}}(\tau)|_0^2 \Big\} + \epsilon_d \int_0^t |\widetilde{\mathsf{w}}_{\mathrm{II}}(\tau) - \widetilde{\mathsf{w}}_{\mathrm{II}}^\star|_{l+1}^2 \, d\tau$$

$$+\frac{1}{\epsilon}\int_0^t |\pi\widetilde{w}(\tau)|_l^2\,d\tau \;\le\; c_1^2\exp\big(c_2(t+M_t\sqrt{t})\big)\Big(|\widetilde{w}_0-\widetilde{w}^\star|_l^2+\frac{1}{\epsilon}|\pi\,\widetilde{w}_0|_0^2$$
$$+\epsilon_d c_2\int_0^t |g_{\mathrm{II}}(\tau)|_{l-1}^2\,d\tau + c_2\Big\{\int_0^t |f(\tau)|_l\,d\tau\Big\}^2 + c_2\int_0^t |\pi f(\tau)|_0^2\,d\tau\Big), \tag{3.111}$$

$$\frac{1}{\epsilon}\sup_{0\le\tau\le t}|\pi\,\widetilde{w}(\tau)|_{l-1}^2+\frac{1}{\epsilon^2}\int_0^t|\pi\,\widetilde{w}(\tau)|_{l-1}^2\,d\tau+\int_0^t|\partial_t\widetilde{w}(\tau)|_{l-1}^2\,d\tau$$
$$\le c_2\exp\big(c_2(t+M_t\sqrt{t})\big)\Big(|\widetilde{w}_0-\widetilde{w}^\star|_l^2+\frac{1}{\epsilon}|\pi\,\widetilde{w}_0|_{l-1}^2+\epsilon_d\int_0^t|g_{\mathrm{II}}|_{l-1}^2\,d\tau$$
$$+\Big\{\int_0^t|f|_l\,d\tau\Big\}^2+\int_0^t|f|_{l-1}^2\,d\tau\Big). \tag{3.112}$$

Sketch of the Proof The differences with the non stiff estimates (3.33) of Theorem 3.6 are the inclusion of extra terms associated with the fast variables $\pi w/\epsilon$ and $\pi w/\sqrt{\epsilon}$ as well as the coupling with the estimates for time derivatives. The scalar products of the linearized equation governing $\partial^\alpha w$ must be taken with $\partial^\alpha w$ as in Theorem 3.6 but also with $\partial^\alpha\pi w$ and $\partial^\alpha\partial_t w$ and this yields the new terms involving $\pi w/\epsilon$ and $\partial_t w$ in the a priori estimates [53]. □

We now establish a local existence theorem on a time interval $\bar{\tau} > 0$ independent of ϵ_d and ϵ for the system of partial differential equation in normal form. We are interested in convergence results on time intervals including the time origin $t = 0$, we assume in this section that the initial data is *well prepared* or equivalently that πw_0 is small.

Theorem 3.17 *Let $d \ge 1$ and $l \ge l_0 + 1$, be integers with $l_0 = [d/2] + 1$, and let $b > 0$ be given. Let O_0 such that $\overline{O}_0 \subset O_w$, d_1 such that $0 < d_1 < d(\overline{O}_0, \partial O_w)$, and define $O_1 = \{\, w \in O_w;\ d(w, \overline{O}_0) < d_1 \,\}$. There exists $\bar{\tau} > 0$ depending on O_1 and b, and independent on $\epsilon_d \in (0, 1]$ and $\epsilon \in (0, 1]$, such that for any w_0 with*

$$|w_0 - w^\star|_l^2 + \frac{1}{\epsilon}|\pi\,w_0|_{l-1}^2 < b^2, \tag{3.113}$$

and $w_0 \in \overline{O}_0$, there exists a unique local solution w to the system

$$\overline{A}_0(w)\partial_t w+\sum_{i\in\mathfrak{D}}\overline{A}_i(w)\partial_i w-\epsilon_d\sum_{i,j\in\mathfrak{D}}\partial_i\big(\overline{B}_{ij}(w)\partial_j w\big)+\frac{1}{\epsilon}\overline{L}(w)\,w=\epsilon_d\overline{h}(w,\partial_x w_{\mathrm{II}}), \tag{3.114}$$

with initial condition $w(0, x) = w_0(x)$, $x \in \mathbb{R}^d$ such that $w(t, x) \in O_1$ for $t \in [0, \bar{\tau}]$ and $x \in \mathbb{R}^d$ and

$$\mathsf{w}_{\mathrm{I}} - \mathsf{w}_{\mathrm{I}}^{\star} \in C^0\big([0,\bar{\tau}], H^l\big) \cap C^1\big([0,\bar{\tau}], H^{l-1}\big),$$
$$\mathsf{w}_{\mathrm{II}} - \mathsf{w}_{\mathrm{II}}^{\star} \in C^0\big([0,\bar{\tau}], H^l\big) \cap C^1\big([0,\bar{\tau}], H^{l-2}\big) \cap L^2\big((0,\bar{\tau}), H^{l+1}\big).$$

In addition, there exists $C > 0$ which only depend on $\mathcal{O}_1$ and b, such that

$$\sup_{0\le\tau\le\bar{\tau}} \Big(|\mathsf{w}(\tau) - \mathsf{w}^{\star}|_l^2 + \frac{1}{\epsilon}|\pi\,\mathsf{w}(\tau)|_{l-1}^2 \Big) + \epsilon_d \int_0^{\bar{\tau}} |\mathsf{w}_{\mathrm{II}}(\tau) - \mathsf{w}_{\mathrm{II}}^{\star}|_{l+1}^2 \, d\tau$$
$$+ \frac{1}{\epsilon}\int_0^{\bar{\tau}} |\pi\,\mathsf{w}(\tau)|_l^2 \, d\tau + \frac{1}{\epsilon^2}\int_0^{\bar{\tau}} |\pi\,\mathsf{w}(\tau)|_{l-1}^2 \, d\tau + \int_0^{\bar{\tau}} |\partial_t \mathsf{w}(\tau)|_{l-1}^2 \, d\tau$$
$$\le C\Big(|\mathsf{w}_0 - \mathsf{w}^{\star}|_l^2 + \frac{1}{\epsilon}|\pi\,\mathsf{w}_0|_{l-1}^2 \Big). \tag{3.115}$$

Sketch of the Proof The proof is similar to the nonstiff case once the linearized a priori estimates are available. An invariant set is found for the map $\mathsf{w} \mapsto \widetilde{\mathsf{w}}$ and the successive iterates shown to be convergent towards the solution using a lower norm. The convergence of the iterates is then established successively on subintervals small enough in order to have a local contraction property [53]. A more elegant proof is to note that there exists an iterated of the map $\mathsf{w} \mapsto \widetilde{\mathsf{w}}$ that is a contraction over the whole time interval [119, 120]. □

We now strengthen the linearized estimates for $\partial_t \mathsf{w}$ obtained with the local existence theorem under stronger assumptions. Such extra estimates for the time derivatives $\partial_t \mathsf{w} \in C^0\big([0,\bar{\tau}], H^{l-2}\big)$ and $\pi\,\partial_t \mathsf{w}/\epsilon \in L^2\big((0,\bar{\tau}), H^{l-3}\big)$, uniformly in $\epsilon_d, \epsilon \in (0,1]$ when $l \ge [d/2]+4$ are relevant when the initial time derivative $\partial_t \mathsf{w}_0$ at $t = 0$ is close to the equilibrium manifold and will be needed in the convergence analysis of the fast relaxation limit.

Theorem 3.18 *Keep the assumptions of Theorem 3.17 and further assume that $l \ge l_0 + 4$. There exists C depending on $\mathcal{O}_1$ and b such that*

$$\sup_{0\le\tau\le\bar{\tau}} \Big(|\partial_t \mathsf{w}(\tau)|_{l-2}^2 + \frac{1}{\epsilon}|\pi\,\partial_t \mathsf{w}(\tau)|_{l-3}^2 \Big) + \epsilon_d \int_0^{\bar{\tau}} |\partial_t \mathsf{w}_{\mathrm{II}}(\tau)|_{l-1}^2 \, d\tau$$
$$+ \frac{1}{\epsilon}\int_0^{\bar{\tau}} |\pi\,\partial_t \mathsf{w}(\tau)|_{l-2}^2 \, d\tau + \frac{1}{\epsilon^2}\int_0^{\bar{\tau}} |\pi\,\partial_t \mathsf{w}(\tau)|_{l-3}^2 \, d\tau + \int_0^{\bar{\tau}} |\partial_t^2 \mathsf{w}(\tau)|_{l-3}^2 \, d\tau$$
$$\le C\Big(|\mathsf{w}_0 - \mathsf{w}^{\star}|_l^2 + \frac{1}{\epsilon}|\pi\,\mathsf{w}_0|_{l-1}^2 + |\partial_t \mathsf{w}_0|_{l-2}^2 + \frac{1}{\epsilon}|\pi\,\partial_t \mathsf{w}_0|_{l-3}^2 \Big). \tag{3.116}$$

Corollary 3.1 *Keeping the assumptions of Theorem 3.18, there exists C only depending on $\mathcal{O}_1$ and b such that*

$$\frac{1}{\epsilon^2} \sup_{0\le\tau\le t} |\pi\,\mathsf{w}(\tau)|_{l-2}^2 \le C\Big(|\mathsf{w}_0 - \mathsf{w}^{\star}|_l^2 + \frac{1}{\epsilon}|\pi\,\mathsf{w}_0|_{l-1}^2 + |\partial_t \mathsf{w}_0|_{l-2}^2 + \frac{1}{\epsilon}|\pi\,\partial_t \mathsf{w}_0|_{l-3}^2 \Big). \tag{3.117}$$

Moreover, if w *and* w' *corresponds to two initial conditions* w_0 *and* w_0' *and if* $|\mathsf{w}_0 - \mathsf{w}^\star|_l^2 + \frac{1}{\epsilon}|\pi\,\mathsf{w}_0|_{l-1}^2 + |\partial_t\mathsf{w}_0|_{l-2}^2 + \frac{1}{\epsilon}|\pi\,\partial_t\mathsf{w}_0|_{l-3}^2$ *remains bounded by some constant only depending on* O_1 *and b, letting* $\delta\mathsf{w} = \mathsf{w} - \mathsf{w}'$ *and* $\delta\mathsf{w}_0 = \mathsf{w}_0 - \mathsf{w}_0'$*, then there exists C only depending on* O_1 *and b such that*

$$\sup_{0\le\tau\le\bar\tau}\Big(|\delta\mathsf{w}(\tau)|_{l-1}^2+\frac{1}{\epsilon}|\pi\,\delta\mathsf{w}(\tau)|_{l-2}^2\Big)+\epsilon_d\int_0^{\bar\tau}|\delta\mathsf{w}_{\mathrm{II}}(\tau)|_l^2\,d\tau+\frac{1}{\epsilon}\int_0^{\bar\tau}|\pi\,\delta\mathsf{w}(\tau)|_{l-1}^2\,d\tau$$
$$+\frac{1}{\epsilon^2}\int_0^{\bar\tau}|\pi\,\delta\mathsf{w}(\tau)|_{l-2}^2\,d\tau+\int_0^{\bar\tau}|\partial_t\delta\mathsf{w}(\tau)|_{l-2}^2\,d\tau\le C\Big(|\delta\mathsf{w}_0|_{l-1}^2+\frac{1}{\epsilon}|\pi\,\delta\mathsf{w}_0|_{l-2}^2\Big). \tag{3.118}$$

These estimates are obtained under the assumption $l \ge l_0 + 4$ by considering the hyperbolic-parabolic equation satisfied by the time derivative $\partial_t\mathsf{w}$ [53].

3.3.6 Asymptotics for Well Prepared Data

We consider the linear operator $\varphi = [\mathsf{e}_1, \ldots, \mathsf{e}_{d+2}]$ having columns $\mathsf{e}_1, \ldots, \mathsf{e}_{d+2}$ in such a way that $\varphi\mathsf{w} = (\rho, \boldsymbol{v}, -1/T)^t$. The slow components of w are thus $\varphi\mathsf{w}$ and our goal is to estimate the difference between $\varphi\mathsf{w}$ and the equilibrium normal variable $\mathsf{w}_\mathrm{e} = (\rho_\mathrm{e}, \boldsymbol{v}_\mathrm{e}, -1/T_\mathrm{e})^t$ in terms of ϵ_d and ϵ. To this aim, we estimate to what extend the equilibrium projection $\varphi\mathsf{w}$ of the normal variable w out of equilibrium satisfies the equations of the normal variable w_e at thermodynamic equilibrium. We then use a stability theorem for the equations governing fluids at thermodynamic equilibrium. By combining these results, we rigorously establish that the difference $\varphi\mathsf{w} - \mathsf{w}_\mathrm{e}$ is $O\big(\epsilon(\epsilon + \epsilon_d)\big)$. This shows that w_e is a second order approximation of $\varphi\mathsf{w}$ in the fast relaxation limit and yields a rigorous justification of the volume viscosity terms appearing in the equilibrium fluid model.

We thus have to estimate the residual h defined by

$$\overline{\mathsf{A}}_0^\mathrm{e}(\varphi\mathsf{w})\partial_t(\varphi\mathsf{w}) + \sum_{i\in\mathfrak{D}}\overline{\mathsf{A}}_i^\mathrm{e}(\varphi\mathsf{w})\partial_i(\varphi\mathsf{w}) - \epsilon_d\sum_{i,j\in\mathfrak{D}}\partial_i\big(\overline{\mathsf{B}}_{ij}^\mathrm{e}(\varphi\mathsf{w})\partial_j(\varphi\mathsf{w})\big)$$
$$- \epsilon_d\overline{\mathsf{h}}_\mathrm{e}\big(\varphi\mathsf{w}, \partial_\mathsf{x}(\varphi\mathsf{w})_{\mathrm{II}}\big) = \mathsf{h}, \tag{3.119}$$

that is the ‘default to equilibrium residual’ of the projection $\varphi\mathsf{w}$. Using estimates for h we will deduce estimates for the difference $\varphi\mathsf{w} - \mathsf{w}_\mathrm{e}$. We begin by expressing h in terms of fluid properties and denote by θ the reduced temperature difference

$$\theta = \frac{T_\mathrm{tr} - T_\mathrm{in}}{\epsilon} = -T_\mathrm{tr}T_\mathrm{in}\frac{\pi\,\mathsf{w}}{\epsilon}, \tag{3.120}$$

which essentially corresponds to the fast variable $\pi \mathsf{w}/\epsilon$. The a priori estimates for hyperbolic-parabolic systems with stiff sources already obtained yield in particular various estimates of θ uniformly with respect to the parameters $\epsilon, \epsilon_d \in (0,1]$. The difference $\phi(T_{\text{tr}}, T_{\text{in}}) - \phi(T,T)$ for a smooth function of the translational and internal temperatures is estimated in the following lemma.

Lemma 3.5 *Let ϕ be a C^γ function $\phi(T_{\text{tr}}, T_{\text{in}})$ of both temperatures T_{tr} and T_{in}. Then*

$$\phi(T_{\text{tr}}, T_{\text{in}}) - \phi(T,T) = \epsilon \overline{\mathsf{h}}_\phi(\theta, T_{\text{tr}}, T_{\text{in}}), \tag{3.121}$$

where $\overline{\mathsf{h}}_\phi$ is a $C^{\gamma-1}$ function of T_{tr} and T_{in} and is proportional to θ.

This simply results from Taylor expansions of the function ϕ and of the definition of θ. We may now use this lemma in order to rewrite in a convenient form the various transport coefficients appearing in the governing equations. Keeping in mind that the viscosity η is given by $\eta = \epsilon_d \bar{\eta}$ from (3.68), and using Lemma 3.5 with $\phi = \bar{\eta}$, we obtain that $\bar{\eta}(T_{\text{tr}}, T_{\text{in}}) - \bar{\eta}_{\text{e}}(T) = \epsilon \overline{\mathsf{h}}_{\bar{\eta}}$ where $\bar{\eta}_{\text{e}}(T) = \bar{\eta}(T,T)$. Further defining the total conductivity as $\lambda = \lambda_{\text{tr,tr}} + \lambda_{\text{tr,in}} + \lambda_{\text{in,tr}} + \lambda_{\text{in,in}}$, we have from (3.68) that $\lambda = \epsilon_d \bar{\lambda}$ where $\bar{\lambda} = \bar{\lambda}_{\text{tr,tr}} + \bar{\lambda}_{\text{tr,in}} + \bar{\lambda}_{\text{in,tr}} + \bar{\lambda}_{\text{in,in}}$, and using Lemma 3.5 with $\phi = \bar{\lambda}$, yields $\bar{\lambda}(T_{\text{tr}}, T_{\text{in}}) - \bar{\lambda}_{\text{e}}(T) = \epsilon \overline{\mathsf{h}}_{\bar{\lambda}}$ where $\bar{\lambda}_{\text{e}}(T) = \bar{\lambda}(T,T)$. Finally, the volume viscosity may be written $\kappa = \epsilon \bar{\kappa}$ and using Lemma 3.5 with $\phi = \bar{\kappa}$ yields $\bar{\kappa}(T_{\text{tr}}, T_{\text{in}}) - \bar{\kappa}_{\text{e}}(T) = \epsilon \overline{\mathsf{h}}_{\bar{\kappa}}$ where $\bar{\kappa}_{\text{e}}(T) = \bar{\kappa}(T,T)$. The transport coefficients are thus in the form $\kappa(T_{\text{tr}}, T_{\text{in}}) - \kappa_{\text{e}}(T) = \epsilon^2 \overline{\mathsf{h}}_{\bar{\kappa}}$, $\eta(T_{\text{tr}}, T_{\text{in}}) - \eta_{\text{e}}(T) = \epsilon\epsilon_d \overline{\mathsf{h}}_{\bar{\eta}}$, $\lambda(T_{\text{tr}}, T_{\text{in}}) - \lambda_{\text{e}}(T) = \epsilon\epsilon_d \overline{\mathsf{h}}_{\bar{\lambda}}$ with $\kappa_{\text{e}}(T) = \epsilon \bar{\kappa}_{\text{e}}(T)$, $\eta_{\text{e}}(T) = \epsilon_d \bar{\eta}_{\text{e}}(T)$, and $\lambda_{\text{e}}(T) = \epsilon \bar{\lambda}_{\text{e}}(T)$. We further define the reduced transport fluxes $\overline{\boldsymbol{Q}}_{\text{tr}}$, $\overline{\boldsymbol{Q}}_{\text{in}}$, $\overline{\boldsymbol{Q}} = \overline{\boldsymbol{Q}}_{\text{tr}} + \overline{\boldsymbol{Q}}_{\text{in}}$, and $\overline{\boldsymbol{\Pi}}$ as the fluxes evaluated with the reduced transport coefficients. We then obtain that

$$\begin{aligned}\rho\mathfrak{r}(T_{\text{tr}} - T) = -\kappa_{\text{e}}(T)\nabla\cdot\boldsymbol{v} - \epsilon^2\Big(\overline{\mathsf{h}}_{\bar{\kappa}}\nabla\cdot\boldsymbol{v} + \frac{\bar{\kappa}}{\mathfrak{r}T_{\text{tr}}}(\partial_t\theta + \boldsymbol{v}\cdot\nabla\theta)\Big)\\ -\epsilon\epsilon_d\frac{\bar{\kappa}}{p}\Big(\overline{\boldsymbol{\Pi}}:\nabla\boldsymbol{v} + \nabla\cdot\overline{\boldsymbol{Q}}_{\text{tr}} - \frac{c_{\text{v,tr}}}{c_{\text{in}}}\nabla\cdot\overline{\boldsymbol{Q}}_{\text{in}}\Big),\end{aligned} \tag{3.122}$$

and we are ready to investigate the structure of the residual h.

Proposition 3.4 *The residual h may be written in the form*

$$\mathsf{h} = \epsilon^2 \mathsf{h}_r + \epsilon\epsilon_d \mathsf{h}_d, \tag{3.123}$$

where $\mathsf{h}_r = \big((\partial_{\mathsf{w}_\text{e}}\mathsf{v}_\text{e})(\varphi\mathsf{w})\big)^t \sum_{i\in\mathfrak{D}} \partial_i \mathsf{h}_{r,i}$, $\mathsf{h}_d = \big((\partial_{\mathsf{w}_\text{e}}\mathsf{v}_\text{e})(\varphi\mathsf{w})\big)^t \sum_{i\in\mathfrak{D}} \partial_i \mathsf{h}_{d,i}$, and denoting by $\boldsymbol{\xi} = (\xi_1, \dots, \xi_d)^t$ an arbitrary vector of $\mathbb{R}^d$, the components $\mathsf{h}_{r,i}$ and $\mathsf{h}_{d,i}$ are given by

$$\sum_{i\in\mathfrak{D}} \xi_i \mathsf{h}_{r,i} = \big(0,\ a\boldsymbol{\xi},\ a\boldsymbol{v}\cdot\boldsymbol{\xi}\big)^t, \tag{3.124}$$

$$\sum_{i\in\mathfrak{D}} \xi_i \mathsf{h}_{d,i} = \big(0,\ \overline{h}_{\bar\eta} \mathsf{S}\cdot\boldsymbol{\xi} + b\boldsymbol{\xi},\ \overline{h}_{\bar\eta}\langle \mathsf{S}\boldsymbol{v}, \boldsymbol{\xi}\rangle + b\boldsymbol{v}\cdot\boldsymbol{\xi} + \overline{h}_{\bar\lambda}\boldsymbol{\nabla}T\cdot\boldsymbol{\xi} + \boldsymbol{c}\big)^t, \tag{3.125}$$

where $\mathsf{S} = \boldsymbol{\nabla v} + (\boldsymbol{\nabla v})^t - \frac{2}{3}(\boldsymbol{\nabla}\cdot\boldsymbol{v})\boldsymbol{I}$ *is the rate of strain tensor and where*

$$a = \overline{h}_{\bar\kappa}\boldsymbol{\nabla}\cdot\boldsymbol{v} + (\bar\kappa/\mathfrak{r}T_{\rm tr})(\partial_t\theta + \boldsymbol{v}\cdot\boldsymbol{\nabla}\theta), \tag{3.126}$$

$$b = (\bar\kappa/p)\big(\overline{\boldsymbol{\Pi}}{:}\boldsymbol{\nabla v} + \boldsymbol{\nabla}\cdot\overline{\boldsymbol{Q}}_{\rm tr} - (c_{\rm v,tr}/c_{\rm in})\boldsymbol{\nabla}\cdot\overline{\boldsymbol{Q}}_{\rm in}\big), \tag{3.127}$$

$$c_i = (\bar\lambda_{\rm tr,tr} + \bar\lambda_{\rm in,tr})\partial_i(\theta\widetilde{c}_{\rm in}/\widetilde{c}_{\rm v}) - (\bar\lambda_{\rm tr,in} + \bar\lambda_{\rm in,in})\partial_i(\theta c_{\rm v,tr}/\widetilde{c}_{\rm v}). \tag{3.128}$$

This results from lengthy calculations that are technically not difficult [53]. We may now estimate the residuals h_r and h_d from (3.123) in the functional spaces $L^2\big((0,\bar\tau), H^{l-4}\big)$ and $L^2\big((0,\bar\tau), H^{l-3}\big)$, respectively, uniformly with respect to the parameters $\epsilon, \epsilon_d \in (0,1]$, as a direct application of the estimates from previous section.

Theorem 3.19 *Assume that* $l \geq l_0 + 4$ *and that* w_0 *is such that*

$$|\mathsf{w}_0 - \mathsf{w}^\star|_l^2 + \frac{1}{\epsilon}|\pi\,\mathsf{w}_0|_{l-1}^2 + |\partial_t\mathsf{w}_0|_{l-2}^2 + \frac{1}{\epsilon}|\pi\,\partial_t\mathsf{w}_0|_{l-3}^2$$

is uniformly bounded independently of $\epsilon \in (0,1]$. *Then the residual* h_r *belongs to* $C^0\big([0,\bar\tau], H^{l-3}\big)$ *and is bounded in* $L^2\big((0,\bar\tau), H^{l-4}\big)$ *independently of* $\epsilon, \epsilon_d \in (0,1]$, *and the residual* h_d *belongs to* $C^0\big([0,\bar\tau], H^{l-3}\big)$ *and is bounded in* $L^2\big((0,\bar\tau), H^{l-3}\big)$ *independently of* $\epsilon, \epsilon_d \in (0,1]$.

We now combine the estimate of the residual h with the stability theorem at equilibrium in order to obtain a convergence theorem.

Theorem 3.20 *Let* $d \geq 1$, $l \geq l_0 + 4$, $l_0 = [d/2] + 1$, *be integers and let* $b > 0$ *be given. Let* O_0 *be given such that* $\overline{O}_0 \subset O_{\mathsf{w}}$, d_1 *such that* $0 < d_1 < d(\overline{O}_0, \partial O_{\mathsf{w}})$, *and define* $O_1 = \{\, \mathsf{w} \in O_{\mathsf{w}};\ d(\mathsf{w}, \overline{O}_0) < d_1 \,\}$. *There exists* $\bar\tau > 0$ *depending on* O_1 *and* b, *and independent on* $\epsilon_d \in (0,1]$ *and* $\epsilon \in (0,1]$, *such that for any* $\mathsf{w}_0 \in \overline{O}_0$ *with*

$$|\mathsf{w}_0 - \mathsf{w}^\star|_l^2 + \frac{1}{\epsilon}|\pi\,\mathsf{w}_0|_{l-1}^2 + |\partial_t\mathsf{w}_0|_{l-2}^2 + \frac{1}{\epsilon}|\pi\,\partial_t\mathsf{w}_0|_{l-3}^2 < b^2, \tag{3.129}$$

there exists a unique solution of the out of equilibrium system such that estimates (3.115) *and* (3.118) *holds, as well as estimates* (3.116), *and furthermore, there exists a unique solution of the equilibrium system starting from* $\varphi\mathsf{w}_0$. *Moreover there exists a constant* C *depending on* O_1 *and* b *and independent of* $\epsilon, \epsilon_d \in [0,1)$ *such that*

$$\sup_{\tau\in[0,\bar{\tau}]} |\varphi \mathsf{w} - \mathsf{w}_{\mathrm{e}}|_{l-4} \leq C\epsilon(\epsilon + \epsilon_d). \tag{3.130}$$

Proof We may use the estimates of the residual term h in Theorem 3.19 and apply Theorem 3.8 with $\mathsf{f} = \mathsf{h}$, $\mathsf{g} = 0$, and use the different estimates (3.44) to complete the proof. □

In the special case $\epsilon_d = \epsilon$, these estimates show that the two term Chapman-Enskog expansion u_{e} established in [52], which includes the $O(\epsilon)$ volume viscosity terms, is effectively of second order accuracy. The loss of four derivatives in the estimate (3.130) arise from terms in the form $\boldsymbol{\nabla}\partial_t(\pi w/\epsilon)$ in the residual h, all other terms being easier to control. There is a first loss of derivative due to the fact that we need to estimate $\pi w/\epsilon$ and not $\pi w/\sqrt{\epsilon}$ as established in (3.115). There are next two extra losses of derivatives since we need to control the time derivative $\partial_t(\pi w/\epsilon)$ of $\pi w/\epsilon$ and we are dealing with second-order systems of partial differential equations. There is finally a fourth loss since we need to control the first-order spatial derivative $\boldsymbol{\nabla}\partial_t(\pi w/\epsilon)$ of $\partial_t(\pi w/\epsilon)$.

3.3.7 Asymptotics for Ill Prepared Data

In the situation of ill prepared data, it is necessary to take into account the existence of *initial layers*. In order to simplify the analysis it is assumed that

$$\epsilon_d = \epsilon.$$

The solution to the Cauchy problem is denoted by u_ϵ with the subscript ϵ and a similar notation is used for other variables like for instance w_ϵ for the normal variable. In order to distinguish between initial conditions and zeroth order expansion terms, the initial data will be denoted with the $_{\mathrm{data}}$ subscript in this section.

A multiscale asymptotic expansion need to be introduced using a convenient variable $\mathcal{w}_\epsilon$ and the difference between the expansion and the nonequilibrium solution need to be estimated. We rely on results obtained for a class of hyperbolic-parabolic systems that are shown to apply to the particular situation of two-temperature fluids [60]. We introduce the linear operators $\Pi_{\mathrm{s}} = [\mathsf{e}_1, \mathsf{e}_2, \dots, \mathsf{e}_{d+2}]$ and $\Pi_{\mathrm{r}} = [\mathsf{e}_{d+3}]$, where the subscripts r and s are mnemonics for rapid and slow, as well as the *quasinormal* variable $\mathcal{w}_\epsilon$ defined by

$$\mathcal{w}_\epsilon = (u_\epsilon, q_\epsilon)^t, \qquad u_\epsilon = \Pi_{\mathrm{s}}^t \mathsf{u}_\epsilon, \qquad q_\epsilon = \Pi_{\mathrm{r}}^t \mathsf{v}_\epsilon. \tag{3.131}$$

In other words, u_ϵ is the conservative slow variable and q_ϵ the fast symmetrized variable given by

$$u_\epsilon = \left(\rho,\ \rho \boldsymbol{v}, \rho(e_{\mathrm{tr}} + e_{\mathrm{in}} + \tfrac{1}{2}\rho|\boldsymbol{v}|^2)\right) \qquad q_\epsilon = \frac{1}{\mathrm{r}}\Big(\frac{1}{T_{\mathrm{tr}}} - \frac{1}{T_{\mathrm{tr}}}\Big). \tag{3.132}$$

The variable w_ϵ is termed quasinormal since it shares many similarities with the normal variables w_ϵ and it is easily established that the map $\mathsf{u} \mapsto w$ is a $C^{\gamma-1}$ diffeomorphism from the open set O_{u} onto an open set O_w of $\mathbb{R}^{\mathsf{n}}$. The variable w_ϵ is especially convenient in order to build a multiscale expansion as $\varepsilon \to 0$. The initial value of the quasinormal variable is denoted by w_{data}.

We introduce the multiscale—indeed multitime—expansion of w_ϵ

$$w_\varepsilon(\mathsf{x}, t) = w_0(\mathsf{x}, t) + \varepsilon w_1(\mathsf{x}, t) + w_0^{\mathrm{il}}(\mathsf{x}, t/\varepsilon) + \varepsilon w_1^{\mathrm{il}}(\mathsf{x}, t/\varepsilon) + O(\varepsilon^2), \tag{3.133}$$

Such a composite expansion (3.133) includes an *outer expansion*

$$w_0(\mathsf{x}, t) + \varepsilon w_1(\mathsf{x}, t) + O(\varepsilon^2), \tag{3.134}$$

associated with the standard time t, where w_0 and w_1 denote the zeroth and first order terms. Such an outer expansion (3.134), however, cannot generally satisfy the prescribed initial value w_{data}. It is thus necessary to add a fast evolving *initial layer corrector* in the form

$$w_0^{\mathrm{il}}(\mathsf{x}, t/\varepsilon) + \varepsilon w_1^{\mathrm{il}}(\mathsf{x}, t/\varepsilon) + O(\varepsilon^2),$$

associated with the fast time $\tau = t/\varepsilon$, where w_0^{il} and w_1^{il} denote the zeroth and first order terms. The initial layer correctors w_0^{il} and w_1^{il} have to decay exponentially to zero as $\tau = t/\epsilon$ goes to infinity from general matching principles of asymptotic expansions [59]. The complete approximation of w_ε is then a composite expansion since it involves both the standard time t and the fast time $\tau = t/\varepsilon$. As a formal solution with initial value w_{data}, the composite expansion must also take the prescribed initial value

$$w_{\mathrm{data}}(\mathsf{x}) = w_0(\mathsf{x}, 0) + \varepsilon w_1(\mathsf{x}, 0) + w_0^{\mathrm{il}}(\mathsf{x}, 0) + \varepsilon w_1^{\mathrm{il}}(\mathsf{x}, 0) + O(\varepsilon^2).$$

Naturally assuming that the initial condition w_{data} has an asymptotic expansion $w_{\mathrm{data}} = w_{\mathrm{data}\,0} + \varepsilon w_{\mathrm{data}\,1} + O(\varepsilon^2)$, we deduce that

$$w_0(\mathsf{x}, 0) + w_0^{\mathrm{il}}(\mathsf{x}, 0) = w_{\mathrm{data}\,0}(\mathsf{x}), \qquad w_1(\mathsf{x}, 0) + w_1^{\mathrm{il}}(\mathsf{x}, 0) = w_{\mathrm{data}\,1}(\mathsf{x}). \tag{3.135}$$

The proper initial values for w_0 and w_1 as well as w_0^{il} and w_1^{il} have then to be identified by using the initial value relations and the principles of matched asymptotic expansions [59].

Using the straightforward notation $w_\varepsilon = (u_\varepsilon, q_\varepsilon)^t$, $w_0 = (u_0, q_0)^t$, $w_1 = (u_1, q_1)^t$, $w_0^{\mathrm{il}} = (u_0^{\mathrm{il}}, q_0^{\mathrm{il}})^t$, and $w_1^{\mathrm{il}} = (u_1^{\mathrm{il}}, q_1^{\mathrm{il}})^t$, the projections u_ε and q_ε of the composite expansions are denoted correspondingly by

$$\begin{cases} u_\varepsilon = u_0(\mathsf{x},t) + \varepsilon u_1(\mathsf{x},t) + u_0^{\mathrm{il}}(\mathsf{x},t/\varepsilon) + \varepsilon u_1^{\mathrm{il}}(\mathsf{x},t/\varepsilon) + O(\varepsilon^2), \\ q_\varepsilon = q_0(\mathsf{x},t) + \varepsilon u_1(\mathsf{x},t) + q_0^{\mathrm{il}}(\mathsf{x},t/\varepsilon) + \varepsilon q_1^{\mathrm{il}}(\mathsf{x},t/\varepsilon) + O(\varepsilon^2). \end{cases}$$

From (3.135) we have at time $t = 0$ the relations

$$u_0(\mathsf{x},0) + u_0^{\mathrm{il}}(\mathsf{x},0) = u_{\mathrm{data}\,0}(\mathsf{x}), \qquad u_1(\mathsf{x},0) + u_1^{\mathrm{il}}(\mathsf{x},0) = u_{\mathrm{data}\,1}(\mathsf{x}), \tag{3.136}$$

$$q_0(\mathsf{x},0) + q_0^{\mathrm{il}}(\mathsf{x},0) = q_{\mathrm{data}\,0}(\mathsf{x}), \qquad q_1(\mathsf{x},0) + q_1^{\mathrm{il}}(\mathsf{x},0) = q_{\mathrm{data}\,1}(\mathsf{x}), \tag{3.137}$$

where $\mathcal{w}_{\mathrm{data}\,0} = (u_{\mathrm{data}\,0}, q_{\mathrm{data}\,0})^t$ and $\mathcal{w}_{\mathrm{data}\,1} = (u_{\mathrm{data}\,1}, q_{\mathrm{data}\,1})^t$.

Considering the zeroth order expansion coefficients $\mathcal{w}_0 = (u_0, q_0)^t$, the hyperbolic equation governing u_0 is found in the form

$$\partial_t u_0 + \sum_{i\in\mathfrak{D}} f_i(u_0, 0) = 0,$$

where $f_j = F_j^{\mathrm{s}}\big(\mathsf{u}(\mathcal{w})\big)$ denotes the slow convective fluxe $F_j^{\mathrm{s}} = \Pi_{\mathrm{s}} F_j$ in the jth direction as function of the quasinormal variable $\mathcal{w}$. The initial condition for u_0 is found to be $u_0(\cdot,0) = u_{\mathrm{data}\,0}$ and the fast zeroth order term is uniformly zero

$$q_0 = 0.$$

The zeroth order slow component $u_0 = \big(\rho_0, \rho_0\boldsymbol{v}_0, \rho_0(e(T_0) + \frac{1}{2}|\boldsymbol{v}_0|^2)\big)^t$ is the one-temperature compressible gas conservative variable, the corresponding governing equations are the one-temperature Euler equations, and $\mathcal{w}_0$ is at thermodynamic equilibrium. Similar equations are found at first order with a linear hyperbolic equation governing u_1 whereas q_1 is directly expressed as a linear expression in the gradients of u_0 and their initial values are also identified [60]. The corrector q_1 is found to be proportional to $\nabla\boldsymbol{\cdot}\boldsymbol{v}$ in such a way that from $q = q_0 + \epsilon q_1 + O(\epsilon^2)$ we recover that the temperature difference $T_{\mathrm{tr}} - T_{\mathrm{in}}$ is asymptotically proportional to $\nabla\boldsymbol{\cdot}\boldsymbol{v}$.

Moreover, it is found that the zeroth order initial corrector on the slow manifold is zero $u_0^{\mathrm{il}} = 0$ whereas q_0^{il} is governed by the ordinary differential equation

$$\partial_\tau q_0^{\mathrm{il}} = \partial_u^2\sigma(u_{\mathrm{data}\,0}, q_0^{\mathrm{il}})\ \omega_{\mathrm{in}}(u_{\mathrm{data}\,0}, q_0^{\mathrm{il}}),$$

where $\partial_u^2\sigma$ is written here as a function of $\mathcal{w}$, with initial condition $q_0^{\mathrm{il}}(\cdot,0) = q_{\mathrm{data}\,0}$. The relative entropy

$$\sigma^{\mathrm{il}} = \sigma\big(\mathsf{u}(u,q)\big) - \sigma\big(\mathsf{u}(u,0)\big) - \partial_{\mathsf{u}}\sigma\big(\mathsf{u}(u,0)\big)\big(\mathsf{u}(u,q) - \mathsf{u}(u,0)\big),$$

is a Lyapunov function for the ordinary differential equation governing q_0^{il} and such that $\lim_{|q|\to\infty} \sigma^{\mathrm{il}}(u,q) = +\infty$. Similar equations are found for the first initial layer correctors, that is, u_1^{il} is expressed explicitly in terms of zeroth order terms whereas q_1^{il} is also governed by an ordinary differential equation and its initial value is also identified [60].

All the mathematical problems defining the expansion coefficients are shown to be well posed in properly selected Sobolev spaces [60]. More specifically, assuming that $l \geq l_0 + 5$, that the initial data satisfy $\mathcal{w}_{\mathrm{data}\,0} \in H^l$, $\mathcal{w}_{\mathrm{data}\,1} \in H^{l-2}$, and that $\mathcal{O}_{\mathcal{w}}$ is in the form $\mathcal{O}_{\mathcal{w}} = \mathcal{O}_u \times \mathbb{R}^{n_{\mathrm{r}}}$ then there exists $\mathcal{w}_0 \in C^0\big([0,\bar{\tau}], H^l\big)$, $\mathcal{w}_1 \in C^0\big([0,\bar{\tau}], H^{l-2}\big)$, as well as $\mathcal{w}_0^{\mathrm{il}} \in C^0\big([0,\infty), H^l\big)$, $\mathcal{w}_1^{\mathrm{il}} \in C^0\big([0,\infty), H^{l-2}\big)$, with

$$|\mathcal{w}_0^{\mathrm{il}}|_l \leq C\exp(-\delta t/\varepsilon) \qquad |\mathcal{w}_1^{\mathrm{il}}|_{l-2} \leq C\exp(-\delta t/\varepsilon),$$

for some positive $\delta > 0$ so that the initial layers are decreasing exponentially with respect to the fast time. A key point for these existence results are the symmetrized formulations of the hyperbolic equations governing u_0 and u_1. The relative entropy is also a Lyapunov functions for the ordinary differential equations with respect to the fast time τ governing the initial layer correctors ensuring a global in time existence result [60]. We may then use the second order truncated approximation $\mathcal{w}_\varepsilon^{\mathrm{a}}$ defined by

$$\mathcal{w}_\varepsilon^{\mathrm{a}}(\mathsf{x},t) = \mathcal{w}_0(\mathsf{x},t) + \varepsilon\mathcal{w}_1(\mathsf{x},t) + \mathcal{w}_0^{\mathrm{il}}(\mathsf{x},t/\varepsilon) + \varepsilon\mathcal{w}_1^{\mathrm{il}}(\mathsf{x},t/\varepsilon). \tag{3.138}$$

This four term expansion $\mathcal{w}_\varepsilon^{\mathrm{a}}$ is well defined over some time interval independent of ε and may be used to establish an existence theorem for the out of equilibrium systems [60]. For technical reasons it is also necessary to add an extra term $\mathcal{w}_2' = (0, q_2)^t$ that only includes the fast component of the second term $\mathcal{w}_2$ but these subtleties are omitted [60]. We now establish that the general results from [60] may be applied to the two-temperature nonequilibrium gas model.

Theorem 3.21 *Properties* $(\mathrm{U}_1 - \mathrm{U}_5)$ *required to apply the general results of [60] hold for the system modeling gas out of thermodynamic equilibrium with the mathematical entropy* $\sigma = -\rho(s_{\mathrm{tr}} + s_{\mathrm{in}})/\mathfrak{r}$. *The source term* Ω *is quasilinear in terms of* $\mathcal{w}$ *with* $\Omega = (0,0,0,\omega_{\mathrm{in}})^t = -\mathcal{L}\mathcal{w}$, $\omega_{\mathrm{in}} = -\mathcal{L}_{\mathrm{rr}} q$, $\mathcal{L} = \mathrm{diag}(0,0,0,\mathcal{L}_{\mathrm{rr}})$ *and* $\mathcal{L}_{\mathrm{rr}} = \rho\mathfrak{r}T_{\mathrm{tr}}T_{\mathrm{in}}\frac{c_{\mathrm{in}}}{\tau_{\mathrm{in}}}\, \mathsf{e}_{d+3}\otimes\mathsf{e}_{d+3}$.

Sketch of the Proof The properties $(\mathrm{U}_1 - \mathrm{U}_3)$ of [60] are consequences of the structure of the normal form and the structure of the source term further yields (U_4). The open set $\mathcal{O}_{\mathcal{w}}$ where $\mathcal{w}$ is defined is given by $\mathcal{O}_{\mathcal{w}} = \mathcal{O}_u \times (-\infty,\infty)$ and has the required product form. When the equilibrium temperature T is fixed, we indeed have $\lim_{T_{\mathrm{tr}}\to 0} q = +\infty$ and $\lim_{T_{\mathrm{in}}\to 0} q = -\infty$.

The relative entropy $\eta^{\mathrm{il}} = \sigma(\mathsf{u}) - \sigma(\mathsf{u}_{\mathrm{eq}}) - \partial_{\mathsf{u}}\sigma(\mathsf{u}_{\mathrm{eq}})(\mathsf{u} - \mathsf{u}_{\mathrm{eq}})$ is a natural distance to equilibrium. It is nonnegative by convexity of σ and locally behaves quadratically with respect to q [60]. We have to establish that the relative entropy η^{il} is coercive in the sense that $\lim_{|q|\to\infty} \eta^{\mathrm{il}}(u,q) = \infty$ uniformly with respect to u on compact sets of $\mathcal{O}_u$. This relative entropy may be written $\eta^{\mathrm{il}}(u,q) = \sigma\big(\mathsf{u}(u,q)\big) - \sigma\big(\mathsf{u}(u,0)\big)$ and is evaluated to be

$$\eta^{\mathrm{il}} = \frac{\rho}{\mathrm{r}}\Big(c_{\mathrm{tr}} \log \frac{T}{T_{\mathrm{tr}}} + \int_{T_{\mathrm{in}}}^{T} c_{\mathrm{in}}(\theta)\, d\theta\Big).$$

The relative entropy η^{il} must be investigated as a function of q with u fixed, that is, with ρ, $\boldsymbol{v}$ and T fixed. Using that T is fixed, that $c_{\mathrm{tr}}T_{\mathrm{tr}} + \int_0^{T_{\mathrm{in}}} c_{\mathrm{in}}(\theta)\, d\theta = c_{\mathrm{tr}}T + \int_0^T c_{\mathrm{in}}(\theta)\, d\theta$, and that $\underline{c}_{\mathrm{in}} \leq c_{\mathrm{in}}$ we first obtain the bounds $T_{\mathrm{tr}} \leq \overline{T}$ and $T_{\mathrm{in}} \leq \overline{T}$ where $\overline{T} = \big(c_{\mathrm{tr}}T + \int_0^T c_{\mathrm{in}}(\theta)\, d\theta\big)/\min(c_{\mathrm{tr}}, \underline{c}_{\mathrm{in}})$. Since $\eta^{\mathrm{il}} = \mathrm{Cte} + \rho\big(c_{\mathrm{tr}} \log \frac{\overline{T}}{T_{\mathrm{tr}}} + \int_{T_{\mathrm{in}}}^{\overline{T}} c_{\mathrm{in}}(\theta)\, d\theta\big)/\mathrm{r}$ where Cte denotes a generic constant with respect to q, that is, a quantity only depending on u, we have $\eta^{\mathrm{il}} \geq \mathrm{Cte} - \frac{\rho}{\mathrm{r}} c_{\mathrm{tr}} \log(T_{\mathrm{tr}}) - \frac{\rho}{\mathrm{r}} \underline{c}_{\mathrm{in}} \log(T_{\mathrm{in}})$ in such a way that $\eta^{\mathrm{il}} \geq \mathrm{Cte} - \delta \log(T_{\mathrm{tr}}T_{\mathrm{in}})$ where $\delta = (\rho/\mathrm{r}) \min(c_{\mathrm{tr}}, \underline{c}_{\mathrm{in}})$ is positive and fixed. On the other hand, assuming that $|q| \geq M$ where M is large and positive, we obtain that $q^2 \geq M^2$ and thus

$$\frac{(T_{\mathrm{tr}} - T_{\mathrm{in}})^2}{T_{\mathrm{tr}}^2 T_{\mathrm{in}}^2} = \frac{T_{\mathrm{tr}}^2 + T_{\mathrm{in}}^2 - 2T_{\mathrm{tr}}T_{\mathrm{in}}}{T_{\mathrm{tr}}^2 T_{\mathrm{in}}^2} \geq \mathrm{r}^2 M^2.$$

Letting $\zeta = T_{\mathrm{tr}}T_{\mathrm{in}}$, we have then $\mathrm{r}^2M^2\zeta^2 + 2\zeta - (T_{\mathrm{tr}}^2 + T_{\mathrm{in}}^2) \leq 0$ and thus $\mathrm{r}^2M^2\zeta^2 + 2\zeta - 2\overline{T}^2 \leq 0$. This quadratic equation has only one positive root and this yields $\zeta = T_{\mathrm{tr}}T_{\mathrm{in}} \leq \frac{1}{2\mathrm{r}^2M^2}\Big(\sqrt{1 + 2\mathrm{r}^2M^2\overline{T}^2} - 1\Big)$ and next $T_{\mathrm{tr}}T_{\mathrm{in}} \leq \frac{\sqrt{2}\,\overline{T}}{\mathrm{r}M}$ and finally $\eta^{\mathrm{il}} \geq \mathrm{Cte} + \delta \log M$ in such a way that $\lim_{|q|\to\infty} \eta^{\mathrm{il}} = +\infty$ so that Property (U_5) of [60] is finally established and the proof is complete. □

We may now use the multitime error estimate [60] for the out of equilibrium solution in the situation of ill prepared initial data.

Theorem 3.22 *Assuming that $l \geq l_0+5$, there exists positive constants t_* and $\mathrm{C}(t_*)$ independent of ε such that the out of thermodynamic equilibrium system* (3.53)–(3.56) *has unique solutions in $C^0([0, t_*], H^{l-4})$, the approximated solution $w_\varepsilon^{\mathrm{a}}$ also exists in $C^0([0, t_*], H^{l-4})$, and we have*

$$\sup_{t\in[0,t_*]} |w_\varepsilon(\cdot, t) - w_\varepsilon^{\mathrm{a}}(\cdot, t)|_{l-4} \leq \mathrm{C}(t_*)\varepsilon^2.$$

Sketch of the Proof This directly results from the results established in [60]. The loss of derivatives in the estimates result from the low regularity of the residuals obtained when inserting the approximated solution $w_\varepsilon^{\mathrm{a}}$ in the equations in normal form governing w_ε. □

We thus deduce that the out of equilibrium solution w_ε is close to the multitime expansion $w_\varepsilon^{\mathrm{a}}$. Using the same type of proof, that is, injecting the outer expansion $u_0 + \epsilon u_1$ into the equation governing u_{e}, it is also established that the Hilbert type outer expansion $u_0 + \epsilon u_1$ is close to the Enskog expansion u_{e} with

$$\sup_{t\in[0,t_*]} \left|u_{\mathrm{e}}(t) - \big(u_0(t) + \varepsilon u_1(t)\big)\right|_{l-4} \leq \mathrm{c}\varepsilon^2.$$

As a consequence, we may now estimate the difference between the slow component u_ϵ and the Enskog expansion by simply using the projection on the slow manifold.

Theorem 3.23 *Keeping the assumptions of Theorem 3.22 we have the estimate*

$$\sup_{t\in[0,t_*]} |u_\varepsilon(\cdot,t) - (u_{\mathrm{e}}(\cdot,t) + \varepsilon u_1^{\mathrm{il}}(\cdot,t/\varepsilon))|_{l-4} \leq \mathrm{c}(t_*)\varepsilon^2,$$

where $u_1^{\mathrm{il}}(\cdot,t/\varepsilon))$ is the slow component of the first order initial layer corrector.

We finally deduce that the one-temperature Enskog expansion u_{e} is still a quite good approximation of u_ϵ since the error is $O(\epsilon^2)$ up to the first order $\varepsilon u_1^{\mathrm{il}}$ that is decreasing exponentially with respect to the fast time t/ϵ.

3.4 Diffuse Interface Fluids

We investigate in this section the equations governing diffuse interface fluids also termed Korteweg, capillary, or cohesive fluids [61–74]. Such equations involve nonideal bulk thermodynamic properties [121–125], van der Waals' gradient free energy [61], Korteweg's tensor [62] and Dunn and Serrin's heat flux [63]. Such equations have also been derived from the kinetic theory thereby removing ambiguities associated with thermodynamic methods [71–73]. Such cohesive fluid models have successfully been used for the study of phase change problems including complex liquid-gas interfaces with topological changes [66–70, 74]. Existence theorems of local solutions—weak or strong—have been obtained for such fluids, often in an isothermal framework, in one dimension as well as in several spatial dimensions [75–90].

We present the equations governing single species diffuse interface fluids. The density gradient is added as an extra variable and an augmented system of equations is obtained [76, 79, 81, 84]. These augmented formulations may be related to multifluid hydrodynamic models [126, 127]. The augmented system of equations is recast into a new normal form with symmetric hyperbolic first order terms, symmetric dissipative second order terms and antisymmetric cohesive second order terms. A priori estimates are obtained for such augmented systems of equations in normal form. The time derivatives of the parabolic components are less regular than for standard hyperbolic-parabolic systems and the strongly coupling antisymmetric terms yield new majorizing contributions. Using the augmented system in normal form and a priori estimates, a variant proof of the local existence of strong solutions is established in an Hilbertian framework [90].

We further investigate the asymptotic stability of constant equilibrium states by introducing a new strict stability condition for augmented systems of equations. This condition is found somewhat simpler than the corresponding condition previously introduced for non augmented systems [91–98]. The stability results are also more convenient to formulate by using the augmented variables.

3.4.1 Governing Equations

The equations governing cohesive fluids may be written in the form [66, 67, 71–73]

$$\partial_t \rho + \nabla\cdot(\rho \boldsymbol{v}) = 0, \tag{3.139}$$

$$\partial_t (\rho \boldsymbol{v}) + \nabla\cdot(\rho \boldsymbol{v}\otimes\boldsymbol{v}) + \nabla\cdot\mathcal{P} = 0, \tag{3.140}$$

$$\partial_t \big(\mathcal{E} + \tfrac{1}{2}\rho|\boldsymbol{v}|^2\big) + \nabla\cdot\big(\boldsymbol{v}(\mathcal{E} + \tfrac{1}{2}\rho|\boldsymbol{v}|^2)\big) + \nabla\cdot(\boldsymbol{Q} + \mathcal{P}\cdot\boldsymbol{v}) = 0, \tag{3.141}$$

where ∂_t denotes the time derivative operator, ∇ the spatial differential operator in physical space $\mathbb{R}^d$, d the space dimension, ρ the mass density, $\boldsymbol{v}$ the fluid velocity, $\mathcal{P}$ the total pressure tensor, $\mathcal{E}$ the internal energy, and $\boldsymbol{Q}$ the total heat flux. Vectors and tensors of physical origin in $\mathbb{R}^d$ or $\mathbb{R}^{d,d}$ are denoted by using bold symbols. The transport fluxes $\mathcal{P}$ and $\boldsymbol{Q}$ contain capillary as well as dissipative contributions

$$\mathcal{P} = p\boldsymbol{I} + \varkappa\nabla\rho\otimes\nabla\rho - \rho\nabla\cdot(\varkappa\nabla\rho)\boldsymbol{I} + \boldsymbol{\Pi}, \tag{3.142}$$

$$\boldsymbol{Q} = \varkappa\rho\nabla\cdot\boldsymbol{v}\,\nabla\rho + \boldsymbol{q}, \tag{3.143}$$

where $\varkappa$ denotes the capillarity coefficient, $\boldsymbol{I}$ the d-dimensional unit tensor, $\boldsymbol{\Pi}$ the viscous pressure tensor, and $\boldsymbol{q}$ the Fourier heat flux [66, 67]. In comparison with non cohesive fluids we note that there are first and second order density derivatives in the pressure tensor $\mathcal{P}$ so that the system is of third order as well as first order derivatives of density in the heat flux. The dissipative fluxes $\boldsymbol{\Pi}$ and $\boldsymbol{q}$ are classically of Navier-Stokes-Fourier type [5–7]

$$\boldsymbol{\Pi} = -\,\kappa\nabla\cdot\boldsymbol{v}\,\boldsymbol{I} - \eta\big(\nabla\boldsymbol{v} + \nabla\boldsymbol{v}^t - \tfrac{2}{3}\nabla\cdot\boldsymbol{v}\,\boldsymbol{I}\big), \tag{3.144}$$

$$\boldsymbol{q} = -\,\lambda\nabla T, \tag{3.145}$$

where κ denotes the volume viscosity, η the shear viscosity and λ the thermal conductivity. The transport coefficients κ, η, and λ are obtained from the kinetic theory of dense gases and only depend on ρ and T [71–73].

The free energy per unit volume $\mathcal{F}$ introduced by Van der Waals is in the form

$$\mathcal{F} = \mathcal{F}^{\mathrm{cl}} + \tfrac{1}{2}\varkappa|\nabla\rho|^2, \tag{3.146}$$

where $\mathcal{F}^{\,\mathrm{cl}}$ denotes the classic free energy per unit volume. The superscript ^{cl} is used to denote classic or bulk thermodynamic properties that do not involve gradients. The classic free energy $\mathcal{F}^{\,\mathrm{cl}}$ only depends on the densities ρ and the absolute temperature T whereas the gradient squared term $\frac{1}{2}\varkappa|\boldsymbol{\nabla}\rho|^2$ in $\mathcal{F}$ represents an excess free energy of the interfacial region [61, 62, 64, 66]. As deduced from the kinetic theory of dense gases [71–73], or from experimental measurements [128], the capillary coefficient $\varkappa$ only depend on temperature $\varkappa = \varkappa(T)$ for diffuse interface fluids at variance with the case of quantum fluids [71–73].

Using the thermodynamic relations $\mathbb{D}\mathcal{F}^{\,\mathrm{cl}} = -\mathcal{S}^{\,\mathrm{cl}}\,\mathbb{D}T + g^{\mathrm{cl}}\,\mathbb{D}\rho$, $\partial_T\mathcal{F} = -\mathcal{S}$, and $\partial_\rho\mathcal{F} = g$, the entropy per unit volume $\mathcal{S}$ and the Gibbs function per unit mass g are found in the form

$$\mathcal{S} = \mathcal{S}^{\,\mathrm{cl}} - \tfrac{1}{2}\partial_T\varkappa|\boldsymbol{\nabla}\rho|^2, \qquad g = g^{\mathrm{cl}}. \tag{3.147}$$

The equality of the entropy $\mathcal{S}$ with the classic entropy $\mathcal{S}^{\,\mathrm{cl}}$ when $\varkappa$ is independent of temperature is in agreement with van der Waals theory [61, 62, 64] and the Gibbs function per unit volume $\mathcal{G}$ reads $\mathcal{G} = \rho g = \mathcal{G}^{\,\mathrm{cl}}$. The energy per unit volume $\mathcal{E} = \mathcal{F} + T\mathcal{S}$ and pressure $p = \mathcal{G} - \mathcal{F}$ are further obtained as

$$\mathcal{E} = \mathcal{E}^{\,\mathrm{cl}} + \tfrac{1}{2}(\varkappa - T\partial_T\varkappa)|\boldsymbol{\nabla}\rho|^2, \qquad p = p^{\mathrm{cl}} - \tfrac{1}{2}\varkappa|\boldsymbol{\nabla}\rho|^2, \tag{3.148}$$

as well as the enthalpy per unit volume $\mathcal{H} = \mathcal{E} + p = \mathcal{H}^{\,\mathrm{cl}} - \frac{1}{2}T\partial_T\varkappa|\boldsymbol{\nabla}\rho|^2$ with $\mathcal{H}^{\,\mathrm{cl}} = \mathcal{E}^{\,\mathrm{cl}} + p^{\mathrm{cl}}$. The corresponding properties per unit mass are defined as usual by $s = \mathcal{S}/\rho$ and $e = \mathcal{E}/\rho$. The generalized volumetric Gibbs relation finally reads

$$T\,\mathbb{D}\mathcal{S} = \mathbb{D}\mathcal{E} - g\,\mathbb{D}\rho - \varkappa\boldsymbol{\nabla}\rho\boldsymbol{\cdot}\mathbb{D}\boldsymbol{\nabla}\rho, \tag{3.149}$$

where $\mathbb{D}$ denotes the differentiation operator. Using Gibbs' relations and the governing Eqs. (3.139)–(3.141) it is next obtained that

$$\begin{aligned}\partial_t\mathcal{S} + \boldsymbol{\nabla}\boldsymbol{\cdot}(\boldsymbol{v}\mathcal{S}) + \boldsymbol{\nabla}\boldsymbol{\cdot}\Big(\frac{\boldsymbol{Q}}{T} - \frac{\varkappa\rho\boldsymbol{\nabla}\boldsymbol{\cdot}\boldsymbol{v}\boldsymbol{\nabla}\rho}{T}\Big) &= -\Big(\boldsymbol{Q} - \varkappa\rho\boldsymbol{\nabla}\boldsymbol{\cdot}\boldsymbol{v}\boldsymbol{\nabla}\rho\Big)\boldsymbol{\cdot}\frac{\boldsymbol{\nabla}T}{T^2}\\ &\quad - \frac{1}{T}\Big(\mathcal{P} - p\boldsymbol{I} - \varkappa\boldsymbol{\nabla}\rho\otimes\boldsymbol{\nabla}\rho + \rho\boldsymbol{\nabla}\boldsymbol{\cdot}(\varkappa\boldsymbol{\nabla}\rho)\boldsymbol{I}\Big){:}\boldsymbol{\nabla}\boldsymbol{v},\end{aligned} \tag{3.150}$$

where $\mathsf{v}{:}\mathsf{w}$ denotes the full contraction between any two tensors v and w. Using the expressions (3.142) and (3.143) for $\mathcal{P}$ and $\boldsymbol{Q}$ as well as (3.144) and (3.145) yields

$$\begin{aligned}\partial_t\mathcal{S} + \boldsymbol{\nabla}\boldsymbol{\cdot}(\boldsymbol{v}\mathcal{S}) + \boldsymbol{\nabla}\boldsymbol{\cdot}\Big(\frac{\boldsymbol{Q}}{T} - \frac{\varkappa\rho\boldsymbol{\nabla}\boldsymbol{\cdot}\boldsymbol{v}\boldsymbol{\nabla}\rho}{T}\Big) &= \frac{\lambda}{T^2}|\boldsymbol{\nabla}T|^2\\ &\quad + \frac{1}{T}\Big(\kappa + \eta\big(\tfrac{2}{d} - \tfrac{2}{3}\big)\Big)(\boldsymbol{\nabla}\boldsymbol{\cdot}\boldsymbol{v})^2 + \frac{\eta}{2T}\big|\boldsymbol{\nabla}\boldsymbol{v} + \boldsymbol{\nabla}\boldsymbol{v}^t - \tfrac{2}{d}\boldsymbol{\nabla}\boldsymbol{\cdot}\boldsymbol{v}\,\boldsymbol{I}\big|^2,\end{aligned} \tag{3.151}$$

and there is no entropy production associated with capillary phenomena.

The mathematical properties of the nonideal classic thermodynamic functions $\mathcal{E}^{\,\mathrm{cl}}$, p^{cl}, and $\mathcal{S}^{\,\mathrm{cl}}$ are now presented [121–125]. The mathematical structure of classic thermodynamics and its construction from state laws has been investigated in [123–125]. The properties of the extended thermodynamic functions $\mathcal{E}$, p, $\mathcal{S}$, the capillarity coefficient $\varkappa$, and the transport coefficients κ, η, λ are addressed in the next section using augmented variables. We denote for convenience by $\mathsf{z}^{\,\mathrm{cl}}$ the variable $\mathsf{z}^{\,\mathrm{cl}} = (\rho, T)^t$ and by $\mathsf{u}^{\,\mathrm{cl}}$ the variable $\mathsf{u}^{\,\mathrm{cl}} = (\rho, \mathcal{E}^{\,\mathrm{cl}})^t$. We may commit the traditional abuse of notation of using the same symbol for a given quantity as function of different state variables. We denote in the following by ∂ the derivation operator with respect to the variable $\mathsf{z}^{\,\mathrm{cl}}$ and the regularity class γ denotes an integer such that $\gamma \geq 3$.

($\mathsf{H}_1^{\mathrm{cl}}$) The thermodynamic functions $\mathcal{E}^{\,\mathrm{cl}}$, p^{cl}, and $\mathcal{S}^{\,\mathrm{cl}}$ are C^γ functions of the variable $\mathsf{z}^{\,\mathrm{cl}} = (\rho, T)^t$ defined over a simply connected nonempty open set $\mathcal{O}_{\mathsf{z}^{\,\mathrm{cl}}} \subset (0, \infty)^2$.

($\mathsf{H}_2^{\mathrm{cl}}$) Assume that ($\mathsf{H}_1^{\mathrm{cl}}$) holds and define $\mathcal{G}^{\,\mathrm{cl}} = \mathcal{E}^{\,\mathrm{cl}} + p^{\mathrm{cl}} - T\mathcal{S}^{\,\mathrm{cl}}$ and $g^{\,\mathrm{cl}} = \mathcal{G}^{\,\mathrm{cl}}/\rho$, then we have the volumetric Gibbs' relation

$$T\,\mathbb{D}\mathcal{S}^{\,\mathrm{cl}} = \mathbb{D}\mathcal{E}^{\,\mathrm{cl}} - g^{\mathrm{cl}}\,\mathbb{D}\rho. \tag{3.152}$$

The first property ($\mathsf{H}_1^{\mathrm{cl}}$) is associated with the smoothness of thermodynamic functions and the regularity class γ denotes an integer such that $\gamma \geq 3$. The second property ($\mathsf{H}_2^{\mathrm{cl}}$) concerns Gibbs relation between the differentials of $\mathcal{E}^{\,\mathrm{cl}}$ and $\mathcal{S}^{\,\mathrm{cl}}$. There is also a natural constraint associated with $\mathcal{G}^{\,\mathrm{cl}}$ since we only consider a volumetric Gibbs differential relation [123]. A fundamental question associated with nonideal thermodynamics is then that of thermodynamic stability. According to the second principle of thermodynamics, the evolution of an isolated system tends to maximize its entropy. The physical entropy of a stable isolated system should thus be a concave function of its volume and internal energy [121]. Whenever it is not the case, the system may lose its homogeneity and split between several phases in order to reach equilibrium. This notably arises with mechanically unstable fluids that split between vapor and liquid phases [121]. From a mathematical point of view, these unstable points are associated with a loss of definiteness of entropy Hessian matrices [121–125].

Proposition 3.5 *Assume that (H_1^{cl}) and (H_2^{cl}) hold and that $\mathsf{z}^{\,\mathrm{cl}} \mapsto \mathsf{u}^{\,\mathrm{cl}}$ is locally invertible. Denoting by $\widetilde{\partial}$ the derivation operator with respect to $\mathsf{u}^{\,\mathrm{cl}}$, the following statements are equivalent :*

(i) $\widetilde{\partial}^2_{\mathsf{u}^{\mathrm{cl}}\mathsf{u}^{\mathrm{cl}}}\mathcal{S}^{\,\mathrm{cl}}$ *is negative definite.*

(ii) $\partial_T\mathcal{E}^{\,\mathrm{cl}} > 0$ *and* $\partial_\rho p^{\mathrm{cl}} > 0$.

Proof For any test vector $\mathsf{x}^{\,\mathrm{cl}} = (\mathsf{x}_\rho, \mathsf{x}_{\mathcal{E}})^t \in \mathbb{R}^2$ we have

$$\langle \widetilde{\partial}^2_{\mathsf{u}^{\mathrm{cl}}\mathsf{u}^{\mathrm{cl}}} \mathcal{S}^{\mathrm{cl}} \mathsf{x}^{\mathrm{cl}}, \mathsf{x}^{\mathrm{cl}} \rangle = -\frac{1}{T^2 \partial_T \mathcal{E}^{\mathrm{cl}}} \big(\mathsf{x}_{\mathcal{E}} - \partial_\rho \mathcal{E}^{\mathrm{cl}} \mathsf{x}_\rho \big)^2 - \frac{1}{\rho T} \partial_\rho p^{\mathrm{cl}} \, \mathsf{x}_\rho^2, \tag{3.153}$$

so that thermodynamic stability is equivalent to $\partial_T \mathcal{E}^{\mathrm{cl}} > 0$ and $\partial_\rho p^{\mathrm{cl}} > 0$. □

The inequality $\partial_T \mathcal{E}^{\mathrm{cl}} > 0$ is the thermal stability condition and the inequality $\partial_\rho p^{\mathrm{cl}} > 0$ the mechanical stability condition [121–125]. Of course for an ideal gas we have $\partial_T \mathcal{E}^{\mathrm{pg}} = \rho c > 0$ and $\partial_\rho p^{\mathrm{pg}} = RT/m > 0$ so that all states are stable. The superscript ^{pg} is used to denote the perfect gas thermodynamic as studied in Sect. 3.2.1. For nonideal fluid considered in this section, we do not assume that the Hessian matrix $\widetilde{\partial}^2_{\mathsf{u}^{\mathrm{cl}}\mathsf{u}^{\mathrm{cl}}} \mathcal{S}^{\mathrm{cl}}$ is negative definite. We indeed anticipate the presence of mechanical instabilities where $\partial_\rho p^{\mathrm{cl}} < 0$ and only thermal stability is assumed. A nonempty open set $O_{\mathsf{z}^{\mathrm{cl}}} \subset (0,\infty)^2$ is said to be increasing with respect to temperature if for any $\mathsf{z}^{\mathrm{cl}} = (\rho, T)^t \in O_{\mathsf{z}^{\mathrm{cl}}}$ we have $\{\rho\} \times [T, \infty) \subset O_{\mathsf{z}^{\mathrm{cl}}}$. In order to integrate with respect to the variable T in the open set $O_{\mathsf{z}^{\mathrm{cl}}}$ it is required that the open set $O_{\mathsf{z}^{\mathrm{cl}}}$ is increasing with respect to temperature. The construction of thermodynamics from state laws further requires the open set $O_{\mathsf{z}^{\mathrm{cl}}}$ to be decreasing with respect to density [123] as well as the compatibility with ideal gases for small densities [123–125], but such properties are not required for existence theorem [124]. We now assume that the fluid is thermally stable and this allows the use of the u^{cl} variable.

($\mathsf{H}_3^{\mathrm{cl}}$) Assuming ($\mathsf{H}_1^{\mathrm{cl}}$), the open set $O_{\mathsf{z}^{\mathrm{cl}}}$ is increasing with respect to temperature and the volumetric specific heat is positive $\partial_T \mathcal{E}^{\mathrm{cl}} > 0$ over $O_{\mathsf{z}^{\mathrm{cl}}}$.

Assuming that ($\mathsf{H}_1^{\mathrm{cl}}$) and ($\mathsf{H}_3^{\mathrm{cl}}$) hold, the map $\mathsf{z}^{\mathrm{cl}} \mapsto \mathsf{u}^{\mathrm{cl}}$ is a C^γ diffeomorphism from $O_{\mathsf{z}^{\mathrm{cl}}}$ onto an open set $O_{\mathsf{u}^{\mathrm{cl}}}$. We will also need in the following an instability indicator m^{cl} in order to control negative values of the derivative $\partial_\rho p^{\mathrm{cl}}$ [90].

Lemma 3.6 *Assume that* (H_1^{cl}) *holds and that* $\delta > 0$ *is given. There exists a* $C^{\gamma-1}(O_{\mathsf{z}^{\mathrm{cl}}})$ *function* m^{cl} *such that* $\mathsf{m}^{\mathrm{cl}} \geq 0$, $\mathsf{m}^{\mathrm{cl}} + \partial_\rho p^{\mathrm{cl}}/\rho T > 0$ *and* $\mathsf{m}^{\mathrm{cl}} = 0$ *if* $\partial_\rho p^{\mathrm{cl}}/\rho T \geq \delta$.

Whenever stability is lost, we will add m^{cl} to $\partial_\rho p^{\mathrm{cl}}/\rho T$ in order to obtain a positive quantity. When a thermodynamical state z^{cl} is stable in the sense that $\partial_\rho p^{\mathrm{cl}}/\rho T \geq \delta$ the instability indicator m^{cl} also vanishes in the neighborhood of z^{cl}. From a physical point of view, classic thermodynamic functions for fluids may first be obtained by extending the thermostatic framework to fluids by using the idea of a local state [2, 121]. A more satisfactory derivation is that from the kinetic theory of gases [5–7] since it does take into account the presence of *macroscopic gradients*. The idealized vision that there exists a local state is replaced by the more satisfactory argument that the gas distribution function is a Maxwellian at zeroth order of Enskog expansion. Extended thermodynamics are further valid in the presence of *microscopic gradients* and steep interfaces and may also be derived from the kinetic theory of dense gases [71–73].

3.4.2 Augmented Formulation

We recast in this section the third order system of equation governing cohesive fluids (3.139)–(3.145) in the form of a *second order* system. A main idea is to add the gradient of density as an extra unknown [76, 79, 81, 84].

We introduce the extra unknown vector [76, 79, 81, 84]

$$\boldsymbol{w} = \boldsymbol{\nabla}\rho. \tag{3.154}$$

and the $\boldsymbol{w}$ governing equation is obtained by applying the differential operator $\boldsymbol{\nabla}$ to the mass conservation Eq. (3.139). The resulting equation may be written in the form

$$\partial_t \boldsymbol{w} + \sum_{i\in\mathfrak{D}} \partial_i \big(\boldsymbol{w}\, v_i + \rho \boldsymbol{\nabla} v_i\big) = 0, \tag{3.155}$$

where $\mathfrak{D} = \{1, \dots, d\}$ denotes the indexing set of spatial directions, v_i the velocity in the ith spatial direction and ∂_i the derivative in the ith spatial direction so that $\boldsymbol{v} = (v_1, \dots, v_d)^t$ and $\boldsymbol{\nabla} = (\partial_1, \dots, \partial_d)^t$. The *augmented* conservative unknown u and natural variable z are then in the form

$$\mathsf{u} = \big(\rho, \boldsymbol{w}, \rho\boldsymbol{v}, \mathcal{E} + \tfrac{1}{2}\rho|\boldsymbol{v}|^2\big)^t, \qquad \mathsf{z} = \big(\rho, \boldsymbol{w}, \boldsymbol{v}, T\big)^t, \tag{3.156}$$

and include both ρ and $\boldsymbol{w}$. The variables u and z are in $\mathbb{R}^{\mathsf{n}}$ where $\mathsf{n} = 2d + 2$ is the augmented number of unknowns.

We may then express the thermodynamic functions $\mathcal{E}$, p, and $\mathcal{S}$ in terms of the augmented natural variable z

$$\mathcal{E} = \mathcal{E}^{\,\mathrm{cl}} + \tfrac{1}{2}(\varkappa - T\partial_T\varkappa)|\boldsymbol{w}|^2, \qquad \mathcal{S} = \mathcal{S}^{\,\mathrm{cl}} - \tfrac{1}{2}\partial_T\varkappa|\boldsymbol{w}|^2, \qquad p = p^{\mathrm{cl}} - \tfrac{1}{2}\varkappa|\boldsymbol{w}|^2,$$

as well as $\mathcal{H} = \mathcal{H}^{\,\mathrm{cl}} - \frac{1}{2}T\partial_T\varkappa|\boldsymbol{w}|^2$ and $g = g^{\mathrm{cl}}$.

We now describe the mathematical properties of the thermodynamic functions $\mathcal{E}$, p, and $\mathcal{S}$ as functions of z or u. We simply denote by ∂ the derivation operator with respect to the variable z. The capillarity coefficient $\varkappa$ only depends on temperature following the kinetic theory of dense gases [71–73]. A nonempty open set $\mathcal{O}_{\mathsf{z}} \subset (0, \infty) \times \mathbb{R}^d \times \mathbb{R}^d \times (0, \infty)$ is said to be increasing with respect to temperature if for any $\mathsf{z} = (\rho, \boldsymbol{w}, \boldsymbol{v}, T)^t \in \mathcal{O}_{\mathsf{z}}$ we have $\{\rho, \boldsymbol{w}, \boldsymbol{v}\}\times[T, \infty) \subset \mathcal{O}_{\mathsf{z}}$. The regularity class γ denotes an integer such that $\gamma \geq 3$.

(H_1) The thermodynamic functions $\mathcal{E}$, p, and $\mathcal{S}$ are C^γ functions of the variable z defined over a simply connected nonempty open set $\mathcal{O}_{\mathsf{z}} \subset (0, \infty) \times \mathbb{R}^d \times \mathbb{R}^d \times (0, \infty)$ and the capillarity coefficient $\varkappa$ is a $C^{\gamma+1}$ function of temperature over $\mathcal{O}_{\mathsf{z}}$. For any $(\rho, T)^t \in \mathcal{O}_{\mathsf{z}^{\,\mathrm{cl}}}$ we have $(\rho, \boldsymbol{0}, \boldsymbol{0}, T)^t \in \mathcal{O}_{\mathsf{z}}$ and for any $(\rho, \boldsymbol{w}, \boldsymbol{v}, T)^t \in \mathcal{O}_{\mathsf{z}}$ we have $(\rho, T)^t \in \mathcal{O}_{\mathsf{z}^{\,\mathrm{cl}}}$.

(H_2) Assuming (H_1), and defining $\mathcal{G} = \mathcal{E} + p - T\mathcal{S}$ and $g = \mathcal{G}/\rho$, we have the volumetric Gibbs' relation over $\mathcal{O}_{\mathsf{z}}$

$$T\,\mathbb{D}\mathcal{S} = \mathbb{D}\mathcal{E} - g\,\mathbb{D}\rho - \varkappa \boldsymbol{w}\boldsymbol{\cdot}\mathbb{D}\boldsymbol{w}. \tag{3.157}$$

(H_3) Assuming (H_1), the open set $\mathcal{O}_{\mathsf{z}}$ is increasing with respect to temperature and the volumetric specific heat is positive $\partial_T\mathcal{E} > 0$ over $\mathcal{O}_{\mathsf{z}}$.

(H_4) The capillarity coefficient is positive $\varkappa > 0$ over $\mathcal{O}_{\mathsf{z}}$.

(H_5) The coefficients κ, η and λ are C^γ functions over $\mathcal{O}_{\mathsf{z}}$. The coefficients η and λ are positive and the coefficient κ is nonnegative over $\mathcal{O}_{\mathsf{z}}$.

These assumptions are assumed to hold whenever the cohesive fluid model is considered. Concerning thermodynamic stability, a result similar to that of Proposition 3.5 is established by taking into account the extra $\boldsymbol{w}$ and $\boldsymbol{v}$ components of the variable z.

Proposition 3.6 *Assume that (H_1) and (H_2) are satisfied and that the map $\mathsf{z} \mapsto \mathsf{u}$ is locally invertible. Denoting by $\widetilde{\partial}$ the derivation operator with respect to u, the following statements are equivalent :*

(i) $\widetilde{\partial}^2_{\mathsf{uu}}\mathcal{S}$ *is negative definite.*

(ii) $\partial_T\mathcal{E} > 0$, $\partial_\rho p > 0$, *and* $\varkappa > 0$.

Proof The proof is similar to that of Proposition 3.5 using the expression

$$\begin{aligned}\langle \widetilde{\partial}^2_{\mathsf{uu}}\mathcal{S}\mathsf{x},\, \mathsf{x}\rangle = &-\frac{1}{T^2\partial_T\mathcal{E}}\Big(\mathsf{x}_{\mathcal{E}} - \partial_\rho\mathcal{E}^{\,\mathrm{cl}}\mathsf{x}_\rho - \boldsymbol{v}\boldsymbol{\cdot}\mathsf{x}_{\boldsymbol{v}} - (\varkappa - T\partial_T\varkappa)\boldsymbol{w}\boldsymbol{\cdot}\mathsf{x}_{\boldsymbol{w}}\Big)^2 \\ &-\frac{\partial_\rho p}{\rho T}\mathsf{x}_\rho^2 - \frac{\varkappa}{T}\big|\mathsf{x}_{\boldsymbol{w}}\big|^2 - \frac{1}{\rho T}\big|\mathsf{x}_{\boldsymbol{v}} - \boldsymbol{v}\mathsf{x}_\rho\big|^2,\end{aligned} \tag{3.158}$$

where $\mathsf{x} = (\mathsf{x}_T, \mathsf{x}_{\boldsymbol{w}}, \mathsf{x}_{\boldsymbol{v}}, \mathsf{x}_{\mathcal{E}})^t$ denotes an arbitrary vector of $\mathbb{R}^{\mathsf{n}}$. This expression directly yields that stability is equivalent to $\partial_T\mathcal{E} > 0$, $\partial_\rho p > 0$ and $\varkappa > 0$. □

Note that the same mechanical instabilities are obtained in both Propositions 3.6 and 3.5 since $\partial_\rho p = \partial_\rho p^{\mathrm{cl}}$. This is a natural result since vaporization fronts are stabilized by higher order density derivatives and not by first order density gradients $\boldsymbol{w}$. With Proposition 3.6 and under assumptions (H_3) and (H_4) only mechanical instabilities may arise. Assuming that ($\mathsf{H}_1 - \mathsf{H}_3$) hold, then the map $\mathsf{z} \mapsto \mathsf{u}$ is a C^γ diffeomorphism from the open set $\mathcal{O}_{\mathsf{z}}$ onto an open set $\mathcal{O}_{\mathsf{u}}$. Assumptions ($\mathsf{H}_1 - \mathsf{H}_5$) are assumed to hold in the remaining part of the section.

Since the map $\mathsf{z} \mapsto \mathsf{u}$ is a C^γ diffeomorphism, all the thermodynamic functions and transport coefficients are also C^γ functions of the conservative variable u and defined over the corresponding open set $\mathcal{O}_{\mathsf{u}}$. The thermal stability condition $\partial_T\mathcal{E} = \partial_T\mathcal{E}^{\,\mathrm{cl}} - \frac{1}{2}T\partial^2_T\varkappa\,|\boldsymbol{w}|^2 > 0$ naturally introduces limitations on the norm of $\boldsymbol{w}$. However, the order of magnitude of $\varkappa$ shows that such conditions may hold even when $\boldsymbol{w}$ is of the order of the inverse of the range of interaction potential [72, 73]. We do not assume that the Hessian matrix $\widetilde{\partial}^2_{\mathsf{uu}}\mathcal{S}$ is negative definite since we anticipate the presence of mechanical thermodynamic instabilities. We will also need an instability indicator m in order to control negative values of the derivative $\partial_\rho p = \partial_\rho p^{\,\mathrm{cl}}$ and that may be taken to be the instability indicator $\mathsf{m}^{\,\mathrm{cl}}$ of Lemma 3.6.

Lemma 3.7 *Assuming that* (H_1) *holds, that* $\delta > 0$ *is given, and letting* $\mathsf{m}(\mathsf{z}) = \mathsf{m}^{\,\mathrm{cl}}(\rho, T)$, *then* m *is* $C^{\gamma-1}$ *over* $\mathcal{O}_\mathsf{z}$, *and such that* $\mathsf{m} \geq 0$, $\mathsf{m} + \partial_\rho p/\rho T > 0$ *and* $\mathsf{m} = 0$ *if* $\partial_\rho p/\rho T \geq \delta$.

The governing equations of the augmented system are now written as

$$\partial_t \mathsf{u} + \sum_{i\in\mathfrak{D}} \partial_i \big(\mathsf{F}_i + \mathsf{F}_i^{\mathrm{c}} + \mathsf{F}_i^{\mathrm{d}}\big) = 0, \tag{3.159}$$

where F_i is the augmented convective flux, $\mathsf{F}_i^{\mathrm{c}}$ the augmented cohesive flux, and $\mathsf{F}_i^{\mathrm{d}}$ the augmented dissipative flux. The augmented convective F_i and capillary $\mathsf{F}_i^{\mathrm{c}}$ fluxes have been identified by using the Legendre transform of entropy [90] whereas the augmented dissipative fluxes $\mathsf{F}_i^{\mathrm{d}}$ are similar to that of standard fluids [7, 18, 22]

$$\mathsf{F}_i = \Big(\rho v_i,\ \boldsymbol{w} v_i,\ \rho \boldsymbol{v} v_i + (p + \varkappa|\boldsymbol{w}|^2)\boldsymbol{e}_i,\ (\mathcal{E} + \tfrac{1}{2}\rho|\boldsymbol{v}|^2 + p + \varkappa|\boldsymbol{w}|^2) v_i\Big)^t, \tag{3.160}$$

$$\mathsf{F}_i^{\mathrm{c}} = \Big(0,\ \rho \boldsymbol{\nabla} v_i,\ -\rho \boldsymbol{\nabla}(\varkappa w_i),\ \rho\varkappa \boldsymbol{w}\cdot\boldsymbol{\nabla} v_i - \rho \boldsymbol{v}\cdot\boldsymbol{\nabla}(\varkappa w_i)\Big)^t, \tag{3.161}$$

$$\mathsf{F}_i^{\mathrm{d}} = \Big(0,\ 0_{d,1},\ \boldsymbol{\Pi}_i,\ Q_i + \boldsymbol{\Pi}_i\cdot\boldsymbol{v}\Big)^t, \tag{3.162}$$

where $\boldsymbol{e}_i$ is the ith basis vector in the physical space $\mathbb{R}^d$ and $\boldsymbol{\Pi}_i = (\Pi_{i1}, \dots, \Pi_{id})^t$. It is easily checked that the governing equations associated with these fluxes coincide with the original equations for capillary fluids by using calculus identities [90].

From the expressions (3.161) and (3.162), the capillary $\mathsf{F}_i^{\mathrm{c}}$ and dissipative $\mathsf{F}_i^{\mathrm{d}}$ fluxes may also be written in the form

$$\mathsf{F}_i^{\mathrm{d}} = -\sum_{j\in\mathfrak{D}} \mathsf{B}_{ij}^{\mathrm{d}} \partial_j \mathsf{u}, \qquad \mathsf{F}_i^{\mathrm{c}} = -\sum_{j\in\mathfrak{D}} \mathsf{B}_{ij}^{\mathrm{c}} \partial_j \mathsf{u}, \qquad i \in \mathfrak{D}, \tag{3.163}$$

where $\mathsf{B}_{ij}^{\mathrm{d}}$ and $\mathsf{B}_{ij}^{\mathrm{c}}$ are uniquely defined matrices. We may next introduce the jacobian $\mathsf{A}_i = \partial_\mathsf{u} \mathsf{F}_i$ of the convective flux F_i with respect to the conservative variable u and the governing Eq. (3.159) may finally be written in the form of an augmented quasilinear second order system of equations

$$\partial_t \mathsf{u} + \sum_{i\in\mathfrak{D}} \mathsf{A}_i(\mathsf{u})\partial_i \mathsf{u} - \sum_{i,j\in\mathfrak{D}} \partial_i \big(\mathsf{B}_{ij}^{\mathrm{d}}(\mathsf{u})\partial_j \mathsf{u}\big) - \sum_{i,j\in\mathfrak{D}} \partial_i \big(\mathsf{B}_{ij}^{\mathrm{c}}(\mathsf{u})\partial_j \mathsf{u}\big) = 0. \tag{3.164}$$

The coefficient matrices $\mathsf{A}_i = \partial_{\mathsf{u}}\mathsf{F}_i$, $\mathsf{B}^{\mathrm{d}}_{ij}$, and $\mathsf{B}^{\mathrm{c}}_{ij}$, for $i, j \in \mathfrak{D}$, have at least regularity $C^{\gamma-1}$ over the open set $\mathcal{O}_{\mathsf{u}}$. It is then remarkable that the original system (3.139)–(3.141), that involves third order derivatives of density in the momentum equation, has been rewritten in the form of a second order system. Symmetrization of the corresponding augmented system of partial differential equation is investigated in the next section.

3.4.3 A New Hyperbolic-parabolic Structure

We investigate symmetrized forms for augmented systems of Eq. (3.164). The resulting normal form involves new antisymmetric—dispersive—second order terms that strongly couple the hyperbolic and parabolic variables.

The mathematical entropy σ is taken in the form

$$\sigma = -\mathcal{S} = -\mathcal{S}^{\,\mathrm{cl}} + \tfrac{1}{2}\partial_T \varkappa \, |\boldsymbol{w}|^2, \tag{3.165}$$

and includes capillary contributions when $\varkappa$ depends on temperature. The mathematical entropy σ is a C^{γ} function of z or u from ($\mathsf{H}_1 - \mathsf{H}_5$). After some calculus, we obtain the expression of the augmented entropic variable $\mathsf{v} = (\partial_{\mathsf{u}}\sigma)^t$

$$\mathsf{v} = \frac{1}{T}\Big(g - \tfrac{1}{2}|\boldsymbol{v}|^2, \varkappa\, \boldsymbol{w}, \boldsymbol{v}, -1\Big)^t. \tag{3.166}$$

This entropic variable v is formally similar to that of compressible non capillary gases [7, 18, 22] with the addition of the extra term $\varkappa \boldsymbol{w}/T$ associated with the $\boldsymbol{w}$ component of u.

An entropic symmetrized form of the quasilinear system in conservative form (3.164) may be obtained by using the entropic variable v around a stable state. Assuming that z is a stable state $\mathsf{z} \in \mathcal{O}^{\mathrm{st}}_{\mathsf{z}}$ with $\partial_\rho p > 0$, the map $\mathsf{z} \mapsto \mathsf{v}$ is locally invertible so that $\mathsf{u} \mapsto \mathsf{v}$ is also locally invertible. Proceeding as for classic fluids [7, 18, 22], the following symmetrized entropic form is obtained

$$\widetilde{\mathsf{A}}_0(\mathsf{v})\partial_t \mathsf{v} + \sum_{i\in\mathfrak{D}} \widetilde{\mathsf{A}}_i(\mathsf{v})\partial_i \mathsf{v} - \sum_{i,j\in\mathfrak{D}} \partial_i\big(\widetilde{\mathsf{B}}^{\mathrm{d}}_{ij}(\mathsf{v})\partial_j \mathsf{v}\big) - \sum_{i,j\in\mathfrak{D}} \partial_i\big(\widetilde{\mathsf{B}}^{\mathrm{c}}_{ij}(\mathsf{v})\partial_j \mathsf{v}\big) = 0, \tag{3.167}$$

with $\widetilde{\mathsf{A}}_0 = \partial_{\mathsf{v}}\mathsf{u}$, $\widetilde{\mathsf{A}}_i = \mathsf{A}_i\partial_{\mathsf{v}}\mathsf{u}$, $\widetilde{\mathsf{B}}^{\mathrm{d}}_{ij} = \mathsf{B}^{\mathrm{d}}_{ij}\partial_{\mathsf{v}}\mathsf{u}$, and $\widetilde{\mathsf{B}}^{\mathrm{c}}_{ij} = \mathsf{B}^{\mathrm{c}}_{ij}\partial_{\mathsf{v}}\mathsf{u}$, for $i, j \in \mathfrak{D}$. The system coefficients have been obtained in [90] and are omitted. *A major difficulty* with the entropic variable v, notably identified by Benzoni et al. [79], is that the maps $\mathsf{u} \mapsto \mathsf{v}$ and $\mathsf{z} \mapsto \mathsf{v}$ are generally not globally invertible in the presence of mechanical instabilities. This is typically the case in liquid-vapor flows and denoting by l and g the indices of a liquid and its vapor at equilibrium, the classic equilibrium relations are $T_{\mathrm{l}} = T_{\mathrm{g}}$, ${p_{\mathrm{l}}}^{\mathrm{cl}} = {p_{\mathrm{g}}}^{\mathrm{cl}}$, and $g_{\mathrm{l}} = g_{\mathrm{g}}$. These equilibrium conditions between a liquid and its vapor, completed for instance with the conditions $\boldsymbol{w}_{\rm l} = \boldsymbol{w}_{\rm g} = 0$ and $\boldsymbol{v}_{\rm l} = \boldsymbol{v}_{\rm g} = 0$ mean that the entropic variables $\mathsf{v}_{\rm l}$ and $\mathsf{v}_{\rm g}$ coincide even though the corresponding natural variables $\mathsf{z}_{\rm l}$ and $\mathsf{z}_{\rm g}$ differ. In other words, using the entropic variable v is not a good idea for liquid-vapor flows. It is then mandatory to use a normal variable w such that $\mathsf{u} \mapsto \mathsf{w}$ remains invertible. The matrix $\widetilde{\mathsf{A}}_0$ also becomes singular or negative at points where $\partial_\rho p \leq 0$ and this must also be taken into account.

We thus investigate a normal form of the augmented system of equations using the augmented natural variable

$$\mathsf{w} = \big(\rho, \boldsymbol{w}, \boldsymbol{v}, T\big)^t. \tag{3.168}$$

We use as usual the vector and matrix block structure induced by the partitioning of $\mathbb{R}^{\mathsf{n}}$ into $\mathbb{R}^{\mathsf{n}} = \mathbb{R}^{\mathsf{n}_{\rm I}} \times \mathbb{R}^{\mathsf{n}_{\rm II}}$ with $\mathsf{n} = \mathsf{n}_{\rm I} + \mathsf{n}_{\rm II}$, $\mathsf{n}_{\rm I} = \mathsf{n}_{\rm II} = d+1$, and $\mathsf{w} = (\mathsf{w}_{\rm I}, \mathsf{w}_{\rm II})^t$ with $\mathsf{w}_{\rm I} = (\rho, \boldsymbol{w})^t$ and $\mathsf{w}_{\rm II} = (\boldsymbol{v}, T)^t$. It will also be convenient in the following to split the hyperbolic variable $\mathsf{w}_{\rm I}$ as $\mathsf{w}_{\rm I} = (\mathsf{w}_{\rm I'}, \mathsf{w}_{\rm I''})^t$ where $\mathsf{w}_{\rm I'} = \rho$ and $\mathsf{w}_{\rm I''} = \boldsymbol{w}$. We will also denote by $\mathsf{w}_{\rm r}$ the components of w that are more regular, that is, $\mathsf{w}_{\rm r} = (\rho, \boldsymbol{v}, T)^t$ and more generally $\mathsf{w}_{\rm r} = (\mathsf{w}_{\rm I'}, \mathsf{w}_{\rm II})^t$. The normal form of the system is written for convenience in nonconservative form

$$\overline{\mathsf{A}}_0(\mathsf{w})\partial_t \mathsf{w} + \sum_{i\in\mathfrak{D}} \overline{\mathsf{A}}_i(\mathsf{w})\partial_i \mathsf{w} - \sum_{i,j\in\mathfrak{D}} \overline{\mathsf{B}}^{\rm d}_{ij}(\mathsf{w})\partial_i\partial_j \mathsf{w} - \sum_{i,j\in\mathfrak{D}} \overline{\mathsf{B}}^{\rm c}_{ij}(\mathsf{w})\partial_i\partial_j \mathsf{w} = \mathsf{h}(\mathsf{w}, \nabla\mathsf{w}), \tag{3.169}$$

where h has the structure

$$\mathsf{h} = \big(\mathsf{h}_{\rm I}(\mathsf{w}, \nabla\mathsf{w}_{\rm II}), \mathsf{h}_{\rm II}(\mathsf{w}, \nabla\mathsf{w})\big)^t. \tag{3.170}$$

The standard method for deriving normal forms for systems of conservation laws is to substitute $\mathsf{v} = \mathsf{v}(\mathsf{w})$ in the entropic symmetrized form and to multiply on the left by $(\partial_{\mathsf{w}}\mathsf{v})^t$ in order to maintain the symmetry properties of convective and dissipative matrices [7, 22]. The matrix in front of $\partial_t \mathsf{w}$ would then be $(\partial_{\mathsf{z}}\mathsf{v})^t\widetilde{\mathsf{A}}_0\partial_{\mathsf{z}}\mathsf{v}$, that may be rewritten $(\partial_{\mathsf{z}}\mathsf{v})^t\partial_{\mathsf{z}}\mathsf{u}$, and the matrix in front of $\partial_i \mathsf{w}$ would be $(\partial_{\mathsf{z}}\mathsf{v})^t\widetilde{\mathsf{A}}_i\partial_{\mathsf{z}}\mathsf{v}$, that may also be rewritten $(\partial_{\mathsf{z}}\mathsf{v})^t\partial_{\mathsf{z}}\mathsf{F}_i$. This standard procedure, however, must be modified in order to take into account the apparition of singularities due to thermodynamic mechanical instabilities arising when $\partial_\rho p$ vanishes and becomes negative. To this aim, after multiplication on the left by $(\partial_{\mathsf{z}}\mathsf{v})^t$, *we add the total mass conservation equation multiplied by the instability indicator* m of Lemma 3.7 to the first equation.

The matrices $\overline{\mathsf{A}}_0$ and $\overline{\mathsf{A}}_i$ are thus taken in the form

$$\overline{\mathsf{A}}_0 = (\partial_{\mathsf{w}}\mathsf{v})^t\partial_{\mathsf{w}}\mathsf{u} + \mathsf{m}\,\mathsf{e}_1\otimes\mathsf{e}_1, \qquad \overline{\mathsf{A}}_i = (\partial_{\mathsf{w}}\mathsf{v})^t\partial_{\mathsf{w}}\mathsf{F}_i + \mathsf{m}v_i\,\mathsf{e}_1\otimes\mathsf{e}_1, \quad i \in \mathfrak{D},$$

where e_i, $1 \leq i \leq \mathsf{n}$, denotes the canonical basis of $\mathbb{R}^{\mathsf{n}}$. There is also a cancelation of singularities when rewriting $(\partial_{\mathsf{w}}\mathsf{v})^t\partial_{\mathsf{v}}\mathsf{u}\,\partial_{\mathsf{w}}\mathsf{v}$ as $(\partial_{\mathsf{w}}\mathsf{v})^t\partial_{\mathsf{w}}\mathsf{u}$ and $(\partial_{\mathsf{w}}\mathsf{v})^t\partial_{\mathsf{v}}\mathsf{F}_i\,\partial_{\mathsf{w}}\mathsf{v}$ as $(\partial_{\mathsf{w}}\mathsf{v})^t\partial_{\mathsf{w}}\mathsf{F}_i$ and furthermore $\overline{\mathsf{B}}^{\rm d}_{ij} = (\partial_{\mathsf{w}}\mathsf{v})^t\mathsf{B}^{\rm d}_{ij}\partial_{\mathsf{w}}\mathsf{u}$ and $\overline{\mathsf{B}}^{\rm c}_{ij} = (\partial_{\mathsf{w}}\mathsf{v})^t\mathsf{B}^{\rm c}_{ij}\partial_{\mathsf{w}}\mathsf{u}$. The right hand side is correspondingly in the form

$$\mathsf{h} = \sum_{i,j\in\mathfrak{D}} (\partial_{\mathsf{w}}\mathsf{v})^t \partial_i \widehat{\mathsf{B}}^{\mathrm{d}}_{ij} \partial_j \mathsf{w} + \sum_{i,j\in\mathfrak{D}} (\partial_{\mathsf{w}}\mathsf{v})^t \partial_i \widehat{\mathsf{B}}^{\mathrm{c}}_{ij} \partial_j \mathsf{w} - \mathsf{m}\rho \nabla\cdot\boldsymbol{v}\, \mathsf{e}_1, \tag{3.171}$$

where we have written $\mathsf{F}^{\mathrm{d}}_i = -\sum_{j\in\mathfrak{D}} \widehat{\mathsf{B}}^{\mathrm{d}}_{ij}\partial_j \mathsf{w}$ and $\mathsf{F}^{\mathrm{c}}_i = -\sum_{j\in\mathfrak{D}} \widehat{\mathsf{B}}^{\mathrm{c}}_{ij}\partial_j \mathsf{w}$ with $\widehat{\mathsf{B}}^{\mathrm{d}}_{ij} = \mathsf{B}^{\mathrm{d}}_{ij}\partial_{\mathsf{w}}\mathsf{u}$ and $\widehat{\mathsf{B}}^{\mathrm{c}}_{ij} = \mathsf{B}^{\mathrm{c}}_{ij}\partial_{\mathsf{w}}\mathsf{u}$.

The system coefficients have been evaluated in [90] and are omitted. The matrix $\overline{\mathsf{A}}_0$ is symmetric, positive definite, block diagonal $\overline{\mathsf{A}}_0 = \mathrm{diag}(\overline{\mathsf{A}}_0^{\mathrm{I,I}}, \overline{\mathsf{A}}_0^{\mathrm{II,II}})$ and $\overline{\mathsf{A}}_0^{\mathrm{II,II}}$ only depends on the subvariable w_{r}. The matrices $\overline{\mathsf{A}}_i(\mathsf{w})$, $i \in \mathfrak{D}$, are symmetric, the dissipation matrices are such that $(\overline{\mathsf{B}}^{\mathrm{d}}_{ij})^t = \overline{\mathsf{B}}^{\mathrm{d}}_{ji}$, $i, j \in \mathfrak{D}$, have nonzero components only into the right lower $\overline{\mathsf{B}}^{\mathrm{d\,II,II}}_{ij}$ blocks, and $\overline{\mathsf{B}}^{\mathrm{d\,II,II}} = \sum_{i,j\in\mathfrak{D}} \overline{\mathsf{B}}^{\mathrm{d\,II,II}}_{ij}\xi_i\xi_j$ is positive definite for $\boldsymbol{\xi} \in \Sigma^{d-1}$ where Σ^{d-1} is the sphere in d dimension. The matrices $\overline{\mathsf{B}}^{\mathrm{c}}_{ij}$ are such that $(\overline{\mathsf{B}}^{\mathrm{c}}_{ij})^t = -\overline{\mathsf{B}}^{\mathrm{c}}_{ji}$, the blocks $\overline{\mathsf{B}}^{\mathrm{c,I,I}}_{ij}$ vanish $\overline{\mathsf{B}}^{\mathrm{c,I,I}}_{ij} = 0$, and the blocks $\overline{\mathsf{B}}^{\mathrm{c,I,II}}_{ij}$ and $\overline{\mathsf{B}}^{\mathrm{c,II,I}}_{ij}$ only depend on the subvariable w_{r}. The system coefficients $\overline{\mathsf{A}}_0$, $\overline{\mathsf{A}}_i$ $\overline{\mathsf{B}}^{\mathrm{d}}_{ij}$, and $\overline{\mathsf{B}}^{\mathrm{c}}_{ij}$, $i, j \in \mathfrak{D}$, are smooth functions of $\mathsf{w} \subset \mathbb{R}^{\mathsf{n}}$ defined over an open set $\mathcal{O}_{\mathsf{w}}$ with at least C^{l+2} regularity assuming that the regularity class γ is such that $\gamma \geq l + 4$.

The general structure of the system of partial differential Eq. (3.159) may be discussed with the normal form (3.169). The variable $\mathsf{w}_{\mathrm{I}} = (\rho, \boldsymbol{w})^t$ with $\boldsymbol{w} = \nabla\rho$ is first an hyperbolic variable, i.e., when w_{II} is given, w_{I} is governed by a symmetric hyperbolic system of conservation laws. Similarly, $\mathsf{w}_{\mathrm{II}} = (\boldsymbol{v}, T)^t$ is a parabolic variable, i.e., when w_{I} is given, w_{II} is governed by a strongly parabolic system of conservation laws. For such symmetric system, Petrovsky parabolicity is equivalent to strong parabolicity [124]. As a consequence, the variable ρ is not a parabolic variable but rather the couple $(\rho, \nabla\rho)^t$ is an hyperbolic variable. The system (3.169) is a new type of composite hyperbolic-parabolic-dispersive system involving matrices $\overline{\mathsf{B}}^{\mathrm{d}}_{ij}$ with symmetry properties and matrices $\overline{\mathsf{B}}^{\mathrm{c}}_{ij}$ with antisymmetry properties. The matrices $\overline{\mathsf{B}}^{\mathrm{c}}_{ij}$ introduce extra couplings between the hyperbolic and parabolic variables and these coupling blocks $\overline{\mathsf{B}}^{\mathrm{c\,I,II}}_{ij}$ and $\overline{\mathsf{B}}^{\mathrm{c\,II,I}}_{ij}$ require more regularity. Last but not least, it will be important in practice to maintain the constraint $\boldsymbol{w} = \nabla\rho$.

After some calculus, the hyperbolic part $\mathsf{h}_{\mathrm{I}} = (\mathsf{h}_{\mathrm{I}'}, \mathsf{h}_{\mathrm{I}''})^t$ of the right hand side in (3.169) is found to be

$$\mathsf{h}_{\mathrm{I}} = \Big(-\mathsf{m}\rho\nabla\cdot\boldsymbol{v}, -\frac{\varkappa}{T}\sum_{i\in\mathfrak{D}} w_i \nabla v_i \Big)^t, \tag{3.172}$$

where $\boldsymbol{w} = (w_1, \ldots, w_d)^t$. The derivatives of density appearing in $\mathsf{h}_{\mathrm{I}''}$ of h_{I} have notably been written as components of $\boldsymbol{w}$ to ensure that the gradient constraint holds for linearized equations as investigated in the following. On the other hand, the parabolic part is in the general form

$$\mathsf{h}_{\mathrm{II}} = \mathsf{h}_{\mathrm{II}}(\mathsf{w}, \boldsymbol{\nabla}\mathsf{w}), \tag{3.173}$$

and depends on the complete gradient $\boldsymbol{\nabla}\mathsf{w}$. The term $\mathsf{h}_{\mathrm{I}'}$ is linear in the gradient $\boldsymbol{\nabla}\mathsf{w}_{\mathrm{II}}$ and arises from the stabilization procedure whereas $\mathsf{h}_{\mathrm{I}''}$ has been rewritten formally as linear in $\boldsymbol{\nabla}\mathsf{w}_{\mathrm{II}}$ for convenience by using $\boldsymbol{w} = \boldsymbol{\nabla}\rho$ but is intrinsically quadratic. The term h_{II} is quadratic in terms of the gradient $\boldsymbol{\nabla}\mathsf{w}$ with coefficients depending on w that have at least regularity C^{l+2}.

We now investigate how the natural gradient constraint $\boldsymbol{w} = \boldsymbol{\nabla}\rho$ is a consequence of the governing equations. Considering the nonlinear equations in normal form (3.169), we obtain after some calculus that the first equation reads $\partial_t\rho + \boldsymbol{\nabla}\cdot(\rho\boldsymbol{v}) = 0$ where we have used that $\mathsf{m} + \frac{1}{\rho T}\partial_\rho p > 0$. Here m is the instability indicator of Lemma 3.7 and applying the $\boldsymbol{\nabla}$ operator then yields an equation for $\boldsymbol{\nabla}\rho$. Dividing by $\varkappa/T$ the second equation of (3.169) governing $\boldsymbol{w}$ and subtracting the $\boldsymbol{\nabla}\rho$ equation further yields

$$\partial_t(\boldsymbol{w} - \boldsymbol{\nabla}\rho) + \boldsymbol{v}\cdot\boldsymbol{\nabla}(\boldsymbol{w} - \boldsymbol{\nabla}\rho) + (\boldsymbol{w} - \boldsymbol{\nabla}\rho)\boldsymbol{\nabla}\cdot\boldsymbol{v} + (\boldsymbol{\nabla}\boldsymbol{v})^t\cdot(\boldsymbol{w} - \boldsymbol{\nabla}\rho) = 0.$$

This Friedrichs system implies that $\boldsymbol{w} - \boldsymbol{\nabla}\rho = 0$ provided that the solution w is smooth enough, that the constraint $\boldsymbol{w}_0 - \boldsymbol{\nabla}\rho_0 = 0$ holds at the initial time $t = 0$, where $\mathsf{w}(0, \mathsf{x}) = \big(\rho_0(\mathsf{x}), \boldsymbol{w}_0(\mathsf{x}), \boldsymbol{v}_0(\mathsf{x}), T_0(\mathsf{x})\big)^t$ and that the state at infinity $\mathsf{w}^\star$ is naturally such that $\boldsymbol{w}^\star = 0$ [25].

In order to establish existence results for the nonlinear system (3.169), we will also need to use linearized equations. It is then mandatory to ensure that the natural gradient constraint $\boldsymbol{w} = \boldsymbol{\nabla}\rho$ *also holds for solutions of properly selected linearized equations*. To this aim, the linearized unknown $\widetilde{\mathsf{w}}$ is written $\widetilde{\mathsf{w}} = (\widetilde{\mathsf{w}}_{\mathrm{I}}, \widetilde{\mathsf{w}}_{\mathrm{II}})^t = \big(\widetilde{\rho}, \widetilde{\boldsymbol{w}}, \widetilde{\boldsymbol{v}}, \widetilde{T}\big)^t$ and the linearized system taken in the form

$$\overline{\mathsf{A}}_0(\mathsf{w})\partial_t\widetilde{\mathsf{w}} + \sum_{i\in\mathfrak{D}} \overline{\mathsf{A}}_i(\mathsf{w})\partial_i\widetilde{\mathsf{w}} - \sum_{i,j\in\mathfrak{D}} \big(\overline{\mathsf{B}}^{\mathrm{d}}_{ij}(\mathsf{w}) + \overline{\mathsf{B}}^{\mathrm{c}}_{ij}(\mathsf{w})\big)\partial_i\partial_j\widetilde{\mathsf{w}} = \big(\widetilde{\mathsf{h}}_{\mathrm{I}}, \mathsf{h}_{\mathrm{II}}(\mathsf{w}, \boldsymbol{\nabla}\mathsf{w})\big)^t,$$

where the hyperbolic part of the right hand side $\widetilde{\mathsf{h}}_{\mathrm{I}}$ is taken to be

$$\widetilde{\mathsf{h}}_{\mathrm{I}} = \Big(-\mathsf{m}\,\rho\,\boldsymbol{\nabla}\cdot\widetilde{\boldsymbol{v}}, -\sum_{i\in\mathfrak{D}} \frac{\varkappa}{T}\widetilde{w}_i\boldsymbol{\nabla}v_i\Big)^t. \tag{3.174}$$

The linearized equations are thus in the form

$$\overline{\mathsf{A}}_0(\mathsf{w})\partial_t\widetilde{\mathsf{w}} + \sum_{i\in\mathfrak{D}} \overline{\mathsf{A}}_i{}'(\mathsf{w})\partial_i\widetilde{\mathsf{w}} - \sum_{i,j\in\mathfrak{D}} \overline{\mathsf{B}}^{\mathrm{d}}_{ij}(\mathsf{w})\partial_i\partial_j\widetilde{\mathsf{w}} - \sum_{i,j\in\mathfrak{D}} \overline{\mathsf{B}}^{\mathrm{c}}_{ij}(\mathsf{w})\partial_i\partial_j\widetilde{\mathsf{w}}$$
$$+ \overline{\mathsf{L}}(\mathsf{w}, \boldsymbol{\nabla}\mathsf{w}_{\mathrm{II}})\widetilde{\mathsf{w}} = \mathsf{h}'(\mathsf{w}, \boldsymbol{\nabla}\mathsf{w}), \tag{3.175}$$

where $\mathsf{h}' = \big(0_{\mathsf{n}_\mathrm{I},1}, \mathsf{h}_\mathrm{II}(\mathsf{w},\nabla\mathsf{w})\big)^t$, $\overline{\mathsf{A}}_i{}'(\mathsf{w}) = \overline{\mathsf{A}}_i(\mathsf{w}) + \mathsf{m}\rho\mathsf{e}_1\otimes\mathsf{e}_{d+1+i}$ and $\overline{\mathsf{L}}(\mathsf{w},\nabla\mathsf{w}_\mathrm{II}) = \sum_{i\in\mathfrak{D}}\frac{\varkappa}{T}(0,\nabla v_i, 0_{1,\mathsf{n}_\mathrm{I}},0)^t\otimes\mathsf{e}_{i+1}$ with $\mathsf{e}_1,\dots,\mathsf{e}_\mathsf{n}$ denoting the canonical basis of $\mathbb{R}^\mathsf{n}$. The hyperbolic blocks $\overline{\mathsf{A}}_i{}'^{\mathrm{I,I}}$ and $\overline{\mathsf{A}}_i{}^{\mathrm{I,I}}$ coincide and the matrix $\overline{\mathsf{L}}$ is block diagonal $\overline{\mathsf{L}} = \mathrm{diag}(\overline{\mathsf{L}}^{\mathrm{I,I}},\overline{\mathsf{L}}^{\mathrm{II,II}})$ with $\overline{\mathsf{L}}^{\mathrm{I,I}} = \mathrm{diag}\Big(0,\sum_{i\in\mathfrak{D}}\frac{\varkappa}{T}\nabla v_i\otimes\boldsymbol{e}_i\Big)$ and $\overline{\mathsf{L}}^{\mathrm{II,II}} = 0_{d+1,d+1}$.

We now deduce that the first equation of (3.175) governing $\widetilde{\rho}$ is in the form $\partial_t\widetilde{\rho}+\boldsymbol{v}\cdot\nabla\widetilde{\rho}+\rho\nabla\cdot\widetilde{\boldsymbol{v}} = 0$ and the equation governing $\nabla\widetilde{\rho}$ is obtained by applying the ∇ operator. Using the second equation of (3.175) governing $\widetilde{\boldsymbol{w}}$, after simplification by the $\varkappa/T$ factor and subtraction of the $\nabla\widetilde{\rho}$ governing equation, we obtain

$$\partial_t(\widetilde{\boldsymbol{w}}-\nabla\widetilde{\rho}) + \boldsymbol{v}\cdot\nabla(\widetilde{\boldsymbol{w}}-\nabla\widetilde{\rho}) + (\boldsymbol{w}-\nabla\rho)\,\nabla\cdot\widetilde{\boldsymbol{v}} + \nabla\boldsymbol{v}^t\cdot(\widetilde{\boldsymbol{w}}-\nabla\widetilde{\rho}) = 0.$$

This Friedrichs system then implies that $\widetilde{\boldsymbol{w}}-\nabla\widetilde{\rho} = 0$ provided that the constraint $\boldsymbol{w}-\nabla\rho = 0$ holds for w, that the solutions w and $\widetilde{\mathsf{w}}$ are smooth enough, that the constraint $\widetilde{\boldsymbol{w}}_0-\nabla\widetilde{\rho}_0 = 0$ holds at the initial time $t=0$, and that the state at infinity $\widetilde{\mathsf{w}}^\star$ is naturally such that $\widetilde{\boldsymbol{w}}^\star = 0$ [25]. The natural gradient constraint may thus be transmitted from w to $\widetilde{\mathsf{w}}$ by using suitable linearizations like (3.174). The right hand sides are also naturally related by $\sum_{i\in\mathfrak{D}}\big(\overline{\mathsf{A}}_i{}'(\mathsf{w})-\overline{\mathsf{A}}_i(\mathsf{w})\big)\partial_i\mathsf{w} + \overline{\mathsf{L}}(\mathsf{w},\nabla\mathsf{w}_\mathrm{II})\mathsf{w} + \mathsf{h}(\mathsf{w},\nabla\mathsf{w}) = \mathsf{h}'(\mathsf{w},\nabla\mathsf{w})$ for any regular function w.

3.4.4 Local Existence

The equations governing cohesive fluids have been the object of intensive work over the last decades [75–90]. Local existence of strong solutions has been established by Hattori and Li [75], Danchin [77] and Kotschote [82] in the isothermal situation and by Haspot [83, 84] and Kotschote [84] in the non isothermal case. We present in this section a local existence of strong solutions using augmented symmetrized systems taking into account the temperature dependence of the capillarity coefficient [90].

We first need to establish a priori estimates for hyperbolic-parabolic linearized systems in the form

$$\overline{\mathsf{A}}_0(\mathsf{w})\partial_t\widetilde{\mathsf{w}} + \sum_{i\in\mathfrak{D}}\overline{\mathsf{A}}_i{}'(\mathsf{w})\partial_i\widetilde{\mathsf{w}} - \sum_{i,j\in\mathfrak{D}}\big(\overline{\mathsf{B}}^{\mathrm{d}}_{ij}(\mathsf{w})+\overline{\mathsf{B}}^{\mathrm{c}}_{ij}(\mathsf{w})\big)\partial_i\partial_j\widetilde{\mathsf{w}} + \overline{\mathsf{L}}(\mathsf{w},\nabla\mathsf{w}_\mathrm{r})\widetilde{\mathsf{w}} = \mathsf{f}+\mathsf{g}, \tag{3.176}$$

where t denotes the time variable, ∂_i the derivative in the ith spatial direction, w a given function, w_r a subset of w components that have more regularity, and $\widetilde{\mathsf{w}}$ the linearized unknown. The variables w and $\widetilde{\mathsf{w}}$ have $\mathsf{n}\geq 1$ components and $\mathbb{R}^\mathsf{n}$ is decomposed into $\mathbb{R}^\mathsf{n} = \mathbb{R}^{\mathsf{n}_\mathrm{I}}\times\mathbb{R}^{\mathsf{n}_\mathrm{II}}$ with $\mathsf{n} = \mathsf{n}_\mathrm{I}+\mathsf{n}_\mathrm{II}$ and the subvariable w_r is composed of n_r components of w with $1\leq\mathsf{n}_\mathrm{r}\leq\mathsf{n}$. In practical applications the subvariable w_{r} is taken to be $\mathsf{w}_{\mathrm{r}} = (\mathsf{w}_{\mathrm{I}'}, \mathsf{w}_{\mathrm{II}})^t$ for an augmented system but only the general properties of w_{r} are required for the a priori estimates. The bloc structure induced by the partitioning of $\mathbb{R}^{\mathsf{n}}$ is used in the following so that $\widetilde{\mathsf{w}}$ is decomposed into $\widetilde{\mathsf{w}} = (\widetilde{\mathsf{w}}_{\mathrm{I}}, \widetilde{\mathsf{w}}_{\mathrm{II}})^t$ with $\widetilde{\mathsf{w}}_{\mathrm{I}} \in \mathbb{R}^{\mathsf{n}_{\mathrm{I}}}$ and $\widetilde{\mathsf{w}}_{\mathrm{II}} \in \mathbb{R}^{\mathsf{n}_{\mathrm{II}}}$. The integer l_0 denotes the minimum integer for an embedding of the Sobolev space $H^{l_0}(\mathbb{R}^d)$ into the space of bounded continuous functions $l_0 = [d/2] + 1$ and l is an integer such that $l \geq l_0 + 2$. More regularity is required with $l \geq l_0 + 2$ rather than $l \geq l_0 + 1$ like for standard hyperbolic-parabolic systems [7, 18] since the time derivative $\partial_t \mathsf{w}_{\mathrm{II}}$ are naturally obtained to be in $C^0\big((0, \bar{\tau}), H^{l-2}\big)$, due to the antisymmetric coupling terms $\sum_{i,j\in\mathfrak{D}} \overline{\mathsf{B}}{}^{\mathrm{c}}_{ij} \partial_i \partial_j \mathsf{w}$, even if stronger estimates may afterwards be obtained from stronger assumptions.

The following assumptions are made on the Eq. (3.176) and essentially say that it is a linearized normal form. The matrix $\overline{\mathsf{A}}_0$ is symmetric, positive definite, block diagonal $\overline{\mathsf{A}}_0 = \operatorname{diag}(\overline{\mathsf{A}}_0^{\mathrm{I,I}}, \overline{\mathsf{A}}_0^{\mathrm{II,II}})$, and the block $\overline{\mathsf{A}}_0^{\mathrm{II,II}}$ only depends on the subvariable w_{r}. The matrices $\overline{\mathsf{A}}_i{}'(\mathsf{w}) \in \mathbb{R}^{\mathsf{n},\mathsf{n}}$, $i \in \mathfrak{D}$, have their hyperbolic blocks $\overline{\mathsf{A}}_i^{\prime\mathrm{I,I}}$ symmetric, the dissipation matrices are such that $(\overline{\mathsf{B}}{}^{\mathrm{d}}_{ij})^t = \overline{\mathsf{B}}{}^{\mathrm{d}}_{ji}[ji]$, $i, j \in \mathfrak{D}$, have nonzero components only into the right lower $\overline{\mathsf{B}}{}^{\mathrm{d\,II,II}}_{ij}$ blocks, and $\overline{\mathsf{B}}{}^{\mathrm{d\,II,II}} = \sum_{i,j\in\mathfrak{D}} \overline{\mathsf{B}}{}^{\mathrm{d\,II,II}} \xi_i \xi_j$ is positive definite for $\boldsymbol{\xi} \in \Sigma^{d-1}$ where Σ^{d-1} is the sphere in d dimension. The matrices $\overline{\mathsf{B}}{}^{\mathrm{c}}_{ij}$ are such that $(\overline{\mathsf{B}}{}^{\mathrm{c}}_{ij})^t = -\overline{\mathsf{B}}{}^{\mathrm{c}}_{ji}$, the blocks $\overline{\mathsf{B}}{}^{\mathrm{c,I,I}}_{ij}$ vanish $\overline{\mathsf{B}}{}^{\mathrm{c,I,I}}_{ij} = 0$, and the blocks $\overline{\mathsf{B}}{}^{\mathrm{c,I,II}}_{ij}$ and $\overline{\mathsf{B}}{}^{\mathrm{c,II,I}}_{ij}$ only depend on the subvariable w_{r}. The matrix $\overline{\mathsf{L}} = \overline{\mathsf{L}}(\mathsf{w}, \mathfrak{p}_{\mathrm{r}})$ is block diagonal $\overline{\mathsf{L}} = \operatorname{diag}(\overline{\mathsf{L}}^{\mathrm{I,I}}, \overline{\mathsf{L}}^{\mathrm{II,II}})$ and is a linear function of $\mathfrak{p}_{\mathrm{r}} \in \mathbb{R}^{\mathsf{n}_{\mathrm{r}},d}$ so that $\overline{\mathsf{L}}^{\mathrm{I,I}} = \mathfrak{L}^{\mathrm{I,I}}(\mathsf{w})\mathfrak{p}_{\mathrm{r}}$ and $\overline{\mathsf{L}}^{\mathrm{II,II}} = \mathfrak{L}^{\mathrm{II,II}}(\mathsf{w})\mathfrak{p}_{\mathrm{r}}$ where $\mathfrak{L}^{\mathrm{I,I}} \in \mathbb{R}^{\mathsf{n}_{\mathrm{I}},\mathsf{n}_{\mathrm{I}},\mathsf{n}_{\mathrm{r}},d}$ and $\mathfrak{L}^{\mathrm{II,II}} \in \mathbb{R}^{\mathsf{n}_{\mathrm{II}},\mathsf{n}_{\mathrm{II}},\mathsf{n}_{\mathrm{r}},d}$ depend on w. The system coefficients $\overline{\mathsf{A}}_0$, $\overline{\mathsf{A}}_i{}'$ $\overline{\mathsf{B}}{}^{\mathrm{d}}_{ij}$, $\overline{\mathsf{B}}{}^{\mathrm{c}}_{ij}$, $i, j \in \mathfrak{D}$, $\mathfrak{L}^{\mathrm{I,I}}$, and $\mathfrak{L}^{\mathrm{II,II}}$ are smooth functions of $\mathsf{w} \subset \mathbb{R}^{\mathsf{n}}$ defined over an open set O_{w} with at least C^{l+2} regularity. The right hand sides f and g as well as w are given functions of time and space defined over $[0, \bar{\tau}] \times \mathbb{R}^d$ for some given positive time $\bar{\tau} > 0$.

Denoting by $\mathsf{w}^\star \in O_{\mathsf{w}}$ a constant state in the w variable, it is assumed that w is such that

$$\begin{cases} \mathsf{w}_{\mathrm{I}} - \mathsf{w}^\star_{\mathrm{I}} \in C^0\big([0, \bar{\tau}], H^l\big) \cap C^1\big([0, \bar{\tau}], H^{l-2}\big), \\ \mathsf{w}_{\mathrm{II}} - \mathsf{w}^\star_{\mathrm{II}} \in C^0\big([0, \bar{\tau}], H^l\big) \cap C^1\big([0, \bar{\tau}], H^{l-2}\big) \cap L^2\big((0, \bar{\tau}), H^{l+1}\big), \end{cases} \tag{3.177}$$

and the quantities M, M_{t} and M_{r} are defined by

$$M^2 = \sup_{0\leq\tau\leq\bar{\tau}} |\mathsf{w}(\tau) - \mathsf{w}^\star|_l^2, \quad M_{\mathrm{t}}^2 = \int_0^{\bar{\tau}} |\partial_t \mathsf{w}(\tau)|_{l-2}^2 \, d\tau, \quad M_{\mathrm{r}}^2 = \int_0^{\bar{\tau}} |\nabla \mathsf{w}_{\mathrm{r}}(\tau)|_l^2 \, d\tau.$$

The quantities M and M_{t} have been used in the a priori estimates for solutions of hyperbolic-parabolic linearized systems in previous sections. Since $l \geq l_0 + 2$ we may use $|\partial_t \mathsf{w}(\tau)|_{l-2}^2$ in the definition of M_{t} instead of $|\partial_t \mathsf{w}(\tau)|_{l-1}^2$ since $|\partial_t \mathsf{w}(\tau)|_{l-2}^2$

majorizes $|\partial_t \mathsf{w}(\tau)|^2_{L^\infty}$. The nonstandard use of M_r is now required due to the strong coupling arising from the antisymmetric blocks $\overline{\mathsf{B}}_{ij}^{\mathrm{c,I,II}}$ and $\overline{\mathsf{B}}_{ij}^{\mathrm{c,II,I}}$ that only depend on w_r and from the matrix $\overline{\mathsf{L}}$ that depends on w and $\nabla \mathsf{w}_\mathrm{r}$. It is assumed that $\mathcal{O}_0$ is an open set such that $\mathcal{O}_0 \subset \overline{\mathcal{O}}_0 \subset \mathcal{O}_\mathsf{w}$, that a_1 is such that $0 < a_1 < \mathrm{dist}(\overline{\mathcal{O}}_0, \partial\mathcal{O}_\mathsf{w})$, and the open set $\mathcal{O}_1$ is defined by

$$\mathcal{O}_1 = \{\, \mathsf{w} \in \mathcal{O}_\mathsf{w};\ \mathrm{dist}(\mathsf{w}, \overline{\mathcal{O}}_0) < a_1 \,\}. \tag{3.178}$$

It is further assumed that $\mathsf{w}_0(\mathsf{x}) = \mathsf{w}(0, \mathsf{x}) \in \mathcal{O}_0$, and that $\mathsf{w}(t, \mathsf{x}) \in \mathcal{O}_1$, for $(t, \mathsf{x}) \in [0, \bar{\tau}] \times \mathbb{R}^d$ so that the values of w are controlled with the open set $\mathcal{O}_1$.

Let now l' be such that $1 \le l' \le l$ and assume that the right hand sides $\mathsf{f} \in \mathbb{R}^\mathsf{n}$ and $\mathsf{g} \in \mathbb{R}^\mathsf{n}$ are such that

$$\mathsf{f} \in C^0\big([0, \bar{\tau}], H^{l'-1}\big) \cap L^1\big((0, \bar{\tau}), H^{l'}\big), \tag{3.179}$$

$$\mathsf{g} \in C^0\big([0, \bar{\tau}], H^{l'-1}\big), \qquad \mathsf{g}_\mathrm{I} = 0. \tag{3.180}$$

Denoting by $\widetilde{\mathsf{w}}^\star$ a constant state in $\mathbb{R}^\mathsf{n}$ such that $\overline{\mathsf{L}}(\mathsf{w}, \mathfrak{p}_\mathrm{r})\widetilde{\mathsf{w}}^\star = 0$ for any $\mathsf{w} \in \mathcal{O}_\mathsf{w}$ and $\mathfrak{p}_\mathrm{r} \in \mathbb{R}^{\mathsf{n}_\mathrm{r},d}$, a priori estimates of $\widetilde{\mathsf{w}} - \widetilde{\mathsf{w}}^\star$ are obtained in the following theorem for solutions $\widetilde{\mathsf{w}}$ to the linearized Eq. (3.176).

Theorem 3.24 *Assume that the solution $\widetilde{\mathsf{w}}$ of the system* (3.176) *is such that*

$$\begin{aligned} &\widetilde{\mathsf{w}}_\mathrm{I} - \widetilde{\mathsf{w}}_\mathrm{I}^\star \in C^0\big([0, \bar{\tau}], H^{l'}\big) \cap C^1\big([0, \bar{\tau}], H^{l'-2}\big), \\ &\widetilde{\mathsf{w}}_\mathrm{II} - \widetilde{\mathsf{w}}_\mathrm{II}^\star \in C^0\big([0, \bar{\tau}], H^{l'}\big) \cap C^1\big([0, \bar{\tau}], H^{l'-2}\big) \cap L^2\big((0, \bar{\tau}), H^{l'+1}\big), \end{aligned} \tag{3.181}$$

where $1 \le l' \le l$ and denote by $\widetilde{\mathsf{w}}_0$ the initial state $\widetilde{\mathsf{w}}_0(\mathsf{x}) = \widetilde{\mathsf{w}}(0, \mathsf{x})$. Then there exists constants $\mathsf{c}_1(\mathcal{O}_1) \ge 1$ and $\mathsf{c}_2(\mathcal{O}_1, M) \ge 1$, with $\mathsf{c}_2(\mathcal{O}_1, M)$ increasing with M, such that for any $t \in [0, \bar{\tau}]$

$$\begin{aligned} \sup_{0 \le \tau \le t} |\widetilde{\mathsf{w}}(\tau) - \widetilde{\mathsf{w}}^\star|^2_{l'} + \int_0^t |\widetilde{\mathsf{w}}_\mathrm{II}(\tau) - \widetilde{\mathsf{w}}_\mathrm{II}^\star|^2_{l'+1}\, d\tau \le\ & \mathsf{c}_1^2 \exp\big(\mathsf{c}_2\big(t + M_\mathrm{t}\sqrt{t} + M_\mathrm{r}\sqrt{t}\big)\big) \\ & \times \Big(|\widetilde{\mathsf{w}}_0 - \widetilde{\mathsf{w}}^\star|^2_{l'} + \mathsf{c}_2 \Big\{ \int_0^t |\mathsf{f}|_{l'}\, d\tau \Big\}^2 + \mathsf{c}_2 \int_0^t |\mathsf{g}_\mathrm{II}|^2_{l'-1}\, d\tau \Big). \end{aligned} \tag{3.182}$$

These estimates differ from the classic estimates for hyperbolic-parabolic linearized systems in several points. The Sobolev order l associated with w is first larger than $l_0 + 2$, rather than $l_0 + 1$, since the time derivative $\partial_t \mathsf{w}_\mathrm{II}$ is a priori only in H^{l-2} due to the coupling blocks $\overline{\mathsf{B}}_{ij}^{\mathrm{c,II,I}}$ that yield second derivatives of the hyperbolic variable in the parabolic equation. The extra coupling terms $\overline{\mathsf{B}}_{ij}^{\mathrm{c,I,II}}$ and $\overline{\mathsf{B}}_{ij}^{\mathrm{c,II,I}}$ also require more regularity by solely depending on the subvariable w_r. The antisymmetric coupling terms and the source term both yield new contributions like $M_{\mathrm{r}}\sqrt{t}$ in the exponential term of the right hand side of (3.182).

Sketch of the Proof The proof is similar to the standard proof for the convective and dissipative terms whereas the antisymmetric terms systematically need to be decomposed by blocks between the hyperbolic and parabolic variables [90]. The properties that $\overline{\mathsf{B}}_{ij}^{\mathrm{c,I,I}} = 0$ and that the blocks $\overline{\mathsf{A}}_0^{\mathrm{II,II}}\overline{\mathsf{A}}_0^{\mathrm{II,II}}$, $\overline{\mathsf{B}}_{ij}^{\mathrm{c,I,II}}$, and $\overline{\mathsf{B}}_{ij}^{\mathrm{c,II,I}}$ only depending on the subvariable w_{r} are then key points. Existence of solutions for linearized equations is next obtained by first regularizing the antisymmetric strongly coupling terms $\overline{\mathsf{B}}_{ij}^{\mathrm{c}}\partial_i\partial_j\delta\widetilde{\mathsf{w}}$ allowing to uncouple the hyperbolic and parabolic subsystems and then letting the regularizing parameter to go to zero [90]. □

We now establish local existence of solutions for augmented systems including the situation of cohesive fluids. We consider a symmetric-antisymmetric hyperbolic-parabolic-dispersive augmented system in normal form

$$\overline{\mathsf{A}}_0(\mathsf{w})\partial_t\mathsf{w} + \sum_{i\in\mathfrak{D}}\overline{\mathsf{A}}_i(\mathsf{w})\partial_i\mathsf{w} - \sum_{i,j\in\mathfrak{D}}\overline{\mathsf{B}}_{ij}^{\mathrm{d}}(\mathsf{w})\partial_i\partial_j\mathsf{w}$$
$$- \sum_{i,j\in\mathfrak{D}}\overline{\mathsf{B}}_{ij}^{\mathrm{c}}(\mathsf{w})\partial_i\partial_j\mathsf{w} = \mathsf{h}(\mathsf{w}, \nabla\mathsf{w}), \tag{3.183}$$

as obtained in Sect. 3.4.3. The right hand side $\mathsf{h} = (\mathsf{h}_{\mathrm{I}}, \mathsf{h}_{\mathrm{II}})^t$ may more generally be taken in the form

$$\mathsf{h}_{\mathrm{I}} = \sum_{i\in\mathfrak{D}}\overline{\mathrm{M}}_i^{\mathrm{I}}(\mathsf{w})\partial_i\mathsf{w}_{\mathrm{r}} + \sum_{i,j\in\mathfrak{D}}\overline{\mathrm{M}}_{ij}^{\mathrm{I,I}}(\mathsf{w})\partial_i\mathsf{w}_{\mathrm{r}}\partial_j\mathsf{w}_{\mathrm{r}}, \tag{3.184}$$

$$\mathsf{h}_{\mathrm{II}} = \sum_{i\in\mathfrak{D}}\overline{\mathrm{M}}_i^{\mathrm{II}}(\mathsf{w})\partial_i\mathsf{w} + \sum_{i,j\in\mathfrak{D}}\overline{\mathrm{M}}_{ij}^{\mathrm{II,II}}(\mathsf{w})\partial_i\mathsf{w}\partial_j\mathsf{w}, \tag{3.185}$$

where $\overline{\mathrm{M}}_i^{\mathrm{I}} \in \mathbb{R}^{\mathsf{n}_{\mathrm{I}},\mathsf{n}_{\mathrm{r}}}$, $\overline{\mathrm{M}}_{ij}^{\mathrm{I,I}}(\mathsf{w}) \in \mathbb{R}^{\mathsf{n}_{\mathrm{I}},\mathsf{n}_{\mathrm{r}},\mathsf{n}_{\mathrm{r}}}$, $\overline{\mathrm{M}}_i^{\mathrm{II}} \in \mathbb{R}^{\mathsf{n}_{\mathrm{II}},\mathsf{n}}$, and $\overline{\mathrm{M}}_{ij}^{\mathrm{II,II}} \in \mathbb{R}^{\mathsf{n}_{\mathrm{II}},\mathsf{n},\mathsf{n}}$ are linear operators. The hyperbolic components h_{I} thus includes linear and quadratic contributions in terms of $\nabla\mathsf{w}_{\mathrm{r}}$ whereas h_{II} includes linear and quadratic contributions in terms of $\nabla\mathsf{w}$. All the system coefficients $\overline{\mathsf{A}}_0$, $\overline{\mathsf{A}}_i$, $i \in \mathfrak{D}$, $\overline{\mathsf{B}}_{ij}^{\mathrm{d}}$, $\overline{\mathsf{B}}_{ij}^{\mathrm{c}}$, $i, j \in \mathfrak{D}$, $\overline{\mathrm{M}}_i^{\mathrm{I}}$, $\overline{\mathrm{M}}_{ij}^{\mathrm{I,I}}$, $\overline{\mathrm{M}}_i^{\mathrm{II}}$, and $\overline{\mathrm{M}}_{ij}^{\mathrm{II,II}}$ are assumed to have at least regularity $C^{l+2}(O_{\mathsf{w}})$ where O_{w} is a simply connected open set in $\mathbb{R}^d$. We also denote by $\mathsf{w}^\star$ a constant state in the open set O_{w} such that $\mathsf{w}_{\mathrm{I}''}^\star = 0$. In addition, denoting by $\mathsf{w}_0 = (\mathsf{w}_{0\mathrm{I}'}, \mathsf{w}_{0\mathrm{I}''}, \mathsf{w}_{0\mathrm{II}})^t$ the initial condition, it is assumed that whenever $\mathsf{w}_{0\mathrm{I}''} = \nabla\mathsf{w}_{0\mathrm{I}'}$, the regular solutions of (3.183) are such that $\mathsf{w}_{\mathrm{I}''} = \nabla\mathsf{w}_{\mathrm{I}'}$.

Theorem 3.25 *Let $d \geq 1$, $l \geq l_0 + 2$, $l_0 = [d/2] + 1$, be integers and let $b > 0$. Let O_0 be such that $\overline{O}_0 \subset O_{\mathsf{w}}$, a_1 such that $0 < a_1 < \mathrm{dist}(\overline{O}_0, \partial O_{\mathsf{w}})$, and $O_1 = \{\, \mathsf{w} \in O_{\mathsf{w}};\ \mathrm{dist}(\mathsf{w}, \overline{O}_0) < a_1 \,\}$. There exists $\bar{\tau} > 0$ depending on O_1 and b such that for any w_0 with $\mathsf{w}_0 \in \overline{O}_0$, $\mathsf{w}_0 - \mathsf{w}^\star \in H^l$, $\mathsf{w}_{0\mathrm{I}''} = \nabla\mathsf{w}_{0\mathrm{I}'}$ and*

$$|\mathsf{w}_0 - \mathsf{w}^\star|_l^2 < b^2, \tag{3.186}$$

there exists a unique local solution w *to the system* (3.183) *with initial condition* $\mathsf{w}(0, \mathsf{x}) = \mathsf{w}_0(\mathsf{x})$, *such that* $\mathsf{w}(t, \mathsf{x}) \in \mathcal{O}_1$ *for* $(t, \mathsf{x}) \in [0, \bar{\tau}]\times\mathbb{R}^d$, $\mathsf{w}_{\mathrm{I}''} = \nabla \mathsf{w}_{\mathrm{I}'}$, *and*

$$\mathsf{w}_{\mathrm{I}} - \mathsf{w}_{\mathrm{I}}^\star \in C^0\big([0, \bar{\tau}], H^l\big) \cap C^1\big([0, \bar{\tau}], H^{l-2}\big),$$
$$\mathsf{w}_{\mathrm{II}} - \mathsf{w}_{\mathrm{II}}^\star \in C^0\big([0, \bar{\tau}], H^l\big) \cap C^1\big([0, \bar{\tau}], H^{l-2}\big) \cap L^2\big((0, \bar{\tau}), H^{l+1}\big).$$

Moreover, there exists $\mathsf{c}_{\mathrm{loc}} \geq 1$ *only depending on* $\mathcal{O}_1$ *and* b *such that*

$$\sup_{0\leq\tau\leq\bar{\tau}} |\mathsf{w}(\tau) - \mathsf{w}^\star|_l^2 + \int_0^{\bar{\tau}} |\mathsf{w}_{\mathrm{II}}(\tau) - \mathsf{w}_{\mathrm{II}}^\star|_{l+1}^2 d\tau \leq \mathsf{c}_{\mathrm{loc}}^2 |\mathsf{w}_0 - \mathsf{w}^\star|_l^2. \tag{3.187}$$

Sketch of the Proof The proof is globally similar to that of classic hyperbolic-parabolic systems once the linearized estimates are established. The invariant set $\mathsf{X}_{\bar{\tau}}^l\big(\mathcal{O}_1, \overline{M}\big)$ is now defined by $\mathsf{w} - \mathsf{w}^\star \in C^0\big([0, \bar{\tau}], H^l\big)$, $\partial_t \mathsf{w} \in C^0\big([0, \bar{\tau}], H^{l-2}\big)$, $\mathsf{w}_{\mathrm{II}} - \mathsf{w}_{\mathrm{II}}^\star \in L^2\big((0, \bar{\tau}), H^{l+1}\big)$, $\mathsf{w}(t, \mathsf{x}) \in \mathcal{O}_1$, $\mathsf{w}_{\mathrm{I}''} = \nabla \mathsf{w}_{\mathrm{I}'}$, with

$$\sup_{0\leq\tau\leq\bar{\tau}} |\mathsf{w}(\tau) - \mathsf{w}^\star|_l^2 + \int_0^{\bar{\tau}} |\mathsf{w}_{\mathrm{II}}(\tau) - \mathsf{w}_{\mathrm{II}}^\star|_{l+1}^2 \, d\tau \leq \overline{M}^2$$

$$\int_0^{\bar{\tau}} |\partial_t \mathsf{w}(\tau)|_{l-2}^2 \, d\tau \leq \overline{M}^2 \qquad \int_0^{\bar{\tau}} |\nabla \mathsf{w}_{\mathrm{r}}(\tau)|_l^2 \, d\tau \leq \overline{M}^2.$$

Using the linearized estimates, the set $\mathsf{X}_{\bar{\tau}}^l\big(\mathcal{O}_1, \overline{M}\big)$ is shown to be invariant by the map $\mathsf{w} \mapsto \widetilde{\mathsf{w}}$ for $\overline{M} = 2\mathsf{c}_1(\mathcal{O}_1)b$ and $\bar{\tau}$ small enough. The successive approximations $\{\mathsf{w}^k\}_{k\geq 0}$ are then built by letting $\mathsf{w}^0 = \mathsf{w}^\star$ and $\mathsf{w}^{k+1} = \widetilde{\mathsf{w}}^k$ for $k \geq 0$ and are well defined. The sequence $\{\mathsf{w}^k\}_{k\geq 0}$ is then convergent over $[0, \bar{\tau}]$ for the norm

$$\sup_{0\leq\tau\leq\bar{\tau}} |\delta\widetilde{\mathsf{w}}(\tau)|_{l-2}^2 + \int_0^{\bar{\tau}} |\delta\widetilde{\mathsf{w}}_{\mathrm{II}}(\tau)|_{l-1}^2 \, d\tau$$

as obtained by using the a priori estimates for linearized equations for $\mathsf{w}^{k+1} - \mathsf{w}^k$.

We thus have $\mathsf{w}^k \to \overline{\mathsf{w}} \in C^0\big([0, \bar{\tau}], H^{l-2}\big)$ that is a solution (fixed point) and we already have $\overline{\mathsf{w}} \in L^\infty\big((0, \bar{\tau}), H^l\big)$ and $\overline{\mathsf{w}}_{\mathrm{II}} - \mathsf{w}_{\mathrm{II}}^\star \in L^2\big((0, \bar{\tau}), H^{l+1}\big)$. In order to establish that $\overline{\mathsf{w}} \in C^0\big((0, \bar{\tau}), H^l\big)$ we introduce a mollifying sequence of functions $\mathfrak{a}_\epsilon = \epsilon^{-d}\mathfrak{a}(\mathbf{r}/\epsilon)$ where $0 < \epsilon \leq 1$ and $\mathfrak{a}$ is a smooth function positive in the ball $|\mathbf{r}| < 1$ of $\mathbb{R}^d$ and zero otherwise with its integral equal to unity $\int \mathfrak{a} \, d\mathbf{r} = 1$. The corresponding mollifying convolution operators are then denoted by $\mathsf{R}_\epsilon \phi(\mathbf{r}) = \int \mathfrak{a}_\epsilon(\mathbf{r} - \widehat{\mathbf{r}})\phi(\widehat{\mathbf{r}}) \, d\widehat{\mathbf{r}}$ for $\phi \in L^1(\mathbb{R}^d)$ and $0 < \epsilon \leq 1$. It is then established that $\mathsf{w}^\delta = \mathsf{R}_\delta \overline{\mathsf{w}}$ form a Cauchy sequence in $C^0\big([0, \bar{\tau}], H^l\big)$ for δ going to zero and this completes the proof. □

Theorem 3.25 may then be directly applied to the diffuse interface fluid equations in normal form (3.169) under the assumptions $(\mathsf{H}_1 - \mathsf{H}_5)$. The unknown vector reads $\mathsf{w} = (\rho, \boldsymbol{w}, \boldsymbol{v}, T)^t$ and has $\mathsf{n} = 2d + 2$ components with $\mathsf{n}_{\mathrm{I}} = d + 1$ hyperbolic components and $\mathsf{n}_{\mathrm{II}} = d + 1$ parabolic components given by

$$\mathsf{w}_{\mathrm{I}} = (\rho, \boldsymbol{w})^t, \qquad \mathsf{w}_{\mathrm{II}} = (\boldsymbol{v}, T)^t,$$

and the subvariable w_{r} is given by $\mathsf{w}_{\mathrm{r}} = (\rho, \boldsymbol{v}, T)^t$ with $\mathsf{n}_{\mathrm{r}} = 1 + \mathsf{n}_{\mathrm{II}}$. Letting $\mathcal{O}_{\mathsf{w}}$ to be the open set $\mathcal{O}_{\mathsf{z}}$ of (H_1), we select $\mathsf{w}^\star = (\rho^\star, \boldsymbol{0}, \boldsymbol{v}^\star, T^\star)^t \in \mathcal{O}_{\mathsf{w}}$, an open set $\mathcal{O}_0$ with $\overline{\mathcal{O}}_0 \subset \mathcal{O}_{\mathsf{w}}$ and $\mathsf{w}^\star \in \mathcal{O}_0$, a_1 such that $0 < a_1 < \mathrm{dist}(\overline{\mathcal{O}}_0, \partial\mathcal{O}_{\mathsf{w}})$, and we define $\mathcal{O}_1 = \{\mathsf{w} \in \mathcal{O}_{\mathsf{w}}; \mathrm{dist}(\mathsf{w}, \overline{\mathcal{O}}_0) < a_1\}$. The regularity class of thermodynamic functions γ taken to be such that $\gamma - 2 \geq l + 2$ so that all the system coefficients $\overline{\mathsf{A}}_0, \overline{\mathsf{A}}_i, i \in \mathfrak{D}, \overline{\mathsf{B}}^{\mathrm{d}}_{ij}, \overline{\mathsf{B}}^{\mathrm{c}}_{ij}, i, j \in \mathfrak{D}$, have at least regularity C^{l+2} over $\mathcal{O}_{\mathsf{w}}$. Moreover there exists a linearized version that enforces that gradient constraint. The matrix $\overline{\mathsf{L}} = \overline{\mathsf{L}}(\mathsf{w}, \boldsymbol{\nabla}\mathsf{w}_{\mathrm{r}})$ is block diagonal $\overline{\mathsf{L}} = \mathrm{diag}(\overline{\mathsf{L}}^{\mathrm{I,I}}, \overline{\mathsf{L}}^{\mathrm{II,II}})$ with $\overline{\mathsf{L}}^{\mathrm{II,II}} = 0$ and $\overline{\mathsf{L}}^{\mathrm{I,I}}$ is a linear function of $\boldsymbol{\nabla}\mathsf{w}_{\mathrm{r}}$ so that $\overline{\mathsf{L}}^{\mathrm{I,I}} = \mathfrak{L}^{\mathrm{I,I}}(\mathsf{w})\boldsymbol{\nabla}\mathsf{w}_{\mathrm{r}}$ and the right hand sides are in the form (3.184) and (3.185). The systems coefficients $\mathfrak{L}^{\mathrm{I,I}}$, $\overline{\mathrm{M}}_i'^{\mathrm{I}}$, $\overline{\mathrm{M}}_{ij}'^{\mathrm{I,I}}$, $\overline{\mathrm{M}}_i'^{\mathrm{II}}$, and $\overline{\mathrm{M}}_{ij}'^{\mathrm{II,II}}$ also have at least regularity $C^{l+2}(\mathcal{O}_{\mathsf{w}})$.

Application of Theorem 3.25 then yields the existence result for diffuse interface fluids [90]. For $d \geq 1$, $l \geq l_0 + 2$, and $b > 0$, there exists $\bar{\tau} > 0$ depending on $\mathcal{O}_1$ and b such that for any w_0 with $\mathsf{w}_0 \in \overline{\mathcal{O}}_0$, $\mathsf{w}_0 - \mathsf{w}^\star \in H^l$, with $\boldsymbol{w}_0 = \boldsymbol{\nabla}\rho_0$ and $|\mathsf{w}_0 - \mathsf{w}^\star|_l^2 < b^2$ there exists a unique local solution w to the system (3.169) with initial condition $\mathsf{w}(0, \mathsf{x}) = \mathsf{w}_0(\mathsf{x})$, $\mathsf{w}(t, \mathsf{x}) \in \mathcal{O}_1$ for $(t, \mathsf{x}) \in [0, \bar{\tau}]\times\mathbb{R}^d$, $\boldsymbol{w} = \boldsymbol{\nabla}\rho$ and with

$$\rho - \rho^\star \in C^0\big([0, \bar{\tau}], H^{l+1}\big), \tag{3.188}$$

$$\boldsymbol{v} - \boldsymbol{v}^\star \in C^0\big([0, \bar{\tau}], H^l\big) \cap L^2\big((0, \bar{\tau}), H^{l+1}\big) \tag{3.189}$$

$$T - T^\star \in C^0\big([0, \bar{\tau}], H^l\big) \cap L^2\big((0, \bar{\tau}), H^{l+1}\big). \tag{3.190}$$

The time derivatives are also such that $\partial_t\rho \in C^0\big([0, \bar{\tau}], H^{l-1}\big) \cap L^2\big((0, \bar{\tau}), H^l\big)$, $\partial_t\boldsymbol{v} \in C^0\big([0, \bar{\tau}], H^{l-2}\big)$ and $\partial_t T \in C^0\big([0, \bar{\tau}], H^{l-2}\big)$.

3.4.5 Asymptotic Stability

Strict dissipativity conditions for linearized systems of arbitrary order around constant equilibrium states have been investigated by Pego [91] and Humpherys [93]. Pego studied the stability of shock profiles and Humpherys has established the general equivalence between strict stability, the genuine stability condition, and the existence of compensating matrices for symbol symmetrizable systems [93]. Application of Humpherys results to one-dimensional isothermal Korteweg fluids has been conducted by Plaza and Valachos [98] who strengthened the assumptions on compensating matrices in order to derive a priori estimates. Strict dissipativity conditions for third order systems with constant coefficients have then been thoroughly investigated by Kawashima et al. [96, 97]. The systems of equations being not symmetric, the complex symmetry properties used in these analyses are obtained after linearization, taking the Fourier transform, using projection operators and mixing first and third order derivative terms in an intricate fashion. We introduce in this section a simpler formulation of strict dissipativity condition by using the augmented formulation.

We consider an augmented system in normal form linearized around an equilibrium state $\mathsf{w}^\star$

$$\overline{\mathsf{A}}_0^\star \partial_t \mathsf{w} + \sum_{i\in\mathfrak{D}} \overline{\mathsf{A}}_i^\star \partial_i \mathsf{w} - \sum_{i,j\in\mathfrak{D}} \overline{\mathsf{B}}_{ij}^{\mathrm{d}\star} \partial_i \partial_j \mathsf{w} - \sum_{i,j\in\mathfrak{D}} \overline{\mathsf{B}}_{ij}^{\mathrm{c}\star} \partial_i \partial_j \mathsf{w} + \overline{\mathsf{L}}^\star \mathsf{w} = 0, \tag{3.191}$$

where $\overline{\mathsf{A}}_0^\star = \overline{\mathsf{A}}_0(\mathsf{w}^\star)$, $\overline{\mathsf{A}}_i^\star = \overline{\mathsf{A}}_i(\mathsf{w}^\star)$, $\overline{\mathsf{B}}_{ij}^{\mathrm{d}\star} = \overline{\mathsf{B}}_{ij}^{\mathrm{d}}(\mathsf{w}^\star)$, $\overline{\mathsf{B}}_{ij}^{\mathrm{c}\star} = \overline{\mathsf{B}}_{ij}^{\mathrm{c}}(\mathsf{w}^\star)$, and $\overline{\mathsf{L}}^\star = \overline{\mathsf{L}}(\mathsf{w}^\star)$. We assume also that the gradient constraint is satisfied by the solution, i.e., that $\mathsf{w}_{\mathrm{I}''} = \boldsymbol{\nabla}\mathsf{w}_{\mathrm{I}'}$. We introduce a new dissipative conditions for *augmented* systems directly written by using the Fourier transform of the linearized equations.

The new strict stability condition associated with (3.191) is written for convenience by using compensating matrices K^j, $j \in \mathfrak{D}$, in each spatial direction, and involves both convective and capillary matrices.

(D) The matrices K^j, $j \in \mathfrak{D}$, are such that the products $\mathsf{K}^j\overline{\mathsf{A}}_0(\mathsf{w}^\star)$ are skew-symmetric for any $j \in \mathfrak{D}$

$$\big(\mathsf{K}^j\overline{\mathsf{A}}_0^\star\big)^t = -\mathsf{K}^j\overline{\mathsf{A}}_0^\star, \qquad j \in \mathfrak{D}. \tag{3.192}$$

Moreover there exists $\alpha > 0$ and $C > 0$ such that for any $\xi = (\xi_1, \dots, \xi_d)^t$ and any $\mathsf{x} = (\mathsf{x}_{\mathrm{I}}, \mathsf{x}_{\mathrm{II}})^t = (\mathsf{x}_{\mathrm{I}'}, \mathsf{x}_{\mathrm{I}''}, \mathsf{x}_{\mathrm{II}})^t$ with $\mathsf{x}_{\mathrm{I}''} = \xi\mathsf{x}_{\mathrm{I}'}$ we have

$$\sum_{i,j\in\mathfrak{D}} \big\langle \mathsf{x}, \xi_i\xi_j \mathsf{K}_j \overline{\mathsf{A}}_i^\star \mathsf{x}\big\rangle + \Big\langle \mathsf{x}, \sum_{i,j,j'\in\mathfrak{D}} \xi_i\xi_j\xi_{j'} (\mathsf{K}_j \overline{\mathsf{B}}_{ij'}^{\mathrm{c}\star})\mathsf{x}\Big\rangle$$
$$\geq \alpha(|\xi|^2 + |\xi|^4)|\mathsf{x}_{\mathrm{I}'}|^2 - C|\xi|^2|\mathsf{x}_{\mathrm{II}}|^2. \tag{3.193}$$

There is a link between the Fourier variable ξ and the test vector x in D since we consider an augmented system whose solutions satisfy the gradient constraint. Both convective and cohesive matrices are also involved in the new dissipativity condition (3.193). This condition may also naturally be split between convective and dispersive conditions by using scaling properties of derivatives. More specifically, (3.193) may be replaced by the two inequalities

$$\Big\langle \mathsf{x}, \sum_{i,j\in\mathfrak{D}} \xi_i\xi_j \mathsf{K}_j \overline{\mathsf{A}}_i^{\star} \mathsf{x}\Big\rangle \geq \alpha|\xi|^2|\mathsf{x}_{\mathrm{I}'}|^2 - C|\xi|^2|\mathsf{x}_{\mathrm{II}}|^2, \tag{3.194}$$

$$\sum_{i,j,j',l\in\mathfrak{D}} \xi_j\xi_l (\mathsf{K}_j\overline{\mathsf{B}}_{ij'}^{\mathrm{c}\star})_{1,1+l}\xi_i\xi_{j'} \geq \alpha|\xi|^4|\mathsf{x}_{\mathrm{I}'}|^2. \tag{3.195}$$

The first condition (3.194) differs from the traditional strict dissipativity condition since the matrix $\sum_{i,j\in\mathfrak{D}} \xi_i\xi_j \mathsf{K}_j\overline{\mathsf{A}}_i^{\star} + \sum_{i,j\in\mathfrak{D}} \xi_i\xi_j \overline{\mathsf{B}}_{ij}^{\star}$ is not positive definite, but only positive semidefinite, and positive definite on the linear subspace $(\mathsf{e}_{1+1},\ldots,\mathsf{e}_{1+d})^{\perp}$. In the second inequality (3.195) the only nonzero contributions in the scalar product involving the matrices $\mathsf{K}_j\overline{\mathsf{B}}_{ij'}^{\mathrm{c}\star}$ are obtained with the $\mathsf{x}_{\mathrm{I}'}$ on the left and with its Fourier derivatives $\mathsf{x}_{\mathrm{I}''} = \xi_l\mathsf{x}_{\mathrm{I}'}$ on the right. The coefficients $(\mathsf{K}_j\overline{\mathsf{B}}_{ij'}^{\mathrm{c}\star})_{k,k'}$ are indeed zero except when $(k,k') = (1, 1+l)$, $l \in \mathfrak{D}$ and we have $(\mathsf{K}_j\overline{\mathsf{B}}_{ij'}^{\mathrm{c}\star})_{1,1+l} = \alpha\,\delta_{ij}\delta_{j'l}$.

From a practical point of view, under the new condition of strict dissipativity, there exists $\alpha > 0$ and $\mathsf{c} > 0$ so that whenever $\mathsf{w}_{\mathrm{I}''} = \nabla\mathsf{w}_{\mathrm{I}'}$ and w is regular enough we have

$$\sum_{i,j\in\mathfrak{D}} \int \langle\partial_j\mathsf{w},\, \mathsf{K}^j\overline{\mathsf{A}}_i^{\star}\partial_i\mathsf{w}\rangle d\mathsf{x} - \sum_{i,j,j'\in\mathfrak{D}} \int \langle\partial_j\mathsf{w},\, \mathsf{K}^j\overline{\mathsf{B}}_{ij'}^{\mathrm{c}\star}\partial_i\partial_{j'}\mathsf{w}\rangle d\mathsf{x}$$
$$\geq \alpha\Big(\int |\nabla\mathsf{w}_{\mathrm{I}'}|^2 d\mathsf{x} + \int |\Delta\mathsf{w}_{\mathrm{I}'}|^2 d\mathsf{x}\Big) - \mathsf{c}\int |\nabla\mathsf{w}_{\mathrm{II}}|^2 d\mathsf{x}, \tag{3.196}$$

and this inequality may replace the traditional strict dissipativity estimates for non augmented systems without capillarity [18].

We now consider an augmented system in conservative form with a mathematical entropy σ as well as a normal form as in the previous sections. We use the conservative unknown u, the entropic variable v and the normal variable w. We consider a constant state written $\mathsf{u}^\star$, $\mathsf{v}^\star$ and $\mathsf{w}^\star$ in the various variables. We assume that it is a thermodynamically stable state in the sense that the mathematical entropy hessian matrix $\partial^2_{\mathsf{uu}}\sigma$ is positive definite at $\mathsf{u}^\star$. Anticipating that we stay in a neighborhood of $\mathsf{u}^\star$, the instability indicator m may then taken to be zero. We denote again with the $\star$ superscript any function evaluated at the state $\mathsf{w}^\star$ so that in particular $\overline{\mathsf{A}}_0^{\star} = \overline{\mathsf{A}}_0(\mathsf{w}^\star)$, $\overline{\mathsf{A}}_i^{\star} = \overline{\mathsf{A}}_i(\mathsf{w}^\star)$, $\overline{\mathsf{B}}_{ij}^{\mathrm{d}\star} = \overline{\mathsf{B}}_{ij}^{\mathrm{d}}(\mathsf{w}^\star)$, and $\overline{\mathsf{B}}_{ij}^{\mathrm{c}\star} = \overline{\mathsf{B}}_{ij}^{\mathrm{c}}(\mathsf{w}^\star)$. The first step in order to obtain global existence results is to establish new a priori estimates that are independent of length intervals. The system of partial differential equations in normal form is assumed to be strictly dissipative at $\mathsf{w}^\star$ in the sense of D.

A priori estimates of solutions over time intervals of arbitrary length are investigated when w remains close to the constant equilibrium state $\mathsf{w}^\star$. It is assumed that for some time interval $\bar{\tau}$

$$\mathsf{w}_{\mathrm{I}} - \mathsf{w}_{\mathrm{I}}^{\star} \in C^0\big([0, \bar{\tau}], H^l\big) \cap C^1\big([0, \bar{\tau}], H^{l-2}\big), \qquad \partial_{\mathsf{x}} \mathsf{w}_{\mathrm{I}} \in L^2\big((0, \bar{\tau}), H^{l-1}\big), \tag{3.197}$$

$$\mathsf{w}_{\mathrm{II}} - \mathsf{w}_{\mathrm{II}}^{\star} \in C^0\big([0, \bar{\tau}], H^l\big) \cap C^1\big([0, \bar{\tau}], H^{l-2}\big), \qquad \partial_{\mathsf{x}} \mathsf{w}_{\mathrm{II}} \in L^2\big((0, \bar{\tau}), H^l\big), \tag{3.198}$$

and $N_l(t)$ is defined for $0 \leq t \leq \bar{\tau}$ to be

$$N_l^2(t) = \sup_{0 \leq \tau \leq t} |\mathsf{w}(\tau) - \mathsf{w}^{\star}|_l^2 + \int_0^t \big(|\nabla \mathsf{w}_{\mathrm{I}}(\tau)|_{l-1}^2 + |\nabla \mathsf{w}_{\mathrm{II}}(\tau)|_l^2 \big)\, d\tau. \tag{3.199}$$

The norm N_l^2 will play a key role in order to obtain a priori estimates independent of the length of time intervals $\bar{\tau}$.

Lemma 3.8 *Let* w *with* $\mathsf{w} - \mathsf{w}^{\star} \in C^0\big([0, \bar{\tau}], H^l\big)$ *such that* (3.197) *and* (3.198) *hold. There exists* $b_{\overline{\sigma}}$ *such that*

$$N_l(t) \leq b_{\overline{\sigma}} \implies \mathsf{w}(\tau, \mathsf{x}) \in \mathcal{O}_{\overline{\sigma}}, \quad 0 \leq \tau \leq t, \quad \mathsf{x} \in \mathbb{R}^d. \tag{3.200}$$

Proof This property (3.200) is a consequence of the assumption $l \geq l_0$. □

We now establish a priori estimates with constants independent of the time interval $[0, \bar{\tau}]$.

Theorem 3.26 *Assume that* w *is a solution of* (3.183) *over* $[0, \bar{\tau}] \times \mathbb{R}^d$ *with* (3.197), (3.198) *satisfying notably the gradient constraint. There exists constants* $b_N \leq b_{\overline{\sigma}}$ *and* $\mathsf{c}_N \geq 1$, *independent of* $\bar{\tau}$, *such that*

$$N_l(\bar{\tau}) \leq b_N \quad \implies \quad N_l(\bar{\tau}) \leq \mathsf{c}_N |\mathsf{w}_0 - \mathsf{w}^{\star}|_l. \tag{3.201}$$

Proof It is sufficient to establish that there exists constants $b' \leq b_{\overline{\sigma}}$, c', and c'' such that $N_l(t) \leq b'$ implies

$$N_l^2(t) \leq \mathsf{c}' |\mathsf{w}_0 - \mathsf{w}^{\star}|_l^2 + \mathsf{c}'' N_l^3(t),$$

since it directly yields (3.201). It is also sufficient to consider smooth solutions of (3.183) since one may use mollifiers and convolution operators. The proof is then similar to that of the non augmented system when using the new strict dissipativity condition [18]. □

Global existence of solution may now be obtained by combining local existence and the estimates of Theorem 3.26 [18].

Theorem 3.27 *Let* $d \geq 1$ *and* $l \geq [d/2] + 2$ *be integers. There exists* $\bar{b} > 0$ *small enough such that if* w_0 *satisfies* $\mathsf{w}_0 - \mathsf{w}^{\star} \in H^l(\mathbb{R}^d)$ *and*

$$|\mathsf{w}_0 - \mathsf{w}^{\star}|_l^2 < \bar{b}^2,$$

there exists a unique global solution to the Cauchy problem

$$\overline{A}_0 \partial_t \mathsf{w} + \sum_{i\in\mathfrak{D}} \overline{A}_i \partial_i \mathsf{w} - \sum_{i,j\in\mathfrak{D}} \partial_i (\overline{B}_{ij} \partial_j \mathsf{w}) + \overline{L}\mathsf{w} = \sum_{i,j\in\mathfrak{D}} \overline{\mathrm{M}}_{ij}(\mathsf{w})\, \partial_i \mathsf{w}\, \partial_j \mathsf{w},$$

with initial condition $\mathsf{w}(0, \mathsf{x}) = \mathsf{w}_0(\mathsf{x})$ *and*

$$\begin{aligned} &\mathsf{w}_{\mathrm{I}} - \mathsf{w}_{\mathrm{I}}^{\star} \in C^0\big([0,\infty), H^l\big) \cap C^1\big([0,\infty), H^{l-1}\big),\ \ \partial_{\mathsf{x}} \mathsf{w}_{\mathrm{I}} \in L^2\big((0,\infty), H^{l-1}\big),\\ &\mathsf{w}_{\mathrm{II}} - \mathsf{w}_{\mathrm{II}}^{\star} \in C^0\big([0,\infty), H^l\big) \cap C^1\big([0,\infty), H^{l-2}\big),\ \partial_{\mathsf{x}} \mathsf{w}_{\mathrm{II}} \in L^2\big((0,\infty), H^l\big). \end{aligned}$$

Furthermore, there exists a constant $\bar{c}$ *such that* w *satisfies the estimate*

$$|\mathsf{w}(t) - \mathsf{w}^{\star}|_l^2 + \int_0^t |\partial_{\mathsf{x}} \mathsf{w}_{\mathrm{I}}|_{l-1}^2\, d\tau + \int_0^t |\partial_{\mathsf{x}} \mathsf{w}_{\mathrm{II}}|_l^2\, d\tau \le \bar{c}^2 \Big(|\mathsf{w}_0 - \mathsf{w}^{\star}|_l^2 \Big), \tag{3.202}$$

and $\sup_{\mathsf{x}\in\mathbb{R}^d} |\mathsf{w}(t,\mathsf{x}) - \mathsf{w}^{\star}|$ *goes to zero as* $t \to \infty$.

Sketch of the Proof The theorem results from a combination of the local existence theorem and the a priori estimates independent of the length of the time interval as in the non augmented situation. □

The resulting estimates are similar to the estimates of hyperbolic-parabolic type obtained by Kawashima et al. for non-augmented higher order linearized hyperbolic-parabolic-dispersive systems [96, 97]. The estimates are found to be similar but the augmented formulation allows a more convenient formulation with simpler and natural symmetrizability assumptions. Stronger estimates of parabolic types have also been obtained by Kawashima et al. [96, 97] with the density ρ estimated in $L^2\big((0,\bar{\tau}), H^{l+2}\big)$ instead of $C^0\big([0,\bar{\tau}], H^{l+1}\big)$. Such estimates may also be investigated with the augmented formulation and will be the object of future studies.

3.5 Conclusion and Perspectives

A synthesis of mathematical fluid models derived form the kinetic theory of gases has been presented. The mathematical models essentially lead to symmetrized hyperbolic-parabolic systems of various forms. In the multicomponent flow situation, a general hyperbolic-parabolic structure has been obtained. For the two-temperature model the system further includes stiff source terms, whereas for diffuse interface fluids the system also includes extra antisymmetric—dispersive—terms. The local existence of strong solutions and the asymptotic stability of equilibrium states have been obtained and reveal the mathematical quality of such fluid models.

Concerning multicomponent flows investigated in Sect. 3.2, various extensions are of high scientific interest. A first extension would be to consider initial-boundary value problems [129] with the possibility of inflow or outflow conditions, heat losses, surface reactions with complex heterogeneous chemistry, species surface diffusion as well as heat surface conduction [13, 130]. Various numerical analysis theoretical results could also be extended to the case of mixtures like convergence results of Petrov-Galerkin Streamline–Diffusion finite element techniques [20, 131]. The notion of higher order entropy may also be generalized to the situation of multicomponent flows [132–135] as well as the singular limit of small Mach number flow [136, 137]. Multiphase flows with sprays governed by Boltzmann type equations [8], or derived multifluid sectional models for droplets [138], may also be investigated mathematically.

More general nonequilibrium systems than the two-temperature model investigated in Sect. 3.3 and derived form the kinetic theory may also be introduced and lead to hyperbolic-parabolic equation with stiff source terms as well as to the apparition of new bulk viscosity effects. The—ultimate—nonequilibrium state to state models where each quantum state of a given species is independent may also be investigated. Relaxation issues between all these thermodynamic nonequilibrium systems are of high scientific interest including rigorous mathematical analyses that have not yet been conducted.

Diffuse interface fluid models may also be generalized to the multicomponent situation and this may include Cahn-Hilliard models. Stronger parabolic type estimates as well as decay estimates may also be investigated by using augmented systems in normal form. Boundary conditions at solid walls should also be investigated taking into account detailed interaction with the wall atoms.

References

1. J. Meixner, Zur Thermodynamik der irreversiblen Prozesse in Gasen mit chemisch reagierenden, dissoziierenden und anregbaren Komponenten. Ann. Phys. **publis 43**, 244–270 (1943)
2. S.R. de Groot, P. Mazur, *Non-Equilibrium Thermodynamics* (Dover publications, Inc. New York, 1984)
3. J. Keizer, *Statistical Thermodynamics of Nonequilibrium Processes* (Springer-Verlag, New York, 1987)
4. G. Ludwig, M. Heil, Boundary layer theory with dissociation and ionization, in *Advances in Applied Mechanics*, vol. VI (Academic Press, New York, 1960), pp. 39–118
5. S. Chapman, T.G. Cowling, *The Mathematical Theory of Non-Uniform Gases* (Cambridge University Press, Cambridge, 1970)
6. J.H. Ferziger, H.G. Kaper, *Mathematical Theory of Transport Processes in Gases* (North Holland Pub. Co., Amsterdam, 1972)
7. V. Giovangigli, *Multicomponent Flow Modeling* (Birkhäuser, Boston, 1999)
8. F.A. Williams, *Combustion Theory*, 2nd edn. (The Benjamin/Cummings Pub. Co. Inc., Menlo park, 1985)
9. A. Ern, V. Giovangigli, Thermal diffusion effects in hydrogen-air flames and methane-air flames. Comb. Theor. Mod. **2**, 349–372 (1997)

10. T. Poinsot, D. Veynante, *Theoretical and Numerical Combustion* (Edwards, Philadelphia, 2005)
11. D.E. Rosner, *Transport Processes in Chemically Reacting Flow Systems* (Butterworths, Boston, 1986)
12. A. Ern, V. Giovangigli, M. Smooke, Numerical study of a three-dimensional chemical vapor deposition reactor with detailed chemistry. J. Comput. Phys. **126**, 21–39 (1996)
13. R.J. Kee, M.E. Coltrin, P. Glarborg, *Chemically Reacting Flow* (Wiley, Hoboken, 2003)
14. J.D. Anderson Jr., *Hypersonics and High Temperature Gas Dynamics* (McGraw-Hill Book Company, New York, 1989)
15. B. Laboudigue, V. Giovangigli, S. Candel, Numerical solution of a free-boundary problem in hypersonic flow theory: nonequilibrium viscous shock layers. J. Comput. Phys. **102**, 297–309 (1992)
16. S. Godunov, An interesting class of quasilinear systems. Sov. Math. Dokl. **2**, 947–949 (1961)
17. K.O. Friedrichs, P.D. Lax, Systems of conservation laws with a convex extension. Proc. Natl. Acad. Sci. USA **68**, 1686–1688 (1971)
18. S. Kawashima, *Systems of Hyperbolic-Parabolic Composite Type, with Application to the Equations of Magnetohydrodynamics*. Doctoral Thesis, Kyoto University, 1984
19. Y. Shizuta, S. Kawashima, Systems of equations of hyperbolic-parabolic type with applications to the discrete boltzmann equation. Hokkaido Math. J. **14**, 249–275 (1985)
20. T.J.R. Hughes, L.P. Franca, M. Mallet, A new finite element formulation for computational fluid dynamics: I. Symmetric forms of the compressible euler and navier-stokes equations and the second law of thermodynamics. Comput. Meth. Appl. Mech. Eng. **54**, 223–234 (1986)
21. T. Ruggeri, Thermodynamics and symmetric hyperbolic systems. Rend. Sem. Mat. Univ. Torino. Hyperbolic Equ. 167–183 (1987)
22. S. Kawashima, Y. Shizuta, On the normal form of the symmetric hyperbolic-parabolic systems associated with the conservation laws. Tôhoku Math. J. **40**, 449–464 (1988)
23. E. Godlewski, P.A. Raviart, *Numerical Approximation of Hyperbolic Systems of Conservation Laws* (Springer-Verlag, Heidelberg, 1996)
24. C. Dafermos, *Hyperbolic Conservation Laws in Continuum Physics* (Springer-Verlag, Heidelberg, 2000)
25. S. Benzoni, D. Serre, *Multi-Dimensional Hyperbolic Partial Differential Equations. First Order Systems and Applications* (Oxford Mathematical Monographs, Oxford, 2007)
26. A.I. Vol'Pert, S.I. Hudjaev, On the cauchy problem for composite systems of nonlinear differential equations. Math USSR Sbornik. **16**, 517–544 (1972)
27. S. Kawashima, Large-time behavior of solutions to hyperbolic-parabolic systems of conservations laws and applications. Proc. Roy. Soc. Edinb. **106A**, 169–1944 (1987)
28. V. Giovangigli, M. Massot, Les Mélanges Gazeux Réactifs, (I) symétrisation et existence locale. C. R. Acad. Sci. Paris Série I **323**, 1153–1158 (1996)
29. V. Giovangigli, M. Massot, Les Mélanges Gazeux Réactifs, (II) stabilité asymptotique des etats d'equilibre. C. R. Acad. Sci. Paris Série I **323**, 1207–1212 (1996)
30. V. Giovangigli, M. Massot, The local cauchy problem for multicomponent reactive flows in full vibrational nonequilibrium. Math. Meth. Appl. Sci. **21**, 1415–1439 (1998)
31. V. Giovangigli, M. Massot, Asymptotic stability of equilibrium states for multicomponent reactive flows. Math. Mod. Meth. Appl. Sci. **8**, 251–297 (1998)
32. V. Giovangigli, B. Graille, Asymptotic stability of equilibrium states for ambipolar plasmas. Math. Mod. Meth. Appl. Sci. **14**, 1361–1399 (2004)
33. E. Feireisl, H. Petzeltová, K. Trivisa, Multicomponent reactive flows: global-in-time existence for large data. Commun. Pure Appl. Anal. **7**, 1017–1047 (2008)
34. Y.S. Kwon, K. Trivisa, Stability and large-time behavior for multicomponent reactive flows. Nonlinearity **22**, 2443–2471 (2009)
35. D. Hoff, Asymptotic behavior of solutions to a model for the flow of a reacting fluid. Archiv. Ration. Mech. Anal. **196**, 951–979 (2010)
36. V. Giovangigli, M. Massot, Entropic structure of multicomponent reactive flows with partial equilibrium reduced chemistry. Math. Meth. Appl. Sci. **27**, 739–768 (2004)

37. K. Beauchard, E. Zuazua, Large time asymptotics for partially dissipative hyperbolic systems. Arch. Rat. Mech. Anal. **199**, 177–227 (2011)
38. V. Giovangigli, Solutions for models of chemically reacting mixtures, in *Handbook of Mathematical Analysis in Mechanics of Viscous Fluids*, ed. by Y. Giga, A. Novotný (Springer-Verlag, Berlin, 2018), pp. 2979–3030
39. A. Novotný, M. Pokorný, Weak solutions for some compressible multicomponent fluid models. Archiv. Ration. Mech. Anal. **47**, 3747–3797 (2020)
40. R. Brun, *Transport et Relaxation dans les Ecoulements Gazeux* (Masson, Paris, 1986)
41. V.M. Zhdanov, *Transport Processes in Multicomponent Plasmas* (Taylor and Francis, London, 2002)
42. G. Colonna, I. Armenise, D. Bruno, M. Capitelli, Reduction of state-to-state kinetics to macroscopic models in hypersonic flows. J. Thermophys. Heat Transf. **20**, 477–486 (2006)
43. E. Nagnibeda, E. Kustova, *Non-Equilibrium Reacting Gas Flows* (Springer Verlag, Berlin, 2009)
44. D. Bruno, V. Giovangigli, Relaxation of internal temperature and volume viscosity. Phys. Fluids **23**, 093104 (2011)
45. D. Bruno, V. Giovangigli, Erratum: "Relaxation of internal temperature and volume viscosity". Phys. Fluids **25**, 039902 (2013)
46. D. Bruno, F. Esposito, V. Giovangigli, Relaxation of rotational-vibrational energy and volume viscosity in H-H_2 mixtures. J. Chem. Phys. **138**, 084302 (2013)
47. T.P. Liu, Hyperbolic conservation laws with relaxation. Commun. Math. Physics **108**, 153–175 (1987)
48. G.Q. Chen, C.D. Levermore, T.P. Liu, Hyperbolic conservation laws with stiff relaxation terms and entropy. Commun. Pure Appl. Math. **47**, 787–830 (1994)
49. D. Donatelli, P. Marcati, Convergence of singular limits for multi-d semilinear hyperbolic systems to parabolic systems. Trans. Am. Soc. **356**, 2093–2121 (2004)
50. E. Feireisl, A. Novotný, The low mach number limit for the full Navier-Stokes-fourier system. Arch. Ration. Mech. Anal. **186**, 77–107 (2007)
51. Z.B. Yang, W.A. Yong, Validity of the chapman-enskog expansion for a class of hyperbolic relaxation systems. J. Differ. Equ. **258**, 2745–2766 (2015)
52. V. Giovangigli, W.A. Yong, Volume viscosity and fast internal energy relaxation: symmetrization and Chapman-Enskog expansion. Kin. Rel. Mod. **8**, 79–116 (2015)
53. V. Giovangigli, W.A. Yong, Volume viscosity and fast internal energy relaxation: error estimates. Nonlinear Anal. Real World Appl. **43**, 213–244 (2018)
54. V. Giovangigli, W.A. Yong, Asymptotic stability and relaxation for fast chemistry fluids. Nonlinear Anal. **159**, 208–263 (2017)
55. G. Billet, V. Giovangigli, G. de Gassowski, Impact of volume viscosity on a shock/hydrogen bubble interaction. Combustion Theory Modell. **12**, 221–248 (2008)
56. E. Kustova, On the role of bulk viscosity and relaxation pressure in nonequilibrium flows, in *26th International RGD Symposium, AIP Conference Proceedings*, ed. by T. Abe, vol. 1024 (2009), pp. 807–812
57. D. Bruno, V. Giovangigli, Internal energy relaxation processes and bulk viscosities in fluids. MDPI Fluids **7**, 356 (2022)
58. R.E. O'Malley Jr., *Introduction to Singular Perturbations* (Academic Press, New York, 1974)
59. W. Eckhaus, *Matching Principles and Composite Expansions*. Lecture Notes in Mathematics, vol. 594 (Springer-Verlag, Berlin, 1977) , pp. 146–177
60. V. Giovangigli, Z.B. Yang, W.A. Yong, Relaxation limit and initial-layer for a class of hyperbolic-parabolic systems. SIAM J. Math. Anal. **50**, 4655–4697 (2018)
61. J.D. van der Waals, Thermodynamische theorie der capillariteit in de onderstelling van continue. Dichtheidsverandering Verhand. Kon. Akad. V Wetensch. Amst. (1893) (in Dutch). English translation in J. Stat. Phys. **20**, 197–244 (1979)
62. D.J. Korteweg, Sur la Forme que Prennent les Equations du Mouvement Fluide si l'on tient Compte de Forces Capillaires Causées par les Variations de Densité Considérables mais Continues et sur la Théorie de la Capillarité dans l'Hypothèse d'une Variations Continue de la Densité. Arch. Neerl. Sci Exactes **6**, 1–20 (1901)

63. J.E. Dunn, J. Serrin, On the thermomechanics of interstitial working. Archive Ration. Mech. Anal. **133**, 95–133 (1985)
64. J.S. Rowlinson, B. Widom, *Molecular Theory of Capillarity* (Dover publications, Mineola, 1989/2002)
65. S. Gavrilyuk, S. Shugrin, Media with equations of state that depend on derivatives. J. Appl. Mech. Tech. Phys. **37**, 179–189 (1996)
66. D.M. Anderson, G.B. McFadden, A.A. Wheeler, Diffuse interface methods in fluid mechanics. Ann. Rev. Fluid Mech. **30**, 139–165 (1998)
67. D. Jamet, Diffuse interface models in fluid mechanics. Semantic Scholar Corpus Id: 18437499 (2005)
68. P. Gaillard, V. Giovangigli, L. Matuszewski, A diffuse interface Lox/Hydrogen transcritical flame model. Combust. Theory Model. **20**, 486–520 (2016)
69. P. Gaillard, V. Giovangigli, L. Matuszewski, Erratum for: a diffuse interface Lox/Hydrogen transcritical flame model. https://hal.science/hal-04198003
70. P. Barbante, A. Frezzotti, A comparison of models for the evaporation of a Lennard-Jones fluid. Eur. J. Mech. B. Fluids **64**, 69–80 (2017)
71. V. Giovangigli, Kinetic derivation of diffuse-interface fluid models. Phys. Rev. E **102**, 012110 (2020)
72. V. Giovangigli, Kinetic derivation of cahn-hilliard fluid models. Phys. Rev. E **104**, 054109 (2021)
73. V. Giovangigli, A synthesis of the kinetic derivation of Cahn-Hilliard equations. 32nd RGD conference. AIP Conf. Proc. **2996**, 040002 (2024)
74. D. Nayigizente, S. Ducruix, T. Schmitt, Development of an interface thickening method for the direct numerical simulation of compressible liquid-vapor flows in the framework of the second gradient theory. Phys. Fluids **33**, 052119 (2021)
75. H. Hattori, D. Li, Solutions for two-dimensional system for materials of korteweg type. J. Partial Differ. Equ. **9**, 323–342 (1996)
76. S. Gavrilyuk, H. Gouin, Symmetric form of governing equations for capillary fluids, in *Trends in Applications of Mathematics to Mechanics (Nice, 1998)*. Monographs and Surveys in Pure and Applied Mathematics, vol. 106 (Chapman & Hall/CRC, Boca Raton, 2000), pp. 306–311
77. R. Danchin, B. Desjardins, Existence of solutions for compressible fluid models of Korteweg type. Ann. IHP Anal. Non linéaire **18**, 97–133 (2001)
78. D. Bresch, B. Desjardins, C.K. Lin, On some compressible fluid models: korteweg, lubrication and shallow water systems. Commun. Part. Differ. Equ. **28**, 843–868 (2003)
79. S. Benzoni-Gavage, R. Danchin, S. Descombes, D. Jamet, Structure of Korteweg models and stability of diffuse interfaces. Interfaces Free Bound. **7**, 371–414 (2005)
80. S. Benzoni-Gavage, R. Danchin, S. Descombes, On the well-posedness of the euler-korteweg model in several space dimensions. Indiana J. Math. **56**, 1499–1579 (2007)
81. D. Bresch, F. Couderc, P. Noble , J.-P. Vila, A generalization of the quantum bohm identity: hyperbolic CFL condition for Euler-Korteweg equations. C. Rend. Acad. Sci. **354**, 39–43 (2008)
82. M. Kotschote, Strong solutions for a compressible fluid model of Korteweg type. Anal. Inst. H. Poincaré **25**, 679–696 (2008)
83. B. Haspot, Existence of strong solutions for nonisothermal Korteweg system. Ann. Math. Blaise Pascal **16**, 431–481 (2009)
84. M. Kotschote, Dynamics of compressible non-isothermal fluids of non-Newtonian Korteweg type. SIAM J. Math. Anal. **44**, 74–101 (2012)
85. D. Donatelli, E. Feireisl, P. Marcati, Well/Ill posedness for the Euler-Korteweg-Poisson system and related problems. Commun. Partial Differ. Equ. **40**, 1314–1335 (2014)
86. J. Giesselmann, A. Tzavaras, Stability properties of the Euler-Korteweg system with non-monotone pressures. App. Anal. **96**, 1528–1546 (2017)
87. T. Dębiec, P. Gwiazda, A. Świerczewska-Gwiazda, A. Tzavaras, Conservation of energy for the Euler-Korteweg equations. Cal. Var. **57**, 160 (2018)

88. D. Bresch, M. Gisclon, I. Lacroix-Violet, Navier-Stokes-Korteweg and Euler-Korteweg systems: application to quantum fluids. Arch. Rat. Mech. Anal. **233**, 975–1025 (2019)
89. J. Gao, Z. Lyu, Z. Yao, Lower bound of decay rate for higher-order derivatives of solution to the compressible fluid models of Korteweg type. Z. Angew. Math. Phys. **71**, 108 (2020)
90. V. Giovangigli, Y. Le Calvez, F. Nabet, Symmetrization and local existence of strong solutions for diffuse interface fluid models. J. Math. Fluid Mech. **25**, 82 (2023)
91. R. Pego, Stable viscosities and shock profiles for systems of conservation laws. Trans. Am. Math. Soc. **282**, 749–763 (1984)
92. H. Hattori, D. Li, Global solutions of a high dimensional system for korteweg materials. J. Math. Anal. **198**, 84–97 (1996)
93. J. Humpherys, Admissibility of viscous-dispersive systems. J. Hyp. Differ. Equ. **2**, 963–974 (2005)
94. Y. Wang, Z. Tan, Optimal decay rates for the compressible fluid models of Korteweg type. J. Math. Anal. App. **379**, 256–271 (2011)
95. M. Kotschote, Existence and time-asymptotics of global strong solutions to dynamic korteweg models. Indiana J. Math. **63**, 21–51 (2014)
96. S. Kawashima, Y. Shibata, J. Xiu, The Lp energy methods and decay for the compressible Navier-Stokes equations with capillarity. J. Math. Pure App. **154**, 146–184 (2021)
97. S. Kawashima, Y. Shibata, J. Xiu, Dissipative structure for symmetric hyperbolic-parabolic systems with Korteweg-Type dispersion. Commun. Partial Differ. Equ. **47**, 378–400 (2022)
98. R.G. Plaza, and J.M. Valdovinos, Dissipative structure of one-dimensional isothermal compressible fluids of Korteweg type. J. Math. Anal. App. **514**, 126336 (2022)
99. V. Giovangigli, B. Tran, Mathematical analysis of a Saint-Venant model with variable temperature. Math. Mod. Meth. Appl. Sci. **20**, 1–47 (2010)
100. V. Giovangigli, Plane flames with multicomponent transport and complex chemistry. Math. Mod. Meth. Appl. Sci. **9**, 337–378 (1999)
101. E. Zatorska, On the steady flow of a multicomponent, compressible, chemically reacting gas. Nonlinearity **24**, 3267–3278 (2011)
102. E. Zatorska, On the flow of chemically reacting gaseous mixture. J. Differ. Equ. **253**, 3471–3500 (2012)
103. X. Chen, A. Jüngel, Analysis of an incompressible Navier-Stokes-Maxwell-Stefan system. Commun. Math. Phys. **340**, 471–497 (2015)
104. V. Giovangigli, M. Pokorný, E. Zatorska, On steady flow of reactive gaseous mixture. Analysis **35**, 319–341 (2015)
105. M. Marion, R. Temam, Global existence for fully nonlinear reaction-diffusion systems describing multicomponent reactive flows. J. Math. Pures App. **105**, 102–138 (2015)
106. E. Zatorska, Mixtures: sequential stability of variational entropy solutions. J. Math. Fluid Dyn. **17**, 437–461 (2015)
107. P.B. Mucha, M. Pokorný, E. Zatorska, Heat-conducting, compressible mixtures with multicomponent diffusion: construction of a weak solution. SIAM J. Math. Anal. **47**, 3747–3797 (2015)
108. V. Giovangigli, Mass conservation and singular multicomponent diffusion algorithms. IMPACT Comput. Sci. Eng. **2**, 73–97 (1990)
109. V. Giovangigli, Convergent iterative methods for multicomponent diffusion. IMPACT Comput. Sci. Eng. **3**, 244–276 (1991)
110. A. Ern, V. Giovangigli, *Multicomponent Transport Algorithms*. Lecture Notes in Physics, Series Monographs, vol. m 24 (Springer, Berlin, 1994)
111. A. Ern, V. Giovangigli, The structure of transport linear systems in dilute isotropic gas mixtures. Phys. Rev. E **53**, 485–492 (1996)
112. A. Ern, V. Giovangigli, Projected iterative algorithms with application to multicomponent transport. Linear Algebra Appl. **250**, 289–315 (1997)
113. V. Giovangigli, Multicomponent transport algorithms for partially ionized plasmas. J. Comput. Phys. **229**, 4117–4142 (2010)

114. S. Dworkin, M.D. Smooke, V. Giovangigli, The impact of detailed multicomponent transport and thermal diffusion effects on soot formation in ethylene/air flames. Proc. Combin. Inst. **32**, 1165–1172 (2009)
115. R. Marcelin, Sur la Mécanique des Phénomènes Irréversibles. C. R. Acad. Sci. Paris 1052–1055 (1910)
116. A.N. Gorban, V.I. Bykov, G.S. Yablonski, *Essays on Chemical Relaxation* (in Russian) (Nauka, Novosibirsk, 1986)
117. A. Ern, V. Giovangigli, The Kinetic Equilibrium Regime. Physica-A **260**, 49–72 (1998)
118. V. Giovangigli, L. Matuszewski, Structure of entropies in dissipative multicomponent fluids. Kin. Rel. Mod. **6**, 373–406 (2013)
119. A. Kolmogorov, S.V. Fomin, *Introduction to Functional Analysis*, vol. 1 (Graylock press, New York, 1957)
120. V.W. Bryant, A remark on a fixed-point theorem for iterated mappings. Am. Math. Monthly **75**, 399–400 (1968)
121. E.A. Guggenheim, *Thermodynamics* (North-Holland, Amsterdam, 1967)
122. V. Giovangigli, L. Matuszewski, F. Dupoirieux, Detailed modeling of planar transcritical H_2-O_2-N_2 flames. Combustion Theory Modelling **15**, 141–182 (2011)
123. V. Giovangigli, L. Matuszewski, Supercritical fluid thermodynamics from equations of state. Physica D **241**, 649–670 (2012)
124. V. Giovangigli, L. Matuszewski, Mathematical modeling of supercritical multicomponent reactive fluids. Math. Mod. Meth. App. Sci. **23**, 2193–2251 (2013)
125. V. Giovangigli, Y. Le Calvez, G. Ribert, Multicomponent thermodynamics with instabilities and diffuse interfaces fluids. AIMS Math. **9**, 25979–26034 (2024)
126. D. Bresch, V. Giovangigli, E. Zatorska, Two-velocity hydrodynamics in fluid mechanics: part I well posedness for zero mach number systems. J. Math. Pures App. **104**, 762–800 (2015)
127. D. Bresch, B. Desjardins, E. Zatorska, Two-velocity hydrodynamics in fluid mechanics: part II existence of global κ-entropy solutions to the compressible Navier-Stokes system with degenerate viscosities. J. Math. Pures App. **104**, 801–836 (2015)
128. H. Lin, Y.Y. Duan, Q. Min, Gradient theory modeling of surface tension for pure fluids and binary mixtures. Fluid Phase Equil. **254**, 75–90 (2007)
129. A. Matsumura, T. Nishida, The initial boundary value problem for the equations of motion of a viscous and heat-conductive fluids. Commun. Math. Phys. **89**, 445–464 (1983)
130. K. Aoki, V. Giovangigli, Kinetic theory of chemical reactions on a crystal surface. Physica A **565**, 125573 (2021)
131. C. Johnson, A. Szepessy, P. Hansbo, On the convergence of shock-capturing streamline diffusion finite element methods for hyperbolic conservation laws. Math. Comput. **54**, 107–129 (1990)
132. D. Bresch, B. Desjardins, On the existence of global weak solutions to the Navier-Stokes equations for viscous compressible and heat conducting fluids. J. Math. Pure Appl. **87**, 57–90 (2007)
133. V. Giovangigli, Higher order entropies. Arch. Ration. Mech. Anal. **187**, 221–285 (2008)
134. V. Giovangigli, Higher order entropies for compressible fluid models. Math. Mod. Meth. Appl. Sci. **19**, 67–125 (2009)
135. L.C. Evans, A survey of entropy methods for partial differential equations. Bull. AMS **41**, 409–438 (2004)
136. T. Alazard, Low mach number limit of the full Navier-Stokes equations. Arch. Ration. Mech. Anal. **180**, 1–73 (2006)
137. E. Feireisl, A. Novotný, *Singular Limits in Thermodynamics of Viscous Fluids* (Birkhäuser, Basel, 2009)
138. F. Laurent, M. Massot, Multi-fluid modeling of laminar poly-dispersed spray flames: origin, assumptions and comparison of the sectional and sampling methods. Combin. Theor. Mod. **5**, 537–572 (2001)

Part II
Lectures from Kácov School 2019

This part includes two shorter lecture notes based on series of lectures from the EMS school Mathematical Aspects of Fluid Flows held in the Czech Republic, in Kácov in May 2019. The subsequent period was influenced by the pandemic situation; therefore we were able to collect only two shorter contributions, the tradition of these schools was for a few years interrupted and restarted again in 2024. After discussions with the authors we decided to publish these contributions in this volume in the form as they were presented to us at the beginning of the year 2020, without any changes, in particular in the references. They therefore reflect the state of the art in that period.

Chapter 4
Lecture Notes on Intermittent Weak Solutions of the Three-dimensional Navier-Stokes Equations

Tristan Buckmaster, Maria Colombo, and Vlad Vicol

Abstract This summary closely follows the paper Buckmaster et al. (Wild solutions of the Navier-Stokes equations whose singular sets in time have hausdorff dimension strictly less than 1, 2018. ArXiv:1809.00600) by the authors, which contains all the details regarding the specific gluing and convex integration technique used here. The lectures will be supplemented with material from the recent review paper Buckmaster and Vicol (EMS Surv. Math. Sci. 6(1/2):173–263, 2019), which contains the necessary background concerning weak solutions of the Euler and Navier-Stokes equations, their connection to phenomenological theories of fully developed turbulence, and the convex integration methods used to construct them.

4.1 Introduction

The Navier-Stokes equations are the classical model for the motion of a homogenous incompressible viscous fluid. Written in divergence form, the equations read as

$$\partial_t v + \nabla \cdot (v \otimes v) + \nabla p - \Delta v = 0 \tag{4.1a}$$

$$\nabla \cdot v = 0 \tag{4.1b}$$

$$v|_{t=0} = v_0 \tag{4.1c}$$

where v is the velocity field, p is the pressure, and v_0 is a divergence free initial condition. For simplicity, we have set the kinematic viscosity to equal unity. We consider the equations posed on the three-dimensional periodic box $\mathbb{T}^3 = [-\pi, \pi]^3$,

T. Buckmaster · V. Vicol (✉)
Courant Institute of Mathematical Sciences, New York University, New York, NY, USA
e-mail: buckmaster@cims.nyu.edu; vicol@cims.nyu.edu

M. Colombo
École Polytechnique Fédérale de Lausanne, Faculté des Sciences de Base EPFL PH A2 364 (Bâtiment PH), Lausanne, Switzerland
e-mail: maria.colombo@epfl.ch

M. Bulíček et al. (eds.), *New Perspectives in Mathematical Fluid Mechanics*, Nečas Center Series, https://doi.org/10.1007/978-3-032-02319-3_4

with periodic boundary conditions for all functions. The pressure is uniquely given by the formula

$$p = \nabla \cdot (-\Delta)^{-1} \nabla \cdot (v \otimes v) \tag{4.2}$$

once one assumes the normalization $\int_{\mathbb{T}^3} p(x,t)dx = 0$. The initial datum v_0, and hence also the solution $v(\cdot,t)$, are normalized to have zero mean on $\mathbb{T}^3$, i.e. $\int_{\mathbb{T}^3} v(x,t)dx = 0$ for all $t \geq 0$.

The literature concerning the analysis of the incompressible Navier-Stokes equations is too vast to review here. We refer the interested reader to the books [8, 11, 19, 21, 26, 27, 30, 33, 34, 38, 42–44] and will only recall here our main result. There also exist a number of excellent survey articles on the subject of convex integration in fluid dynamics [14, 17, 18, 41]. We refer to interested reader to the recent review [2] for the most recent developments.

First let us recall the notion of *distributional/weak solution* to (4.1) used in these notes (see also [2, 3, 6]):

Definition 4.1 (Continuous in Time Distributional/Weak Solution) A vector field $v \in C^0([0,T); L^2(\mathbb{T}^3))$ is called a continuous in time weak solution of the Cauchy problem for the Navier-Stokes equations (4.1), with incompressible initial datum $v_0 \in L^2(\mathbb{T}^3)$, if for any $t \in [0,T)$ the vector field $v(\cdot,t)$ is weakly divergence free, has zero mean, and

$$\int_{\mathbb{T}^3} v_0 \cdot \varphi(\cdot,0) + \int_0^T \int_{\mathbb{T}^3} v \cdot (\partial_t \varphi + (v \cdot \nabla)\varphi + \Delta\varphi) dx dt = 0$$

holds for any divergence free test function $\varphi \in C_0^\infty([0,T); C^\infty(\mathbb{T}^3))$.

It is known since the work [20] that weak solutions in the sense of (4.1) are the same as *mild* or *Oseen* solutions (cf. [27, Definition 6.5]), meaning that they satisfy

$$v(\cdot,t) = e^{\Delta t} v(\cdot,0) + \int_0^t e^{\Delta(t-s)} \mathbb{P}_H \nabla \cdot (v(\cdot,s) \otimes v(\cdot,s)) ds\,, \tag{4.3}$$

on $[0,T)$. Here and throughout the paper, $\mathbb{P}_H$ denotes the Helmholtz-Leray projector. We emphasize that *solutions in the sense of Definition 4.1 are not necessarily weak solutions in the sense of Leray [28] and Hopf [22]*: first, because we did not assume $v \in L^2((0,T); \dot{H}^1(\mathbb{T}^3))$, and second, because v is not required to obey the energy inequality.

The main result presented in these lectures is:

Theorem 4.1 (Theorem 1.1 in [6]) *There exists $\beta > 0$ such that the following holds. For $T > 0$, let $u^{(1)}, u^{(2)} \in C^0([0,T]; H^3(\mathbb{T}^3))$ be two strong solutions of the Navier-Stokes equations* (4.1) *on* $[0,T]$ *with* $\int_{\mathbb{T}^3} u^{(1)}(0,x)dx = \int_{\mathbb{T}^3} u^{(2)}(0,x)dx$. *There exist*

- *a weak solution* v *of* (4.1) *on* $[0, T]$ *which has the additional regularity*

$$v \in C^0([0, T]; H^\beta(\mathbb{T}^3) \cap W^{1,1+\beta}(\mathbb{T}^3)),$$

and such that

$$v \equiv u^{(1)} \quad on \quad [0, T/3], \qquad and \qquad v \equiv u^{(2)} \quad on \quad [2T/3, T].$$

- *a compact nonempty set of times* $\Sigma_T \subset (0, T]$ *with box-counting (Minkowski) dimension less than* $1 - \beta$ *such that*

$$v \in C^\infty\big(\mathbb{T}^3 \times ((0, T] \setminus \Sigma_T)\big).$$

In particular, the weak solution v *is almost everywhere smooth.*

The above result implies that weak solutions of (4.1) in the sense of Definition 4.1 are not unique, even if the initial datum is strong. Moreover, the above result gives the first example of a weak solution for the Navier-Stokes equations (4.1) whose singular set of times Σ_T is both not empty and has a Hausdorff dimension strictly less than 1.

For the sake of brevity, in this outline we refer to the introductions of [6] and [2] for an adequate historical account, further references, and a list of problems which remain open.

The proof of Theorem 4.1 is based on an *intermittent convex integration scheme*, inspired by the one in [3]. This builds on the long line of works sparked by the groundbreaking result of De Lellis and Székelyhidi Jr. [15], which gave more robust versions of Scheffer's [39] and Schnirelman's [40] famous counterexamples to uniqueness of weak solutions to the three dimensional Euler equations. Within the context of Euler equations, the development of convex integration techniques [1, 4, 5, 13, 16] has culminated in the resolution of Onsager's conjecture by Isett [24] (see also [7, 23]). The main new idea of [6] is to combine the intermittent convex integration scheme of [3] with a variant the gluing step used in [7, 24]. An additional novelty of [6] is the introduction of *intermittent jets* which replace the intermittent Beltrami flows of [3] as the fundamental building blocks on which the convex integration scheme in based. Recently, this intermittent convex integration approach has proven to be very fruitful in constructing wild weak solutions to other classes of supercritical PDEs [9, 12, 31, 32, 35, 36]. We refer to [2] for further references.

4.2 Outline of the Proof

The rest of this note is dedicated to the proof of Theorem 4.1. We follow [6] without making any significant changes: we only outline here the main arguments, and refer to [6] for details. As opposed to [6], where we also discuss the hyperdissipative

Navier-Stokes equation (see also [10, 25, 29, 32]) with dissipation given by $(-\Delta)^{\alpha} v$ for $\alpha \in [1, 5/4)$, here we only discuss the classical Navier-Stokes system, with $\alpha = 1$.

4.2.1 Parameters and Notation

The proof makes use of a number of parameters, listed below. While the meaning of these parameters will only become apparent later on, for the reader's convenience we collect the definitions of all parameters here.

- $b > 1$ is a fixed sufficiently large integer; this parameter is universal.
- $\beta = \beta(b) > 0$ is a fixed sufficiently small parameter such that $\beta b^2 \ll 1$; the implicit constant is universal.
- $\varepsilon_R = \varepsilon_R(b, \beta) \in (0, 1)$ is a sufficiently small parameter, such that $\varepsilon_R b \ll 1$.
- $a \geq a_0 = a_0(b, \beta, \varepsilon_R) \geq 1$ is a parameter which has to be taken sufficiently large (see Proposition 4.1); for convenience we may choose this parameter such that $a^{1/24} \in \mathbb{N}$.
- λ_q is a (Fourier) *frequency parameter* defined for $q \in \mathbb{N}$ by the super-exponentially growing sequence

$$\lambda_q = a^{(b^q)} . \tag{4.4}$$

- δ_q is an *amplitude parameter* defined as

$$\delta_q = \lambda_1^{3\beta} \lambda_q^{-2\beta} . \tag{4.5}$$

  Note that $\delta_1 = \lambda_1^{\beta} = a^{\beta b}$ can be made arbitrarily large by taking a sufficiently large.
- ϑ_q is a *slow temporal cutoff parameter* defined for $q \geq 1$ as

$$\vartheta_q = \lambda_{q-1}^{-7} \delta_q^{1/2} . \tag{4.6}$$

  We do not define ϑ_0.
- τ_q is a *fast temporal cutoff parameter* defined for $q \geq 1$ by

$$\tau_q = \vartheta_q \lambda_{q-1}^{-\varepsilon_R/4} = \lambda_{q-1}^{-7-\varepsilon_R/4} \delta_q^{1/2} . \tag{4.7}$$

  For the special case $q = 0$ we set $\tau_0 := T/15$. Note that by construction we automatically have

$$\vartheta_{q+1} \ll \tau_q \ll \vartheta_q \ll 1 . \tag{4.8}$$

- $r_\perp \ll r_\| \in (\lambda_{q+1}^{-1}, 1)$ are the *lateral* and the *longitudinal* characteristic length scales of the intermittent jets constructed in Sect. 4.4.1. They are defined explicitly as

$$r_\perp = \lambda_{q+1}^{-19/24} \qquad \text{and} \qquad r_\| = \lambda_{q+1}^{-7/12}. \tag{4.9}$$

  Note that we have $\lambda_{q+1} r_\perp = \lambda_{q+1}^{5/24} \in \mathbb{N}$ because it equals an integer power of $a^{\frac{1}{24}} \in \mathbb{N}$.
- $\mu \geq 1$ is a *temporal oscillation* parameter for the intermittent jets defined in Sect. 4.4.1, given by

$$\mu = \frac{\lambda_{q+1} r_\|}{r_\perp} = \lambda_{q+1}^{29/24}. \tag{4.10}$$

Throughout the paper we shall denote by $\mathbb{P}_H$ the Helmholtz projection and by $\mathbb{P}_{\neq 0}$ the projection onto functions with zero mean. That is,

$$\mathbb{P}_H f = f - \nabla(\Delta^{-1}\nabla \cdot f) \qquad \text{and} \qquad \mathbb{P}_{\neq 0} f = f - \fint_{\mathbb{T}^3} f dx.$$

We also recall from [15, Definition 4.2] the *inverse divergence* operator $\mathcal{R}$, acting on vector fields v with zero mean as

$$\mathcal{R}v = \Delta^{-1}\left(\nabla v + (\nabla v)^T\right) - \frac{1}{2}\left(\mathrm{Id} + \nabla\nabla\Delta^{-1}\right)\nabla \cdot \Delta^{-1} v \tag{4.11}$$

Clearly, $\mathcal{R}v(x)$ is a symmetric trace-free matrix for each $x \in \mathbb{T}^3$, and $\mathcal{R}$ is a right inverse of the $\nabla\cdot$ operator. We overload notation and denote $\mathcal{R}v := \mathcal{R}(v - \fint_{\mathbb{T}^3} v dx)$ even when v does not have zero mean.

We use the convention that the divergence operator contracts on a tensor $S = (S^{ij})_{i,j=1}^3$ on the second component, i.e. $(\nabla \cdot S)^i = \partial_j S^{ij}$. The summation convention on repeated indices is used throughout the manuscript. The traceless part of the tensor $a \otimes b$ will be denoted by $a \mathring{\otimes} b$. Unless explicitly stated, all the norms we consider are C^0 in time, i.e. $\|\cdot\|_{H^s}$ or $\|\cdot\|_{L^p}$ is written instead of $\|\cdot\|_{C_t^0 H_x^s}$ or $\|\cdot\|_{C_t^0 L_x^p}$. Lastly, we remark that the implicit constants that appear in the $\lesssim$ symbol are always taken to be independent of q (or any parameter which depends on q), and they are also independent of the parameter a.

4.2.2 Inductive Assumptions on the Relaxed Navier-Stokes Equations

For every integer $q \geq 0$ our goal is to find an approximate solution $(v_q, \mathring{R}_q)$ to the Navier-Stokes-Reynolds system

$$\partial_t v_q + \nabla \cdot (v_q \otimes v_q) + \nabla p_q - \Delta v_q = \nabla \cdot \mathring{R}_q \tag{4.12a}$$
$$\nabla \cdot v_q = 0\,, \tag{4.12b}$$

where $\mathring{R}_q$ is a traceless symmetric 3×3 matrix and the pressure p_q is given by $p_q = \nabla \cdot \Delta^{-1} \nabla \cdot (\mathring{R}_q - v_q \otimes v_q)$. We inductively assume that the Reynolds stress $\mathring{R}_q$ obeys the bounds

$$\left\| \mathring{R}_q \right\|_{L^1(\mathbb{T}^3)} \leq \lambda_q^{-\varepsilon_R} \delta_{q+1} \tag{4.13a}$$
$$\left\| \mathring{R}_q \right\|_{H^3(\mathbb{T}^3)} \leq \lambda_q^7 \tag{4.13b}$$

and that the approximate velocity field v_q satisfies the estimates

$$\left\| v_q \right\|_{L^2(\mathbb{T}^3)} \leq 2\delta_0^{1/2} - \delta_q^{1/2} \tag{4.14a}$$
$$\left\| v_q \right\|_{H^3(\mathbb{T}^3)} \leq \lambda_q^4\,. \tag{4.14b}$$

Additionally, in order to show that the limiting weak solution $v = \lim_{q \to \infty} v_q$ is smooth for almost every time, with every $q \in \mathbb{N}$ we also inductively construct a relatively open *good set* $\mathcal{G}^{(q)}$ and a closed *bad set*

$$\mathcal{B}^{(q)} = [0, T] \setminus \mathcal{G}^{(q)}$$

which obey the following properties:

1. $\mathcal{G}^{(0)} = [0, T/3) \cup (2T/3, T]$.
2. $\mathcal{G}^{(q-1)} \subset \mathcal{G}^{(q)}$ for every $q \geq 1$.
3. $\mathcal{B}^{(q)}$ is a finite union of disjoint closed intervals of length $5\tau_q$.
4. For $q \geq 1$, the bad sets have Lebesgue measures which obey

$$|\mathcal{B}^{(q)}| \leq |\mathcal{B}^{(q-1)}| \frac{10\tau_q}{\vartheta_q}\,. \tag{4.15}$$

5. The velocity fields obey

$$\text{if} \quad t \in \mathcal{G}^{(q')} \quad \text{for some} \quad q' < q, \quad \text{then} \quad v_q(t) = v_{q'}(t)\,. \tag{4.16}$$

6. The residual Reynolds stress obeys

$$\mathring{R}_q(t) = 0 \quad \text{for all} \quad t \in [0, T] \quad \text{such that} \quad \operatorname{dist}(t, \mathcal{G}^{(q)}) \leq \tau_q\,. \tag{4.17}$$

The main ingredient in the proof of Theorem 4.1 is the following proposition, which states that the inductive assumptions listed in this section may be propagated from step q to step $q+1$. We have

Proposition 4.1 *There exists $b \gg 1$, $0 < \beta = \beta(b) \ll 1$, $0 < \varepsilon_R = \varepsilon_R(b, \beta) \ll 1$, and $a_0 = a_0(b, \beta, \varepsilon_R) \gg 1$ such that for every $a \geq a_0$ the following holds. Let $(v_q, \mathring{R}_q)$ be a pair solving the Navier-Stokes-Reynolds system* (4.12) *which satisfies* (4.13a)–(4.14b), *and has a corresponding good set $\mathcal{G}^{(q)}$ with the properties 1–6. Then there exists a second pair $(v_{q+1}, \mathring{R}_{q+1})$ solving* (4.12) *and a corresponding good set $\mathcal{G}^{(q+1)}$ which satisfies* (4.13a)–(4.14b), *respectively 1–6, with q replaced by $q+1$. In addition the velocity increment obeys*

$$\left\| v_{q+1} - v_q \right\|_{L^2} \leq \delta_{q+1}^{1/2} . \tag{4.18}$$

The proof of Proposition 4.1 is the bulk of the paper, and consists of two steps: a gluing step which is detailed in Sect. 4.3, and a convex integration step which is detailed in Sect. 4.4. Assuming for the moment that Proposition 4.1 is true, we show next that it implies the main result in the paper.

4.2.3 *Proof of Theorem 4.1*

Let $b, \beta, \varepsilon_R, a_0$ be as in Proposition 4.1. Let $\eta \colon [0, T] \to [0, 1]$ be a smooth cutoff function such that $\eta = 1$ on $[0, {}^{2T}/_{5}]$ and $\eta = 0$ on $[{}^{3T}/_{5}, T]$, and define

$$\begin{aligned} v_0 &= \eta\, u^{(1)} + (1-\eta)\, u^{(2)} \\ \mathring{R}_0 &= \partial_t \eta\, \mathcal{R}(u^{(1)} - u^{(2)}) - \eta(1-\eta)(u^{(1)} - u^{(2)}) \mathring{\otimes} (u^{(1)} - u^{(2)}) , \end{aligned}$$

where $u^{(1)}$ and $u^{(2)}$ are as in the statement of the theorem. Since $u^{(1)}$ and $u^{(2)}$ have the same average at time 0, the same holds for any $t > 0$. Hence, by construction, the pair $(v_0, \mathring{R}_0)$ solves (4.12), for a suitable pressure p_0. Choosing $a \geq a_0$ sufficiently large, it is not hard to check that $(v_0, \mathring{R}_0)$ satisfies the bounds (4.13a)–(4.14b). From its definition, $\mathring{R}_0$ is supported on the interval $[{}^{2T}/_{5}, {}^{3T}/_{5}]$ and thus properties (3) and (6) hold when $q = 0$.

For $q \geq 1$ we inductively apply Proposition 4.1. From (4.14b), (4.18), and interpolation we have that

$$\begin{aligned} \sum_{q=0}^{\infty} \left\| v_{q+1} - v_q \right\|_{\dot{H}^{\beta'}} &\lesssim \sum_{q=0}^{\infty} \left\| v_{q+1} - v_q \right\|_{L^2}^{1-\beta'/3} \left(\left\| v_{q+1} \right\|_{\dot{H}^3} + \left\| v_q \right\|_{\dot{H}^3} \right)^{\beta'/3} \\ &\lesssim \sum_{q=0}^{\infty} \lambda_{q+1}^{-\frac{\beta(3-\beta')}{6} + \frac{4\beta'}{3}} \lesssim 1 \end{aligned}$$

for $0 \leq \beta' < \frac{3\beta}{8+\beta}$. Hence there exists a strong limit

$$v := \lim_{q\to\infty} v_q \in C^0([0,T]; H^{\beta'}(\mathbb{T}^3))$$

which in view of the $C^0_t L^1_x$ convergence $\mathring{R}_q \to 0$ as $q \to \infty$ is a weak solution of the Navier-Stokes equation in the sense of Definition 4.1. From (1), (5), and the definition of v_0 we have immediately have that $v \equiv u^{(1)}$ on $[0, T/3]$, and $v \equiv u^{(2)}$ on $[2T/3, T]$. Also, we note that for some $\beta'' > 0$ sufficiently small (in terms of β' and β), we have that $v \in C^0(0,T; W^{1,1+\beta''}(\mathbb{T}^3))$. This follows from the maximal regularity of the heat equation, once we note the estimate $\|\mathbb{P}_H(v\otimes v)\|_{L^{1+\beta''}} \lesssim \|v\|^2_{H^{\beta'}}$ (which holds if β' and β'' are chosen suitably).

Lastly, we define the *singular set of times* Σ_T of the limiting solution v as

$$\Sigma_T = \bigcap_{q\geq 0} \mathcal{B}^{(q)}\,. \tag{4.19}$$

Note that for every $t \in (0,T)\setminus\Sigma_T$, we have that there exists q such that $t \in \mathcal{G}^{(q)}$, and from property (5) we deduce that the limiting velocity obeys $v(t) = v_q(t) \in H^3(\mathbb{T}^3)$. Thus, by the short-time parabolic regularization of the Navier-Stokes equations, we deduce that for every t in the open set $(0,T)\setminus\Sigma_T$ we have $v(t) \in C^\infty(\mathbb{T}^3)$. In fact, regularity in time also holds on the complement on the singular set of times from standard arguments. We remark that the set Σ_T is not empty by the Cantor intersection theorem.

To conclude the proof of Theorem 4.1, it remains to show that there exists $\beta''' > 0$ such that Σ_T has box-counting dimension (and hence Hausdorff dimension) less than $1-\beta'''$. To estimate the box-counting dimension of Σ_T, we use that Σ_T is covered by $\mathcal{B}^{(q)}$ for every $q \geq 0$, and $\mathcal{B}^{(q)}$ consists of disjoint intervals of length $5\tau_q$. To count how many such disjoint intervals are necessary, we use (4.15) and the definitions of τ_q and ϑ_q in (4.6) and (4.7) to estimate

$$|\mathcal{B}^{(q)}| \leq |\mathcal{B}^{(0)}| \prod_{q'=1}^{q} \frac{10\tau_{q'}}{\vartheta_{q'}} \leq 10^q T \prod_{q'=0}^{q-1} \lambda_{q'}^{-\varepsilon_R/4} \leq T\lambda_q^{-\frac{\varepsilon_R}{16(b-1)}}\,.$$

Therefore the number of disjoint intervals needed to cover $\mathcal{B}^{(q)}$ is bounded by $T\lambda_q^{-\frac{\varepsilon_R}{16(b-1)}}(5\tau_q)^{-1}$. From (4.4) and (4.19) we conclude that

$$\begin{aligned}\dim_{\text{box}}(\Sigma_T) &\leq \lim_{q\to\infty} \frac{\log(T) - \frac{\varepsilon_R}{16(b-1)}\log\left(\lambda_q\right) - \log(5\tau_q)}{-\log\left(5\tau_q\right)} \\ &\leq 1 - \frac{\varepsilon_R b}{128(b-1)} =: 1-\beta'''\,.\end{aligned}$$

This concludes the proof of the theorem, which holds with β replaced by $\min\{\beta, \beta', \beta'', \beta'''\} > 0$.

4.3 Stability Estimates and Gluing

This section closely follows the exposition in [6, Section 3]. The first step in the proof of Proposition 4.1 is to construct a *good set* $\mathcal{G}^{(q+1)}$ with properties consistent with (1)–(4), and a *glued velocity-stress pair* $(\overline{v}_q, \overline{\mathring{R}}_q)$, which solves (4.12), obeys bounds which are the same as (4.13a)–(4.14b) up to a factor of two, and such that $\overline{\mathring{R}}_q \equiv 0$ on $\mathbb{T}^3 \times \mathcal{G}^{(q+1)}$. For this purpose we first need to recall a few stability properties of the Navier-Stokes equations.

4.3.1 Short Time Smooth Solutions of the Navier-Stokes Equations and Their Stability

The following statement summarizes the classical local in time existence of a strong solution for (4.1), and its parabolic regularization properties [11].

Proposition 4.2 *Consider the Cauchy problem for* (4.1) *with initial datum* $v_0 = v|_{t=t_0} \in H^3(\mathbb{T}^3)$ *of zero mean. There exists a universal constant* $c \in (0, 1]$ *such that if* t_1 *obeys*

$$0 < t_1 - t_0 \leq \frac{c}{\|v_0\|_{H^3} (1 + \|v_0\|_{L^2})}, \tag{4.20}$$

then there exists a unique strong solution to (4.1) *on* $[t_0, t_1]$, *which obeys the estimates*

$$\sup_{t\in[t_0,t_1]} \|v(t)\|_{L^2}^2 + 2\int_{t_0}^{t_1} \|\nabla v(t)\|_{L^2}^2 \, dt \leq \|v_0\|_{L^2}^2 \,. \tag{4.21a}$$

$$\sup_{t\in[t_0,t_1]} \|v(t)\|_{H^3} \leq 2\,\|v_0\|_{H^3} \,. \tag{4.21b}$$

$$\sup_{t\in(t_0,t_1]} |t - t_0|^{\frac{N}{2}+M} \left\|\partial_t^M D^N v(t)\right\|_{H^3} \lesssim \|v_0\|_{H^3} \,, \tag{4.21c}$$

for any $N \geq 0$ *and* $M \in \{0, 1\}$.

Sketch of Proof for Proposition 4.2 Estimate (4.21a) is just the energy inequality, while (4.21b) follows on a time interval which obeys condition (4.20) from the fact that $H^3 \subset W^{1,\infty}$ and appealing to the higher order energy estimate

$$\frac{1}{2}\frac{d}{dt}\|v\|_{\dot H^3}^2 + \|v\|_{\dot H^4}^2 \lesssim \|v\|_{\dot H^3}^2 \|\nabla v\|_{L^\infty} + \|v\|_{\dot H^3} \|\Delta v\|_{L^4}^2 \lesssim \|v\|_{\dot H^3}^3 \,.$$

The higher regularity stated in (4.21c) follows from the mild form of the solution (cf. (4.3)), and maximal regularity of the heat equation. □

Next, we estimate the difference between an approximate solution and an exact solution of the Navier-Stokes equation. The result adapted to our needs is:

Proposition 4.3 *Assuming that $a \geq 1$ is sufficiently large, the following holds. Let $(v_q, \mathring{R}_q)$ be a $C_t^0 H_x^3$ smooth solution of* (4.12) *which obeys the estimates* (4.13a)–(4.14b). *Let $t_0 \in [0, T]$ and define*

$$v_0 := v_q|_{t=t_0} .$$

If $t_1 \in (t_0, T]$ is such that

$$0 < t_1 - t_0 \leq \delta_0^{-1}\lambda_q^{-4} , \tag{4.22}$$

then by (4.14b) *and Proposition 4.2 there exists a unique $C_t^0 H_x^3$ smooth solution v of* (4.1) *on $[t_0, t_1]$, with initial datum v_0. We claim that this solution v obeys the bounds*

$$\left\|v - v_q\right\|_{L^\infty([t_0,t_1];L^2)} \leq |t_1 - t_0|\lambda_q^5 \tag{4.23a}$$

$$\left\|\mathcal{R}(v - v_q)\right\|_{L^\infty([t_0,t_1];L^1)} \leq |t_1 - t_0|\lambda_q^{-\frac{3}{4}\varepsilon_R}\delta_{q+1} . \tag{4.23b}$$

Sketch of Proof for Proposition 4.3 The bounds (4.23a)–(4.23b) follow the available estimates (4.13a)–(4.13b) on $\mathring{R}_q$ and interpolation, upon proving

$$\left\|v(t) - v_q(t)\right\|_{L^2} \leq C|t - t_0| \left\|\nabla \mathring{R}_q\right\|_{L^\infty([t_0,t_1];L^2)} \tag{4.24a}$$

$$\left\|\mathcal{R}v(t) - \mathcal{R}v_q(t)\right\|_{L^p} \leq C|t - t_0| \left\|\mathring{R}_q\right\|_{L^\infty([t_0,t_1];L^p)} \tag{4.24b}$$

for all $t \in (t_0, t_1]$, for any $p \in (1, 2]$ sufficiently close to 1. The constants in (4.24a)–(4.24b) only depend on p. Here we use that a is sufficiently large, to absorb the constants arising in this procedure.

In order to prove (4.24a), we let $u = v_q - v$ and $\bar{p} = p_q - p$. Then $\nabla \cdot u = 0$ and it obeys the equation

$$\partial_t u - \Delta u + v \cdot \nabla u + \nabla \bar{p} = \nabla \cdot \mathring{R}_q - u \cdot \nabla v_q . \tag{4.25}$$

Taking an L^2 inner product of (4.25) with u, we see that $\|u\|_{L^2}$ is exponentially stretched by $\left\|\nabla v_q\right\|_{L^\infty}$ and is linearly forced by $\left\|\nabla \mathring{R}_q\right\|_{L^2}$. Using the Gagliardo-

Nirenberg estimate $\left\|\nabla v_q\right\|_{L^\infty} \lesssim \left\|v_q\right\|_{L^2}^{1/6} \left\|v_q\right\|_{\dot{H}^3}^{5/6}$, the available estimates (4.14a)–(4.14b) on v_q, and using the temporal restriction (4.22), it follows that the exponential stretching term is harmless on $[t_0, t_1]$. The bound (4.24a) follows upon integrating in time, since $u|_{t=t_0} = 0$.

Proving (4.24b) is slightly more delicate. Let us denote $z = \Delta^{-1}\mathrm{curl}\, u$. Note that since $\nabla \cdot u = 0$ we have $\mathrm{curl}\, z = -u$, and using the Calderón-Zygmund inequality we have $\|\mathcal{R}u(t)\|_{L^p} \lesssim \|z(t)\|_{L^p}$. Thus it is sufficient to obtain L^p estimates for $z(t)$. We apply $\Delta^{-1}\mathrm{curl}$ to (4.25) and obtain

$$\partial_t z + v \cdot \nabla z - \Delta z = \Delta^{-1}\mathrm{curl}\,\mathrm{div} \mathring{R}_q + \Delta^{-1}\mathrm{curl}\,\mathrm{div}\,((z \times \nabla)v) + \Delta^{-1}\nabla \mathrm{div}\,((z \cdot \nabla)v) + \Delta^{-1}\mathrm{curl}\,\mathrm{div}\left(((z \times \nabla)v_q)^T\right). \tag{4.26}$$

The left side of (4.26) is a drift-diffusion equation with incompressible drift, so that it does not increase the L^p norm of the solution. All we have to do is to estimate the L^p norm of the right side of (4.26), which may be estimated in L^p by a constant times $\left\|\mathring{R}_q\right\|_{L^p} + \|z\|_{L^p}\left(\|\nabla v\|_{L^\infty} + \left\|\nabla v_q\right\|_{L^\infty}\right)$, since $p > 1$. Again, the time restriction (4.22) shows that the resulting exponential stretching factors are harmless, and (4.24b) follows upon integrating in time. □

With the above lemmas in hand, we next define a family of exact solutions v_i of the Navier-Stokes equations, which do not depart too much from v_q, at least for short time. First, we subdivide the time interval $[0, T]$ into pieces of length ϑ_{q+1}, and we denote

$$t_i = \vartheta_{q+1} i$$

where $i \in \{0, \ldots, n_{q+1}\}$ for a suitable defined integer n_{q+1}. Next, for every $1 \le i \le n_{q+1}$ we define $v_i(x, t)$ to be the unique smooth solution of the Cauchy problem for (4.1) with initial condition equal to v_q at t_{i-1}:

$$\partial_t v_i + \nabla \cdot (v_i \otimes v_i) + \nabla p_i - \Delta v_i = 0 \tag{4.27a}$$

$$\nabla \cdot v_i = 0 \tag{4.27b}$$

$$v_i(t_{i-1}) = v_q(t_{i-1})\,. \tag{4.27c}$$

Taking a (and hence δ_0) sufficiently large, depending on the universal constant c, by (4.6) and (4.14a)–(4.14b) we have

$$3\vartheta_{q+1} \le \frac{1}{\lambda_q^4 \delta_0} \le \frac{c}{4\lambda_q^4 \delta_0^{1/2}} \le \frac{c}{\left\|v_q\right\|_{H^3}(1 + \left\|v_q\right\|_{L^2})},$$

which is consistent with both (4.20) and (4.22). By Propositions 4.2 and 4.3, and recalling that $t_{i+2} = 3\vartheta_{q+1} + t_{i-1}$, we may thus conclude that

$$\sup_{t\in[t_{i-1},t_{i+2}]} \|v_i(t)\|_{L^2} \le 2\delta_0^{1/2} - \delta_q^{1/2} \tag{4.28a}$$

$$\sup_{t\in[t_{i-1},t_{i+2}]} \|v_i(t)\|_{H^3} \le 2\lambda_q^4 \tag{4.28b}$$

$$\sup_{t\in[t_i,t_{i+2}]} \left\|\partial_t^M D^N v_i(t)\right\|_{H^3} \lesssim \lambda_q^4 \vartheta_{q+1}^{-\frac{N}{2}-M} \tag{4.28c}$$

$$\sup_{t\in[t_{i-1},t_{i+2}]} \left\|(v_i - v_q)(t)\right\|_{L^2} \le 4\vartheta_{q+1}\lambda_q^5 \tag{4.28d}$$

$$\sup_{t\in[t_{i-1},t_{i+2}]} \left\|\mathcal{R}(v_i - v_q)(t)\right\|_{L^1} \le 4\vartheta_{q+1}\lambda_q^{-\frac{3}{4}\varepsilon_R}\delta_{q+1} \tag{4.28e}$$

where the implicit constant in (4.28c) depends on $N \geq 0$ and $M \in \{0, 1\}$.

4.3.2 The Glued Velocity, New Good and Bad Sets, and the Glued Stress

We first introduce a C^∞ smooth partition of unity $\{\eta_i\}_{i=0}^{n_{q+1}}$ such that $0 \le \eta_i \le 1$, $\eta_i \equiv 1$ on $[t_i + \tau_{q+1}, t_{i+1}]$, with supports given by

$$\operatorname{supp}(\eta_i) \subset [t_i, t_{i+1} + \tau_{q+1}],$$

with

$$\sum_{i=0}^{n_{q+1}} \eta_i(t) = 1, \qquad \text{for every} \qquad t \in [T/3, 2T/3], \tag{4.29}$$

and such that $\left\|\partial_t^M \eta_i\right\|_{L^\infty} \lesssim \tau_{q+1}^{-M}$, where the implicit constant is independent of τ_{q+1} ϑ_{q+1}, and i. We conclude from these properties that $\eta_i\eta_j = 0$ whenever $|i - j| > 1$, and that $\operatorname{supp}(\eta_i\eta_{i-1}) \subset [t_i, t_i + \tau_{q+1}]$.

Next we define the glued velocity field $\overline{v}_q$ simply as a linear combination of the exact solutions v_i constructed in (4.27), with weights given by the above defined temporal partition of unity

$$\overline{v}_q(x,t) = \sum_i \eta_i(t) v_i(x,t). \tag{4.30}$$

Note that by construction, on the support of η_i, which is by definition a subset of $[t_i, t_{i+2}]$, the functions v_i obey all the properties listed in (4.28). In particular, by (4.28c), we have that $\bar{v}_q$ is C^∞ smooth in space and time, in spite of v_q only being a priori an $C_t^0 H_x^3$ function.

Note trivially that on the temporal set $\cup_i \{t: \eta_i(t) = 1\}$, the glued velocity $\overline{v}_q$ is an exact solution to (4.1). A slightly more subtle observation is the following. If t_{i-1} and t_i both lie within the τ_q neighborhood of the good set $\mathcal{G}^{(q)}$, then due to property (4.17) v_q is already an exact solution of (4.1) on the overlapping region $\operatorname{supp}(\eta_i \eta_{i+1})$. Therefore, by the uniqueness of strong solutions we have $v_i = v_{i+1} = v_q$ on this region, and due to (4.29) we have that $\bar{v}_q = v_q$ is also an exact solution of (4.1) here. In order to identify the regions where $\overline{v}_q$ may not necessarily be an exact solution of (4.1) we define the index set

$$C = \left\{ i \in \{1, \dots, n_{q+1}\} \colon \text{there exists } t \in (t_{i-1}, t_{i+1} + \tau_{q+1}) \text{ with } \mathring{R}_q(t) \neq 0 \right\} \tag{4.31}$$

and then define the bad set at step $q+1$ as

$$\mathcal{B}^{(q+1)} = \bigcup_{i \in C \text{ or } i-1 \in C} [t_i - 2\tau_{q+1}, t_i + 3\tau_{q+1}] \,. \tag{4.32}$$

By the above remark, $\overline{v}_q$ is already exact solution of (4.1) on $\mathcal{G}^{(q+1)} = [0, T] \setminus \mathcal{B}^{(q+1)}$. The fact that $\mathcal{B}^{(q+1)}$ obeys property (3) is automatic from (4.32). Property (4) follows for $\mathcal{B}^{(q+1)}$ by estimating the cardinality of the set C: if $i \in C$, there exists $t \in (t_{i-1}, t_{i+1} + \tau_{q+1})$ such that $\mathring{R}_q(t) \neq 0$, and thus by (4.17) we have $\operatorname{dist}(t, \mathcal{G}^{(q)}) > \tau_q$. Therefore, $\mathcal{B}^{(q)} \supset (t - \tau_q, t + \tau_q) \supset [t_i, t_{i+1}]$. By the pigeonhole principle, $\operatorname{card}(C) \leq \frac{|\mathcal{B}^{(q)}|}{\vartheta_{q+1}}$, which gives the desired bound. Lastly, we prove that (2) holds at level $q+1$. If $t \in \mathcal{G}^{(q)} \cap [t_i - 2\tau_{q+1}, t_i + 3\tau_{q+1}]$ for some $i \in \{1, \dots, n_{q+1}\}$, by (4.17) we know that $\mathring{R}_q(t') = 0$ for all $|t - t'| \leq \tau_q$. Since $\tau_q \geq 2\vartheta_{q+1} + 3\tau_{q+1}$, which holds by (4.6) and (4.7), we obtain that $\mathring{R}_q \equiv 0$ on $[t_{i-2}, t_{i+1} + \tau_{q+1}]$. Hence, by (4.31) we have that $i, i-1 \notin C$. It follows that $t \notin \mathcal{B}^{(q+1)}$ and so $t \in \mathcal{G}^{(q+1)}$. Thus, the postulated properties (2)–(4) hold for the bad set $\mathcal{B}^{(q+1)}$ given by (4.32).

It remains to define the glued stress $\mathring{\overline{R}}_q$. On the sets $[0, T/3]$ and $[2T/3, T]$ we already know that $\bar{v}_q = v_q$ is an exact solution of the Navier-Stokes equation, so on this time interval we set $\mathring{\overline{R}}_q \equiv 0$. For the remaining time interval, one inserts the definition of the glued velocity (cf. (4.30)) into the Navier-Stokes equations, and deduces the definition

$$\begin{aligned} \mathring{\overline{R}}_q = \partial_t \eta_i \mathcal{R}(v_i - v_{i-1}) - \eta_i(1 - \eta_i)(v_i - v_{i-1}) \mathring{\otimes} (v_i - v_{i-1}) \,, \\ \text{for all} \qquad t \in [t_i, t_{i+1}] \cap [T/3, 2T/3] \,. \end{aligned} \tag{4.33}$$

By construction, the pair $(\bar{v}_q, \mathring{\overline{R}}_q)$ defined by (4.30) and (4.33) solves the Navier-Stokes-Reynolds system (4.12) on $[0, T]$. This construction is successful if we can show that $(\bar{v}_q, \mathring{\overline{R}}_q)$ obey roughly the same estimates as $(v_q, \mathring{R}_q)$, but such that $\mathring{\overline{R}}_q$

has a much smaller temporal support, related to the above defined set $\mathcal{B}^{(q+1)}$. These facts are indeed true, and the precise estimates may be summarized as:

Proposition 4.4 (Proposition 2.2 in [6]) *The pair* $(\overline{v}_q, \overline{\mathring{R}}_q)$ *solves* (4.12), *the velocity field* $\overline{v}_q$ *satisfies:*

$$\overline{v}_q \equiv v_q \qquad on \qquad \mathbb{T}^3 \times \mathcal{G}^{(q)}, \tag{4.34a}$$

$$\left\| \bar{v}_q \right\|_{L^2} \leq 2\delta_0^{1/2} - \delta_q^{1/2} \tag{4.34b}$$

$$\left\| \bar{v}_q \right\|_{H^3} \leq 2\lambda_q^4 \tag{4.34c}$$

$$\left\| \bar{v}_q - v_q \right\|_{L^2} \leq \vartheta_{q+1}\lambda_q^6 \leq \frac{1}{4}\delta_{q+1}^{1/2} \tag{4.34d}$$

$$\|\partial_t^M D^N \bar{v}_q\|_{L^\infty(T/3,2T/3;H^3)} \lesssim \tau_{q+1}^{-M}\vartheta_{q+1}^{-\frac{N}{2}}\lambda_q^4 \lesssim \tau_{q+1}^{-M-N}\lambda_q^4 \tag{4.34e}$$

and the stress tensor $\overline{\mathring{R}}_q$ *satisfies:*

$$\overline{\mathring{R}}_q(t) = 0 \quad for\ all \quad t \in [0, T] \quad such\ that \quad \operatorname{dist}(t, \mathcal{G}^{(q+1)}) \leq 2\tau_{q+1} \tag{4.35a}$$

$$\left\| \overline{\mathring{R}}_q \right\|_{L^1} \leq \tau_{q+1}^{-1}\vartheta_{q+1}\lambda_q^{-\varepsilon_R/2}\delta_{q+1} \leq \lambda_q^{-\varepsilon_R/4}\delta_{q+1} \tag{4.35b}$$

$$\left\| \partial_t^M D^N \overline{\mathring{R}}_q \right\|_{H^3} \lesssim \tau_{q+1}^{-M-1}\vartheta_{q+1}^{-\frac{N}{2}}\lambda_q^4 \lesssim \tau_{q+1}^{-M-N-1}\lambda_q^4 \tag{4.35c}$$

for all $M, N \geq 0$.

The above statement is proven in detail in [6, Section 3]. We only note here that (4.34a) follows from the arguments outlined in the paragraph above (4.31), while the bounds (4.34b)–(4.34e) follow from the properties (4.28a)–(4.28d) of the v_i. The bounds (4.35b)–(4.35c) are a direct consequence of the definition (4.33), the properties of the cutoffs η_i, and the stability estimates (4.28d)–(4.28e). Lastly, (4.35a) holds by a careful inspection of the definition of $\mathcal{B}^{(q+1)}$.

4.4 Intermittent Jets and Convex Integration

In this section we construct a perturbation w_{q+1} to be added to the glued velocity $\overline{v}_q$, so that the new velocity field

$$v_{q+1} := w_{q+1} + \overline{v}_q \tag{4.36}$$

and its associated stress $\mathring{R}_{q+1}$ (which we define in Sect. 4.4.3), obey the bounds (4.13a)–(4.14b) and property (4.17) at level $q+1$. Additionally, the velocity increment w_{q+1} will be designed to have support outside a τ_{q+1} neighborhood of $\mathcal{G}^{(q+1)}$, which automatically ensures that properties (5) and (6) will be satisfied. As in [2, 3, 6], the perturbation w_{q+1} consists of three parts: the principal part $w_{q+1}^{(p)}$, the divergence corrector $w_{q+1}^{(c)}$, and the temporal corrector $w_{q+1}^{(c)}$. The principal part $w_{q+1}^{(p)}$ will be constructed as a sum of *intermittent jets* $W_{(\xi)}$ which we define next.

4.4.1 Intermittent Jets

In this section we summarize the properties of the *intermittent jets* defined by the authors of this paper in [6, Section 4.1]. The exposition here follows closely [2, Section 7.4]. As mentioned in the introduction, intermittent jets are the natural extension of the *Mikado flows* constructed in [13], in order to achieve *more than two-dimensional intermittency*. Indeed, the classical Mikado flows are supported in thin periodized tubes, which are thus only two-dimensionally small. On the other hand, the intermittent jets constructed here have spatial supports in thin ellipsoidal regions which can be made almost three-dimensionally small. The cost of this additional thinness property is that the supports of the intermittent jets are moving linearly in time, at a very fast speed.

We first recall a version of Nash's geometric lemma [37, Lemma 1] (see also [13, Lemma 2.4]).

Lemma 4.1 *There exists a set $\Lambda \subset \mathbb{S}^2 \cap \mathbb{Q}^3$ and there exist C^∞ smooth functions $\{\gamma_\xi\}_{\xi\in\Lambda} : \overline{B_{1/2}^{sym}(\mathrm{Id})} \to \mathbb{R}$ such that for every symmetric matrix R satisfying $|R - \mathrm{Id}| \le 1/2$ we may write*

$$R = \sum_{\xi\in\Lambda} \gamma_\xi^2(R)(\xi \otimes \xi) .$$

For $\xi \in \Lambda$, we let $A_\xi \in \mathbb{S}^2 \cap \mathbb{Q}^3$ be an orthogonal vector to ξ, such that $\{\xi, A_\xi, \xi \times A_\xi\} \subset \mathbb{S}^2 \cap \mathbb{Q}^3$ forms an orthonormal basis for $\mathbb{R}^3$. We label by n_* the smallest natural such that

$$\bigcup_{\xi\in\Lambda} \left\{n_*\xi,\ n_*A_\xi,\ n_*\xi \times A_\xi\right\} \subset \mathbb{Z}^3 . \tag{4.37}$$

That is, n_* is the l.c.m. of the denominators of the rational vectors ξ, A_ξ, and $\xi \times A_\xi$. It is convenient here to fix a geometric parameter M, given by

$$M = \sup_{\xi\in\Lambda} \left(\|\gamma_\xi\|_{C^0} + \|\nabla\gamma_\xi\|_{C^0}\right) . \tag{4.38}$$

It is clear that both parameters n_* and M are purely geometric in nature, and in particular they do not depend on q.

In order to define the intermittent jets, we first introduce three cutoff functions: ϕ and Φ which are two-dimensional, and ψ which is one-dimensional. Let $\Phi : \mathbb{R}^2 \to \mathbb{R}^2$ be a smooth function supported in $B_1(0) \subset \mathbb{R}^2$, define $\phi = -\Delta\Phi$ (which automatically has zero mean), and normalize ϕ (and thus Φ) such that $\|\phi\|_{L^2(\mathbb{R}^2)} = |\mathbb{T}^2|^{1/2}$. Similarly, let $\psi : \mathbb{R} \to \mathbb{R}$ be a smooth function supported in $B_1(0) \subset \mathbb{R}$, which we normalize such that $\|\psi\|_{L^2(\mathbb{R})} = |\mathbb{T}|^{1/2}$. We recall the parameters $r_\perp$ and $r_\|$ fixed earlier in (4.9) and introduce rescaled versions of ϕ, Φ, respectively ψ as

$$\phi_{r_\perp}(x_1, x_2) = \frac{1}{r_\perp}\phi\left(\frac{x_1}{r_\perp}, \frac{x_2}{r_\perp}\right), \quad \Phi_{r_\perp}(x_1, x_2) = \frac{1}{r_\perp}\Phi\left(\frac{x_1}{r_\perp}, \frac{x_2}{r_\perp}\right),$$

$$\psi_{r_\|}(x_3) = \frac{1}{r_\|^{1/2}}\psi\left(\frac{x_3}{r_\|}\right). \tag{4.39}$$

With this rescaling we have that $\operatorname{supp}(\phi_{r_\perp}), \operatorname{supp}(\Phi_{r_\perp}) \subset B_{r_\perp}(0) \subset \mathbb{R}^2$, $\operatorname{supp}(\psi_{r_\|}) \subset B_{r_\|}(0) \subset \mathbb{R}$, and we have maintained the normalizations $\|\phi_{r_\perp}\|^2_{L^2(\mathbb{R}^2)} = |\mathbb{T}^2|$ and $\|\psi_{r_\|}\|^2_{L^2(\mathbb{R}^2)} = |\mathbb{T}|$. Note also that $\phi_{r_\perp} = -r_\perp^2 \Delta\Phi_{r_\perp}$ holds. Lastly, we *periodize* $\phi_{r_\perp}$, $\Phi_{r_\perp}$, $\psi_{r_\|}$ and view these functions as $\mathbb{T}^2$, $\mathbb{T}^2$ respectively $\mathbb{T}$-periodic functions.

Recall the temporal speed parameter μ defined in (4.10), and note that by construction we have $\lambda_{q+1} r_\perp \in \mathbb{N}$. Next we define $\left(\mathbb{T}/r_\perp\lambda_{q+1}\right)^3$-periodic functions

$$\psi_{(\xi)}(x,t) := \psi_{r_\|}\left(n_* r_\perp \lambda_{q+1}(x\cdot\xi + \mu t)\right), \tag{4.40a}$$

$$\Phi_{(\xi)}(x) := \Phi_{r_\perp}\left(n_* r_\perp \lambda_{q+1}(x-\alpha_\xi)\cdot A_\xi, n_* r_\perp \lambda_{q+1}(x-\alpha_\xi)\cdot(\xi\times A_\xi)\right) \tag{4.40b}$$

$$\phi_{(\xi)}(x) := \phi_{r_\perp}\left(n_* r_\perp \lambda_{q+1}(x-\alpha_\xi)\cdot A_\xi, n_* r_\perp \lambda_{q+1}(x-\alpha_\xi)\cdot(\xi\times A_\xi)\right). \tag{4.40c}$$

where $\alpha_\xi \in \mathbb{R}^3$ are shifts which ensure that the functions $\{\Phi_{(\xi)}\}_{\xi\in\Lambda}$ have *mutually disjoint support*. It is important to note that by (4.39), the function $\psi_{(\xi)}$ oscillates at frequency proportional to $r_\perp r_\|^{-1}\lambda_{q+1} = \lambda_{q+1}^{19/24} \ll \lambda_{q+1}$, whereas $\phi_{(\xi)}$ and $\Phi_{(\xi)}$ oscillate at frequency proportional to λ_{q+1}. In fact, we have the bounds

$$\left\|\nabla^N \partial_t^M \psi_{(\xi)}\right\|_{L^p} \lesssim r_\|^{1/p-1/2}\left(\frac{r_\perp\lambda_{q+1}}{r_\|}\right)^N \left(\frac{r_\perp\lambda_{q+1}\mu}{r_\|}\right)^M \tag{4.41}$$

$$\left\|\nabla^N \phi_{(\xi)}\right\|_{L^p} + \left\|\nabla^N \Phi_{(\xi)}\right\|_{L^p} \lesssim r_\perp^{2/p-1}\lambda_{q+1}^N \tag{4.42}$$

for all $p \in [1,\infty]$, and $N, M \geq 0$, where the implicit constants are independent of q.

The *intermittent jets* $W_{(\xi)} \colon \mathbb{T}^3 \times \mathbb{R} \to \mathbb{R}^3$ are then defined as the $\left(\mathbb{T}/r_\perp \lambda_{q+1}\right)^3$-periodic functions

$$W_{(\xi)}(x,t) := \xi\, \psi_{(\xi)}(x,t)\, \phi_{(\xi)}(x)\,. \tag{4.43}$$

By the definition of the spatial shifts α_ξ, we have that

$$W_{(\xi)} \otimes W_{(\xi')} \equiv 0 \qquad \text{whenever} \qquad \xi \neq \xi' \in \Lambda\,, \tag{4.44}$$

and as a consequence of the rescaling (4.39), it holds that

$$\fint_{\mathbb{T}^3} W_{(\xi)}(x,t) \otimes W_{(\xi)}(x,t)\, dx = \xi \otimes \xi\,. \tag{4.45}$$

As a consequence, using Lemma 4.1 and the cancelation (4.44), we have

$$\sum_{\xi,\xi' \in \Lambda} \gamma_\xi(R) \gamma_{\xi'}(R) \fint_{\mathbb{T}^3} W_{(\xi)}(x,t) \otimes W_{(\xi')}(x,t)\, dx = R\,, \tag{4.46}$$

for every symmetric matrix R satisfying $|R - \mathrm{Id}| \leq 1/2$. One of the fundamental properties of the intermittent jets is

$$\operatorname{div}\left(W_{(\xi)} \otimes W_{(\xi)}\right) = 2(W_{(\xi)} \cdot \nabla \psi_{(\xi)})\phi_{(\xi)}\xi = \frac{1}{\mu}\phi^2_{(\xi)} \partial_t \psi^2_{(\xi)} \xi = \frac{1}{\mu}\partial_t \left(\phi^2_{(\xi)} \psi^2_{(\xi)} \xi\right), \tag{4.47}$$

an identity which plays a key role in showing that the leading order error in the oscillation stress is cancelled by the time derivative of the temporal corrector.

Note that the intermittent jets $W_{(\xi)}$ defined in (4.43) *are not divergence free*. In order to correct this, we introduce $V_{(\xi)} : \mathbb{T}^3 \times \mathbb{R} \to \mathbb{R}^3$ defined by

$$V_{(\xi)}(x,t) := \frac{1}{n_*^2 \lambda_{q+1}^2} \xi\, \psi_{(\xi)}(x,t)\, \Phi_{(\xi)}(x)\,. \tag{4.48}$$

Using $V_{(\xi)}$, we introduce $W^{(c)}_{(\xi)}$ to be the incompressibility corrector associated to the intermittent jet, defined as

$$W^{(c)}_{(\xi)} := \operatorname{curl}\operatorname{curl} V_{(\xi)} - W_{(\xi)} = \frac{1}{n_*^2 \lambda_{q+1}^2} \nabla \psi_{(\xi)} \times \operatorname{curl}\left(\Phi_{(\xi)} \xi\right). \tag{4.49}$$

Thus, by the definition of $W^{(c)}_{(\xi)}$, we have $\nabla \cdot \left(W_{(\xi)} + W^{(c)}_{(\xi)}\right) = \operatorname{div}\operatorname{curl}(\operatorname{curl} V_{(\xi)}) = 0$.

Lastly, we summarize the bounds obeyed by the intermittent jets, their incompressibility correctors, and their corresponding building blocks. From (4.41)–(4.42) it is not hard to show that for $N, M \geq 0$ and $p \in [1, \infty]$ we have

$$\left\| \nabla^N \partial_t^M W_{(\xi)} \right\|_{L^p} + \frac{r_\|}{r_\perp} \left\| \nabla^N \partial_t^M W_{(\xi)}^{(c)} \right\|_{L^p} + \lambda_{q+1}^2 \left\| \nabla^N \partial_t^M V_{(\xi)} \right\|_{L^p} \lesssim r_\perp^{2/p-1} r_\|^{1/p-1/2} \lambda_{q+1}^N \left(\frac{r_\perp \lambda_{q+1} \mu}{r_\|} \right)^M \tag{4.50}$$

where implicit constants depend on p, N and M. It is convenient at this place to record for future use that

$$r_\perp r_\|^{1/2} = \lambda_{q+1}^{13/12} \gg \lambda_{q+1}, \qquad \frac{r_\perp \lambda_{q+1}}{r_\|} = \lambda_{q+1}^{19/24} \ll \lambda_{q+1}, \qquad \text{and}$$

$$\frac{r_\perp \lambda_{q+1} \mu}{r_\|} = \lambda_{q+1}^2 .$$

4.4.2 *The Perturbation*

In this section we construct the velocity increment w_{q+1}, and show that $v_{q+1} = \bar{v}_q + w_{q+1}$ obeys the inductive assumptions (4.14a)–(4.14b), and the property (5). The velocity increment w_{q+1} consists of a principal corrector $w_{q+1}^{(p)}$, an incompressibility corrector $w_{q+1}^{(c)}$, and a temporal corrector $w_{q+1}^{(t)}$. We construct each one of these separately, and then conclude the section with their estimates.

4.4.2.1 Amplitudes

Let χ be a smooth function such that

$$\chi(z) = \begin{cases} 1 & \text{if} \quad 0 \leq z \leq 1 \\ z & \text{if} \quad z \geq 2 \end{cases}$$

and with $z \leq 2\chi(z) \leq 4z$ for $z \in (1, 2)$. For such a function χ, and with (4.35b) in mind, we define

$$\rho(x, t) = 4\lambda_q^{-\varepsilon_R/4} \delta_{q+1} \chi \left(\frac{\left| \mathring{\bar{R}}_q(x, t) \right|}{\lambda_q^{-\varepsilon_R/4} \delta_{q+1}} \right) .$$

The function ρ inherits a number of properties from $\mathring{\bar{R}}_q$, such as the bound

$$\|\rho\|_{L^p} \le 16\left(\left|\mathbb{T}^3\right|^{1/p} \lambda_q^{-\varepsilon R/4}\delta_{q+1} + \left\|\mathring{R}_q\right\|_{L^p}\right)$$

$$\Rightarrow \quad \|\rho\|_{L^1} \le 16\left(\left|\mathbb{T}^3\right|^{1/p} + 1\right)\lambda_q^{-\varepsilon R/4}\delta_{q+1} \tag{4.51}$$

which follows from (4.35b). Moreover, we have the estimate

$$\|\rho\|_{C^N_{x,t}} \lesssim \mathbf{1}_{\{N=0\}}\lambda_q^{-2\varepsilon R}\tau_{q+1}^{-1}\delta_{q+1} + \mathbf{1}_{\{N\ge 1\}}\tau_{q+1}^{-3N} \lesssim \tau_{q+1}^{-3N-1} \tag{4.52}$$

which follows from the bounds (4.35b)–(4.35c), the interpolation inequality $\left\|\mathring{R}_q\right\|_{C^0} \le C\left\|\mathring{R}_q\right\|_{L^1}^{1/3}\left\|\mathring{R}_q\right\|_{H^3}^{2/3}$, repeated applications of the chain rule (see e.g. [4, Proposition C.1]), and our choice of parameters. However, the most important property of ρ is that pointwise in (x, t) we have the bound

$$\left|\frac{\mathring{R}_q(x,t)}{\rho(x,t)}\right| \le \frac{1}{2} \tag{4.53}$$

which follows from the definition of ρ and of the function χ. Estimate (4.53) allows us to say that on the support of the function ρ, the matrix $\mathrm{Id} - \rho^{-1}\mathring{R}_q$ lies in the ball of radius $1/2$ around the identity matrix, and thus in the domain of the functions γ_ξ from Lemma 4.1.

In order to ensure that properties (5)–(6) will hold, we introduce a smooth temporal cut-off function $\theta\colon [0, T] \to [0, 1]$ which has the following properties:

- $\theta(t) = 1$ for all t such that $\mathrm{dist}(t, \mathcal{G}^{(q+1)}) \ge 2\tau_{q+1}$.
- $\theta(t) = 0$ for all t such that $\mathrm{dist}(t, \mathcal{G}^{(q+1)}) \le \tau_{q+1}$.
- $\|\theta\|_{C^M} \lesssim \tau_{q+1}^{-M}$.

Such a choice for θ is possible since $\mathcal{B}^{(q+1)}$ consists of a finite disjoint union of intervals of length $5\tau_{q+1}$. From the first property above and (4.35a), we conclude that

$$t \in \mathrm{supp}\,(\mathring{R}_q) \quad \Rightarrow \quad \theta(t) = 1\,. \tag{4.54}$$

From the second property above we further obtain that

$$t \in \mathrm{supp}\,(\theta) \quad \Rightarrow \quad \mathrm{dist}(t, \mathcal{G}^{(q+1)}) > \tau_{q+1}\,. \tag{4.55}$$

Lastly, we introduce the amplitude functions

$$a_{(\xi)}(x,t) := \theta(t)\rho(x,t)^{1/2}\,\gamma_\xi\left(\mathrm{Id} - \frac{\mathring{R}_q(x,t)}{\rho(x,t)}\right). \tag{4.56}$$

for every $\xi \in \Lambda$. We immediately deduce from (4.46), (4.56), (4.54), and Lemma 4.1, that have

$$\sum_{\xi,\xi'\in\Lambda} a_{(\xi)}a_{(\xi')} \fint_{\mathbb{T}^3} W_{(\xi)} \otimes W_{(\xi')} dx = \theta^2 \rho \mathrm{Id} - \mathring{\overline{R}}_q , \tag{4.57}$$

which justifies the definition of the amplitude functions $a_{(\xi)}$. From (4.38), (4.51), (4.52) and the chain rule, it follows that the amplitudes $a_{(\xi)}$ obey the bounds

$$\left\|a_{(\xi)}\right\|_{L^2} \leq \|\rho\|_{L^1}^{1/2} \left\|\gamma_\xi\right\|_{C^0(B_{1/2}(\mathrm{Id}))} \leq 4\left(\left|\mathbb{T}^3\right|^{1/2}+1\right)^{1/2} M \lambda_q^{-\varepsilon_R/8} \delta_{q+1}^{1/2} \leq \lambda_q^{-\varepsilon_R/16} \delta_{q+1}^{1/2} \tag{4.58a}$$

$$\left\|a_{(\xi)}\right\|_{C^N_{x,t}} \lesssim \mathbf{1}_{\{N=0\}} M \lambda_q^{-\varepsilon_R} \tau_{q+1}^{-1/2} \delta_{q+1}^{1/2} + \mathbf{1}_{\{N\geq 1\}} \tau_{q+1}^{-3N-1} \lesssim \tau_{q+1}^{-3N-1} \tag{4.58b}$$

for all $N \geq 0$.

4.4.2.2 The Principal Perturbation, the Incompressibility and Temporal Correctors

We define the principal part of the velocity increment as

$$w_{q+1}^{(p)} := \sum_{\xi\in\Lambda} a_{(\xi)} \, W_{(\xi)} . \tag{4.59}$$

In view of (4.44) we have that the summands in (4.59) have mutually disjoint supports. Also, by (4.57) we have that the low-frequency part of the nonlinear term $w_{q+1}^{(p)} \otimes w_{q+1}^{(p)}$ cancels the glued stress $\mathring{\overline{R}}_q$ modulo a term which we will hide in the pressure

$$w_{q+1}^{(p)} \otimes w_{q+1}^{(p)} + \mathring{\overline{R}}_q = \theta^2 \rho \mathrm{Id} + \sum_{\xi\in\Lambda} a_{(\xi)}^2 \mathbb{P}_{\neq 0} \left(W_{(\xi)} \otimes W_{(\xi)}\right) . \tag{4.60}$$

Recalling (4.48), the incompressibility corrector is defined implicitly by the property that

$$w_{q+1}^{(p)} + w_{q+1}^{(c)} = \sum_{\xi\in\Lambda} \operatorname{curl}\operatorname{curl}\left(a_{(\xi)} V_{(\xi)}\right) , \tag{4.61}$$

which clearly ensures $\nabla \cdot \left(w_{q+1}^{(p)} + w_{q+1}^{(c)}\right) \equiv 0$. Explicitly, the incompressibility corrector is given by

$$w_{q+1}^{(c)} := \sum_{\xi \in \Lambda} \operatorname{curl}\left(\nabla a_{(\xi)} \times V_{(\xi)}\right) + \nabla a_{(\xi)} \times \operatorname{curl} V_{(\xi)} + a_{(\xi)} W_{(\xi)}^{(c)} \tag{4.62}$$

which is a useful identity for bounding this term.

Lastly, we introduce a temporal corrector, which is defined by

$$w_{q+1}^{(t)} := -\frac{1}{\mu} \sum_{\xi \in \Lambda} \mathbb{P}_H \mathbb{P}_{\neq 0} \left(a_{(\xi)}^2 \phi_{(\xi)}^2 \psi_{(\xi)}^2 \xi\right). \tag{4.63}$$

We have $\operatorname{div} w_{q+1}^{(t)} = 0$ due to the Leray-Helmholtz projector $\mathbb{P}_H = \mathrm{Id} - \nabla \Delta^{-1} \operatorname{div}$ present in (4.63). The main property of the temporal corrector is the following identity:

$$\begin{aligned}
&\partial_t w_{q+1}^{(t)} + \sum_{\xi \in \Lambda} \mathbb{P}_{\neq 0}\left(a_{(\xi)}^2 \operatorname{div}\left(W_{(\xi)} \otimes W_{(\xi)}\right)\right) \\
&= -\frac{1}{\mu} \sum_{\xi \in \Lambda} \mathbb{P}_H \mathbb{P}_{\neq 0} \partial_t \left(a_{(\xi)}^2 \phi_{(\xi)}^2 \psi_{(\xi)}^2 \xi\right) + \frac{1}{\mu} \sum_{\xi \in \Lambda} \mathbb{P}_{\neq 0}\left(a_{(\xi)}^2 \partial_t \left(\phi_{(\xi)}^2 \psi_{(\xi)}^2 \xi\right)\right) \\
&= \nabla \left(\frac{1}{\mu} \sum_{\xi \in \Lambda} \Delta^{-1} \operatorname{div} \mathbb{P}_{\neq 0} \partial_t \left(a_{(\xi)}^2 \phi_{(\xi)}^2 \psi_{(\xi)}^2 \xi\right)\right) \\
&\quad - \frac{1}{\mu} \sum_{\xi \in \Lambda} \mathbb{P}_{\neq 0}\left(\partial_t a_{(\xi)}^2 \left(\phi_{(\xi)}^2 \psi_{(\xi)}^2 \xi\right)\right)
\end{aligned} \tag{4.64}$$

which follows from the key property (4.47) of the intermittent jets. Identity (4.64) is the key reason why we will be able to estimate the oscillation error in Sect. 4.4.3, and it shows the essential role played by temporal oscillations in our construction. The total velocity increment w_{q+1} is defined as

$$w_{q+1} := w_{q+1}^{(p)} + w_{q+1}^{(c)} + w_{q+1}^{(t)}. \tag{4.65}$$

By construction w_{q+1} is divergence-free and has zero mean on $\mathbb{T}^3$.

4.4.2.3 Bounds and Properties of the Velocity Increment

The only nontrivial estimate is the one on the L^2 norm of $w_{q+1}^{(p)}$, which we need to show is consistent with (4.18). The main insight here is that in view of (4.58b) the amplitudes $a_{(\xi)}$ oscillate at a frequency approximatively $\tau_{q+1}^{-3N} \ll \lambda_q^{24}$, whereas the intermittent jets $W_{(\xi)}$ are $(\mathbb{T}/r_\perp \lambda_{q+1})^3$-periodic, and $\lambda_{q+1} r_\perp = \lambda_{q+1}^{5/24} \gg \lambda_q^{24}$ (since $b \gg 1$). This means that we may appeal to the following L^p de-correlation lemma.

Lemma 4.2 (Lemma 3.6 in [3]) *Fix integers $N, \kappa \geq 1$ and let $\zeta > 1$ be such that*

$$\frac{2\pi\sqrt{3}\zeta}{\kappa} \leq \frac{1}{3} \qquad \text{and} \qquad \zeta^4 \frac{(2\pi\sqrt{3}\zeta)^N}{\kappa^N} \leq 1\,. \tag{4.66}$$

Let $p \in \{1, 2\}$, and let f be a $\mathbb{T}^3$-periodic function such that there exists a constant $C_f > 0$ such that

$$\|D^j f\|_{L^p} \leq C_f \zeta^j\,,$$

holds for all $0 \leq j \leq N + 4$. In addition, let g be a $(\mathbb{T}/\kappa)^3$-periodic function. Then we have that

$$\|fg\|_{L^p} \lesssim C_f \|g\|_{L^p}\,,$$

where the implicit constant is universal.

From Lemma 4.2 with $p = 2$, $f = a_{(\xi)}$ and $g = W_{(\xi)}$, the normalization (4.45), which implies upon taking traces that $\left\|W_{(\xi)}\right\|_{L^2} = \left|\mathbb{T}^3\right|^{1/2}$, and the bound (4.58a), we obtain

$$\left\|w_{q+1}^{(p)}\right\|_{L^2} \leq \sum_{\xi \in \Lambda} \lambda_q^{-\varepsilon_R/16} \delta_{q+1}^{1/2} \left\|W_{(\xi)}\right\|_{L^2} \leq \frac{1}{2}\delta_{q+1}^{1/2}. \tag{4.67}$$

The other estimates for $w_{q+1}^{(p)}$, $w_{q+1}^{(c)}$ and $w_{q+1}^{(t)}$ follow from standard estimates (we do not need to appeal to Lemma 4.2), which we summarize as

$$\left\|w_{q+1}^{(p)}\right\|_{W^{N,p}} \lesssim \tau_{q+1}^{-2} r_\perp^{2/p-1} r_\parallel^{1/p-1/2} \lambda_{q+1}^N \tag{4.68a}$$

$$\left\|w_{q+1}^{(c)}\right\|_{W^{N,p}} + \left\|w_{q+1}^{(t)}\right\|_{W^{N,p}} \lesssim \lambda_{q+1}^{-1/16} r_\perp^{2/p-1} r_\parallel^{1/p-1/2} \lambda_{q+1}^N \tag{4.68b}$$

for $N \geq 0$ and $p > 1$. The above estimates are proven in detail in [6, Section 4], and we omit these computations here. In particular, from the above estimates, combined with (4.50) and (4.58b) we deduce that the total velocity increment given by (4.65) obeys

$$\left\|w_{q+1}\right\|_{L^2} \lesssim \frac{3}{4}\delta_{q+1}^{1/2} \tag{4.69a}$$

$$\left\|w_{q+1}\right\|_{W^{N,p}} \lesssim \tau_{q+1}^{-2} r_\perp^{2/p-1} r_\parallel^{1/p-1/2} \lambda_{q+1}^N \tag{4.69b}$$

$$\left\|w_{q+1}\right\|_{H^3} \lesssim \lambda_{q+1}^3 \tag{4.69c}$$

$$\left\|\partial_t w_{q+1}\right\|_{H^2} \lesssim \lambda_{q+1}^5\,. \tag{4.69d}$$

From (4.36), the bounds (4.34b)–(4.34d), and estimates (4.69) we immediately deduce that (4.14a)–(4.14b) hold with q replaced by $q + 1$, and that (4.18) holds.

Lastly, using (4.34a), (4.35a), and the properties of the cutoff functions $\theta(t)$, we have that (4.16) holds at level $q+1$, as desired.

4.4.3 The New Reynolds Stress

The new Reynolds stress at level $q+1$ and its associated pressure are defined as

$$\mathring{R}_{q+1} = \mathcal{R}\mathbb{P}_H \operatorname{div} \widetilde{R} \qquad \text{and} \qquad p_{q+1} = \widetilde{p} - \Delta^{-1}\operatorname{div}\operatorname{div}\widetilde{R}\,. \tag{4.70}$$

where $\widetilde{R}$ and $\widetilde{p}$ are given by the next proposition:

Proposition 4.5 *There exists a q-independent parameter $p > 1$ with $p-1 \ll 1$, such that the following holds. There exists a traceless symmetric matrix $\widetilde{R}$ and a scalar pressure $\widetilde{p}$, defined implicitly in* (4.74) *below, satisfying*

$$\partial_t v_{q+1} + \operatorname{div}\left(v_{q+1} \otimes v_{q+1}\right) + \nabla \widetilde{p} - \Delta v_{q+1} = \operatorname{div} \widetilde{R}\,, \tag{4.71a}$$

$$\operatorname{div} v_{q+1} = 0\,, \tag{4.71b}$$

such that $\widetilde{R}$ obeys the bound

$$\left\|\widetilde{R}\right\|_{L^p} \lesssim \lambda_{q+1}^{-2\varepsilon_R}\delta_{q+2}\,, \tag{4.72}$$

with an implicit constant that depends on the choice of p, but is independent of q. Moreover, $\widetilde{R}$ has the support property

$$\operatorname{supp}\widetilde{R} \subset \mathbb{T}^3 \times \left\{t \in [0,T] \colon \operatorname{dist}(t, \mathcal{G}^{(q+1)}) > \tau_{q+1}\right\}. \tag{4.73}$$

An immediate consequence of Proposition 4.5 is that the desired inductive estimates (4.13a) and (4.13b) and the support property (4.17) hold for the Reynolds stress $\mathring{R}_{q+1}$ defined in (4.70) above. Indeed, (4.13a) follows from the Hölder inequality $\left\|\mathring{R}_{q+1}\right\|_{L^1} \lesssim \left\|\mathring{R}_{q+1}\right\|_{L^p}$, the Calderon-Zygmund estimate $\|\mathcal{R}\mathbb{P}_H \operatorname{div}\|_{L^p \to L^p} \lesssim 1$, and using $\lambda_{q+1}^{-\varepsilon_R}$ to absorb all the implicit constants. Property (4.17) holds due to (4.73) since the map $\widetilde{R} \mapsto \mathring{R}_{q+1}$ is time-independent. Lastly, (4.13b) follows from the momentum equation in (4.71) which implies

$$\left\|\mathring{R}_{q+1}\right\|_{H^3} \lesssim \|\mathcal{R}\mathbb{P}_H(\operatorname{div}\widetilde{R})\|_{H^3} \lesssim \|\partial_t v_{q+1} + \operatorname{div}(v_{q+1} \otimes v_{q+1}) - \Delta v_{q+1}\|_{L^\infty(T/3,2T/3;H^2)}$$

and the available estimates (4.34b)–(4.34d) on $\overline{v_q}$ and (4.69) for w_{q+1}. We omit these minor details. The proof of Proposition 4.1 is thus concluded once we give the proof of Proposition 4.5.

Using (4.65) and the fact that both $(\overline{v_q}, \overline{\mathring{R}_q})$ and $(v_{q+1}, \widetilde{R})$ solve the Navier-Stokes-Reynolds system, we obtain

$$
\begin{aligned}
\operatorname{div} \widetilde{R} - \nabla \widetilde{p} &= -\Delta w_{q+1} + \partial_t (w^{(p)}_{q+1} + w^{(c)}_{q+1}) + \operatorname{div} (\overline{v_q} \otimes w_{q+1} + w_{q+1} \otimes \overline{v_q}) \\
&\quad + \operatorname{div} \left((w^{(c)}_{q+1} + w^{(t)}_{q+1}) \otimes w_{q+1} + w^{(p)}_{q+1} \otimes (w^{(c)}_{q+1} + w^{(t)}_{q+1}) \right) \\
&\quad + \operatorname{div} (w^{(p)}_{q+1} \otimes w^{(p)}_{q+1} + \mathring{\overline{R}}_q) + \partial_t w^{(t)}_{q+1} \\
&=: \operatorname{div} \left(\widetilde{R}_{\text{linear}} + \widetilde{R}_{\text{corrector}} + \widetilde{R}_{\text{oscillation}} \right) .
\end{aligned}
\tag{4.74}
$$

The linear error and corrector errors are defined by applying $\mathcal{R}$ to the first and second lines of (4.74), while the oscillation error is defined upon first appealing to the identity (4.64), and then applying the inverse divergence operator $\mathcal{R}$ (see (4.77) below). The pressure $\widetilde{p}$ is implicit in (4.74), and it can be defined in a unique way once it is normalized to have zero mean. then the three estimates above imply (4.80).

4.4.3.1 The Linear Error

The linear error is defined from the first line of (4.74). From identity (4.61), the bounds (4.34c), (4.50), (4.58b), (4.69), our choices of parameters, and a few computations, we obtain

$$
\begin{aligned}
\left\| \widetilde{R}_{\text{linear}} \right\|_{L^p} &\lesssim \left\| \mathcal{R} \Delta w_{q+1} \right\|_{L^p} + \left\| \mathcal{R}(\partial_t (w^{(p)}_{q+1} + w^{(c)}_{q+1})) \right\|_{L^p} \\
&\quad + \left\| \mathcal{R} \operatorname{div} (\overline{v}_q \otimes w_{q+1} + w_{q+1} \otimes \overline{v}_q) \right\|_{L^p} \\
&\lesssim \left\| w_{q+1} \right\|_{W^{1,p}} + \sum_i \sum_{\xi \in \Lambda} \left\| \partial_t \operatorname{curl} (a_{(\xi)} V_{(\xi)}) \right\|_{L^p} + \left\| \overline{v}_q \right\|_{L^\infty} \left\| w_{q+1} \right\|_{L^p} \\
&\lesssim \tau_{q+1}^{-5} \lambda_{q+1} \ell_\perp^{2/p-1} \ell_\parallel^{1/p-1/2} .
\end{aligned}
\tag{4.75}
$$

By our choice of p in (4.80), the above estimate is consistent with (4.72).

4.4.3.2 Corrector Error

The corrector error is defined from the second line of (4.74). Appealing to estimates (4.68a)–(4.68b), we have

$$
\begin{aligned}
\left\| \widetilde{R}_{\text{corrector}} \right\|_{L^p} &\le \left\| \mathcal{R} \operatorname{div} \left((w^{(c)}_{q+1} + w^{(t)}_{q+1}) \otimes w_{q+1} + w^{(p)}_{q+1} \otimes (w^{(c)}_{q+1} + w^{(t)}_{q+1}) \right) \right\|_{L^p} \\
&\lesssim \left\| w^{(c)}_{q+1} + w^{(t)}_{q+1} \right\|_{L^{2p}} \left\| w_{q+1} \right\|_{L^{2p}} + \left\| w^{(p)}_{q+1} \right\|_{L^{2p}} \left\| w^{(c)}_{q+1} + w^{(t)}_{q+1} \right\|_{L^{2p}} \\
&\lesssim \tau_{q+1}^{-5} \lambda_{q+1}^{-1} \ell_\perp^{2/p-2} \ell_\parallel^{1/p-5/2} .
\end{aligned}
$$

Again, due to the choice of p in (4.80) this estimate is sufficient for (4.72).

4.4.3.3 Oscillation Error

The oscillation error is defined from the third line in (4.74) so that it obeys

$$\operatorname{div}\left(\widetilde{R}_{\text{oscillation}}\right) = \operatorname{div}\left(w_{q+1}^{(p)} \otimes w_{q+1}^{(p)} + \mathring{\bar{R}}_q\right) + \partial_t w_{q+1}^{(t)} + (\text{a pressure gradient})\,. \tag{4.76}$$

By combining (4.57) and (4.64) we obtain that a suitable definition of the oscillation error is

$$\begin{aligned}\widetilde{R}_{\text{oscillation}} &= \sum_{\xi\in\Lambda} \mathcal{R}\mathbb{P}_{\neq 0}\left(\mathbb{P}_{\neq 0}\left(\xi\phi_{(\xi)}^2\psi_{(\xi)}^2\right)\left((\xi\cdot\nabla)a_{(\xi)}^2\right)\right)\\ &\quad - \frac{1}{\mu}\sum_{\xi\in\Lambda} \mathcal{R}\mathbb{P}_{\neq 0}\left(\left(\xi\phi_{(\xi)}^2\psi_{(\xi)}^2\right)\left(\partial_t a_{(\xi)}^2\right)\right)\end{aligned} \tag{4.77}$$

Bounding the above terms amounts to using (4.58b) and noting that since the function $\phi_{(\xi)}^2\psi_{(\xi)}^2$ is $(\mathbb{T}/r_\perp\lambda_{q+1})^3$-periodic, we have

$$\mathbb{P}_{\neq 0}\left(\phi_{(\xi)}^2\psi_{(\xi)}^2\right) = \mathbb{P}_{\geq r_\perp\lambda_{q+1}/2}\left(\phi_{(\xi)}^2\psi_{(\xi)}^2\right)$$

Since $\tau_{q+1}^{-3} \ll \lambda_{q+1}r_\perp$, we may appeal to the following lemma:

Lemma 4.3 (Lemma B.1 in [3]) *Fix $\kappa \geq 1$, $p \in (1,2]$, and a sufficiently large $L \geq 1$. Let $a \in C^L(\mathbb{T}^3)$ be such that there exists $1 \leq \lambda \leq \kappa$, and $C_a > 0$ with*

$$\left\| D^j a\right\|_{L^\infty} \leq C_a\lambda^j$$

for all $0 \leq j \leq L$. Assume furthermore that $\int_{\mathbb{T}^3} a(x)\mathbb{P}_{\geq\kappa} f(x)dx = 0$. Then we have

$$\left\| |\nabla|^{-1}(a\,\mathbb{P}_{\geq\kappa} f)\right\|_{L^p} \lesssim C_a\left(1+\frac{\lambda^L}{\kappa^{L-2}}\right)\frac{\|f\|_{L^p}}{\kappa} \tag{4.78}$$

for any $f \in L^p(\mathbb{T}^3)$, where the implicit constant depend on p and L.

Applying estimate (4.78) with $\kappa = \lambda_{q+1}r_\perp$, $\lambda = \tau_{q+1}^{-3}$, $a = \xi\cdot\nabla a_{(\xi)}^2$ respectively $a = \partial_t a_{(\xi)}^2$, and $f = \phi_{(\xi)}^2\psi_{(\xi)}^2$, after a few computations we deduce

$$\|\widetilde{R}_{\text{oscillation}}\|_{L^p} \lesssim \tau_{q+1}^{-5}\lambda_{q+1}^{-1}\ell_\perp^{2/p-3}\ell_\parallel^{1/p-1} + \tau_{q+1}^{-6}\mu^{-1}\ell_\perp^{2/p-2}\ell_\parallel^{1/p-1}\,. \tag{4.79}$$

Due to the choice of p in (4.80) below, this estimate is consistent with (4.72).

4.4.3.4 Choice of p and Support Properties

To conclude the proof, we note that our specific choice of the parameters $r_\perp$, $r_\|$ and μ, enable us to choose p sufficiently small so that

$$\begin{aligned}\tau_{q+1}^{-5}\lambda_{q+1}r_\perp^{2/p-1}r_\|^{1/p-1/2} &+ \tau_{q+1}^{-5}\lambda_{q+1}^{-1}r_\perp^{2/p-2}r_\|^{1/p-5/2}\\ &+ \tau_{q+1}^{-6}\lambda_{q+1}^{-1}r_\perp^{2/p-3}r_\|^{1/p-1}\\ &+ \tau_{q+1}^{-6}\mu^{-1}r_\perp^{2/p-2}r_\|^{1/p-1} \lesssim \lambda_{q+1}^{-2\varepsilon_R}\delta_{q+2}\,. \end{aligned} \tag{4.80}$$

holds. To see that (4.80) is indeed possible, we note that when $p = 1$ the left side of (4.80) divided by the right side of (4.80) is bounded by

$$\begin{aligned}&\frac{\tau_{q+1}^{-6}}{\lambda_{q+1}^{-2\varepsilon_R}\delta_{q+2}}\left(\lambda_{q+1}r_\perp r_\|^{1/2} + \lambda_{q+1}^{-1}\left(r_\|^{-3/2} + r_\perp^{-1} + r_\perp r_\|^{-1}\right)\right)\\ &\lesssim \frac{\tau_{q+1}^{-6}}{\lambda_{q+1}^{-2\varepsilon_R}\delta_{q+2}}\lambda_{q+1}^{-1/12} \lesssim \lambda_{q+1}^{-1/24}\end{aligned}$$

by taking ε_R and β sufficiently small in terms of b, and then b sufficiently large. The above bound for $p = 1$ holds by continuity for $p > 1$ but sufficiently close to 1, with a loss bounded by $\lambda_{q+1}^{1/24}$, which thus implies (4.80).

Lastly, from (4.74) we note that

$$\operatorname{supp}\widetilde{R} \subset \operatorname{supp} w_{q+1}^{(p)} \cup \operatorname{supp} w_{q+1}^{(c)} \cup \operatorname{supp} w_{q+1}^{(t)} \cup \operatorname{supp}\mathring{\overline{R}}_q\,.$$

The support of $\mathring{\overline{R}}_q$ is consistent with (4.73) in view of (4.35a). On the other hand, by (4.55) we have that $a_{(\xi)}(t) = 0$ whenever $\operatorname{dist}(t, \mathcal{G}^{(q+1)}) \le \tau_{q+1}$ and thus $w_{q+1}^{(p)}(t) = w_{q+1}^{(c)}(t) = w_{q+1}^{(t)}(t) = 0$ whenever $\operatorname{dist}(t, \mathcal{G}^{(q+1)}) \le \tau_{q+1}$. Therefore, (4.73) holds, which concludes the proof of Proposition 4.5.

Acknowledgments The work of V.V. was partially supported by the NSF grant DMS-1911413.

References

1. T. Buckmaster, Onsager's conjecture almost everywhere in time. Commun. Math. Phys. **333**(3), 1175–1198 (2015)
2. T. Buckmaster, V. Vicol, Convex integration and phenomenologies in turbulence. EMS Surv. Math. Sci. **6**(1/2), 173–263 (2019)
3. T. Buckmaster, V. Vicol, Nonuniqueness of weak solutions to the Navier-Stokes equation. Ann. Math. **189**(1), 101–144 (2019)

4. T. Buckmaster, C. De Lellis, P. Isett, L. Székelyhidi Jr., Anomalous dissipation for $1/5$-Hölder Euler flows. Ann. Math. **182**(1), 127–172 (2015)
5. T. Buckmaster, C. De Lellis, L. Székelyhidi Jr., Dissipative Euler flows with Onsager-critical spatial regularity. Commun. Pure Appl. Math. **69**(9), 1613–1670 (2016)
6. T. Buckmaster, M. Colombo, V. Vicol, Wild solutions of the Navier-Stokes equations whose singular sets in time have hausdorff dimension strictly less than 1 (2018). ArXiv:1809.00600
7. T. Buckmaster, C. De Lellis, L. Székelyhidi Jr., V. Vicol, Onsager's conjecture for admissible weak solutions. Commun. Pure Appl. Math. **72**(2), 227–448 (2019)
8. J.-Y. Chemin, *Perfect Incompressible Fluids*, vol. 14 (Oxford University Press, Oxford, 1998)
9. A. Cheskidov, X. Luo, Stationary and discontinuous weak solutions of the Navier-Stokes equations (2019). ArXiv:1901.07485
10. M. Colombo, C. De Lellis, A. Massaccesi, The generalized Caffarelli-Kohn-Nirenberg theorem for the hyperdissipative Navier-Stokes system. Commun. Pure Appl. Math. **73**(3), 609–663 (2020)
11. P. Constantin, C. Foias, *Navier-Stokes Equations*. Chicago Lectures in Mathematics (University of Chicago Press, Chicago, 1988)
12. M. Dai, Non-uniqueness of Leray-Hopf weak solutions of the 3d Hall-MHD system (2018). ArXiv:1812.11311
13. S. Daneri, L. Székelyhidi Jr., Non-uniqueness and h-principle for Hölder-continuous weak solutions of the Euler equations. Arch. Ration. Mech. Anal. **224**(2), 471–514 (2017)
14. C. De Lellis, L. Székelyhidi Jr., The h-principle and the equations of fluid dynamics. Bull. Am. Math. Soc. (N.S.) **49**(3), 347–375 (2012)
15. C. De Lellis, L. Székelyhidi Jr., Dissipative continuous Euler flows. Invent. Math. **193**(2), 377–407 (2013)
16. C. De Lellis, L. Székelyhidi Jr., Dissipative Euler flows and Onsager's conjecture. J. Eur. Math. Soc. **16**(7), 1467–1505 (2014)
17. C. De Lellis, L. Székelyhidi Jr., High dimensionality and h-principle in PDE. Bull. Am. Math. Soc. **54**(2), 247–282 (2017)
18. C. De Lellis, L. Székelyhidi Jr., On turbulence and geometry: from Nash to Onsager. Not. AMS **05**, 677–685 (2019)
19. C. Doering, J. Gibbon, *Applied Analysis of the Navier-Stokes Equations*, vol. 12 (Cambridge University Press, Cambridge, 1995)
20. E. Fabes, B. Jones, N. Rivière, The initial value problem for the Navier-Stokes equations with data in L^p. Arch. Ration. Mech. Anal. **45**, 222–240 (1972)
21. C. Foias, O. Manley, R. Rosa, R. Temam, *Navier-Stokes Equations and Turbulence*. Encyclopedia of Mathematics and Its Applications, vol. 83 (Cambridge University Press, Cambridge, 2001)
22. E. Hopf, Über die Anfangswertaufgabe für die hydrodynamischen Grundgleichungen. Math. Nachr. **4**, 213–231 (1951)
23. P. Isett, On the endpoint regularity in Onsager's conjecture (2017). ArXiv:1706.01549
24. P. Isett, A proof of Onsager's conjecture. Ann. Math. **188**(3), 871–963 (2018)
25. N. Katz, N. Pavlović, A cheap Caffarelli–Kohn–Nirenberg inequality for the Navier–Stokes equation with hyper-dissipation. Geom. Funct. Anal. **12**(2), 355–379 (2002)
26. P. Lemarié-Rieusset, *Recent Developments in the Navier-Stokes Problem*. Chapman & Hall/CRC Research Notes in Mathematics, vol. 431 (Chapman & Hall/CRC, Boca Raton, 2002)
27. P. Lemarié-Rieusset, *The Navier-Stokes Problem in the 21st Century* (CRC Press, Boca Raton, 2016)
28. J. Leray, Sur le mouvement d'un liquide visqueux emplissant l'espace. Acta Math. **63**(1), 193–248 (1934)
29. J.-L. Lions, *Quelques Méthodes de Résolution des Problèmes aux Limites Non linéaires*, vol. 1 (Dunod/Gauthier-Villars, Paris, 1969)
30. P.-L. Lions, *Mathematical Topics in Fluid Mechanics*, vol. 1 (The Clarendon Press/Oxford University Press, New York, 1996)

31. X. Luo, Stationary solutions and nonuniqueness of weak solutions for the Navier-Stokes equations in high dimensions. Arch. Ration. Mech. Anal. **233**, 701–747 (2019)
32. T. Luo, E. Titi, Non-uniqueness of weak solutions to hyperviscous Navier-Stokes equations - on sharpness of J.-L. Lions exponent (2018). ArXiv:1808.07595
33. A.J. Majda, A.L. Bertozzi, *Vorticity and Incompressible Flow*. Cambridge Texts in Applied Mathematics, vol. 27 (Cambridge University Press, Cambridge, 2002)
34. C. Marchioro, M. Pulvirenti, *Mathematical Theory of Incompressible Nonviscous Fluids*, vol. 96 (Springer Science & Business Media, Berlin, 2012)
35. S. Modena, L. Székelyhidi Jr., Non-uniqueness for the transport equation with Sobolev vector fields. Ann. PDE **4**, 18 (2018)
36. S. Modena, L. Székelyhidi Jr., Non-renormalized solutions to the continuity equation. Calc. Var. **58**, 208 (2019)
37. J. Nash, C^1 isometric imbeddings. Ann. Math. 383–396 (1954)
38. J. Robinson, J. Rodrigo, W. Sadowski, Classical theory. *The Three-dimensional Navier-Stokes Equations*. Cambridge Studies in Advanced Mathematics, vol. 157 (Cambridge University Press, Cambridge, 2016)
39. V. Scheffer, An inviscid flow with compact support in space-time. J. Geom. Anal. **3**(4), 343–401 (1993)
40. A. Shnirelman, Weak solutions with decreasing energy of incompressible Euler equations. Commun. Math. Phys. **210**(3), 541–603 (2000)
41. L. Székelyhidi Jr., From isometric embeddings to turbulence. HCDTE Lecture Notes. Part II. Nonlinear hyperbolic PDEs. Dispersive Trans. Equ. **7**, 63 (2012)
42. R. Temam, *Navier-Stokes Equations and Nonlinear Functional Analysis*, vol. 66 (SIAM, Philadelphia, 1995)
43. R. Temam, Theory and numerical analysis. *Navier-Stokes Equations*, Reprint of the 1984 edn. (AMS Chelsea Publishing, Providence, 2001)
44. T.-P. Tsai, *Lectures on Navier-Stokes Equations*, vol. 192 (American Mathematical Society, Providence, 2018)

Chapter 5
Conserved Quantities and Regularity in Fluid Dynamics

Emil Wiedemann

Abstract Conserved or dissipated quantities, like energy or entropy, are at the heart of the study of many classes of time-dependent PDEs in connection with fluid mechanics. This is the case, for instance, for the Euler and Navier-Stokes equations, for systems of conservation laws, and for transport equations. In all these cases, a formally conserved quantity may no longer be constant in time for a weak solution at low regularity. The delicate interplay between regularity and conservation of the respective quantity relates to renormalisation in the DiPerna-Lions theory of transport and continuity equations, and to Onsager's conjecture in the realm of ideal incompressible fluids. We will review the classical commutator methods of DiPerna-Lions and Constantin-E-Titi, and then proceed to more recent results.

5.1 Introduction

Quite unsurprisingly, conserved or dissipated quantities play a fundamental role in about any evolution differential equation related to continuum mechanics. On the analytical level, these quantities are often (as in the Navier-Stokes equations) the only source of a priori estimates to yield sufficient compactness for the existence of weak solutions. In the theory of transport equations, conserved quantities allow to show uniqueness and stability through the renormalisation theory of DiPerna-Lions [20] discussed below.

On the physical level, quantities that are formally shown to be conserved can, in fact, be observed to be dissipated. This kind of anomalous dissipation occurs, e.g., in the incompressible Euler equations at low regularity due to turbulent energy transfer to high scales, as predicted by Onsager [41] based on Kolmogorov's phenomenological theory of turbulence. Denoting by u the fluid velocity, at low regularity one may thus observe that the kinetic energy

E. Wiedemann (✉)
Department of Mathematics, Chair of Analysis, Friedrich-Alexander-Universität, Erlangen-Nürnberg, Germany
e-mail: emil.wiedemann@fau.de

M. Bulíček et al. (eds.), *New Perspectives in Mathematical Fluid Mechanics*, Nečas Center Series, https://doi.org/10.1007/978-3-032-02319-3_5

$$\frac{1}{2}\int |u(x,t)|^2 dx \tag{5.1}$$

is actually decreasing in time, although by formal computation (as it is justified for smooth solutions) it would be constant. In fact, denoting by p the pressure, energy conservation formally holds in the stronger local sense

$$\partial_t \frac{|u|^2}{2} + \operatorname{div}\left(\left(\frac{|u|^2}{2} + p\right) u\right) = 0, \tag{5.2}$$

which implies the total conservation of energy upon integration in time and using the Divergence Theorem.

The past decades have seen increasingly sophisticated constructions giving rigorous examples of weak solutions to the Euler equations with non-conserved energy [8, 15–17, 35, 43–45], culminating in a complete proof of Onsager's Conjecture on the threshold regularity up to which anomalous dissipation can occur. This regularity is essentially at $C^{1/3}$ (the Hölder space with exponent $1/3$), and below we will at least show that $1/3$ is an upper bound.

A more classical phenomenon in hyperbolic conservation laws is the formation of shock waves, which also leads to a decrease of the (mathematical) entropy. For instance, the inviscid Burgers equation

$$\partial_t u + \partial_x(u^2) = 0 \tag{5.3}$$

is easily seen to satisfy the equality $\partial_t \frac{u^2}{2} + \partial_x \frac{2u^3}{3} = 0$, and thus to conserve the quantity $\frac{1}{2}\int u^2 dx$, as long as the solution remains smooth, but it is equally easy to explicitly find an example of a weak solution that becomes discontinuous in finite time and for which $\frac{1}{2}\int u^2 dx$ is strictly decreasing in time. More precisely, given the Lipschitz initial datum

$$u_0(x) := \begin{cases} 1 & \text{if } x \le 0, \\ 1 - x & \text{if } 0 \le x \le 1, \\ 0 & \text{if } x > 1, \end{cases}$$

for any time $t < 1$ a solution is given by

$$u(x,t) = \begin{cases} 1 & \text{if } x \le t, \\ \frac{1-x}{1-t} & \text{if } t \le x \le 1, \\ 0 & \text{if } x > 1, \end{cases}$$

which can be extended to $t \ge 1$ by

$$u(x,t) = \begin{cases} 1 & \text{if } x < \frac{t+1}{2}, \\ 0 & \text{if } x > \frac{t+1}{2}, \end{cases}$$

and it holds true that, after $t = 1$, there are smooth test functions $\phi \geq 0$ such that

$$\int \int \left(\partial_t \phi \frac{u^2}{2} + \partial_x \phi \frac{2u^3}{3} \right) dx dt > 0,$$

which means $\frac{1}{2} \int u^2 dx$ is not conserved. In the context of scalar conservation laws, one sometimes calls a convex function of a solution an *entropy* (which in some situations can be interpreted as minus the thermodynamic entropy), and we see that in the given example, the entropy conservation is violated in accordance with the Second Law of Thermodynamics.

But even when a certain quantity is seen to be dissipative already on the formal level, it is still important to know whether it satisfies a predicted balance as an equality or not. For example, the incompressible Navier-Stokes equations formally satisfy

$$\frac{1}{2} \partial_t \int |u|^2 dx + \nu \int_0^t \int |\nabla u|^2 dx ds = \frac{1}{2} \int |u_0|^2 dx, \tag{5.4}$$

where $\nu > 0$ is the viscosity of the modelled fluid; but it can only be shown to hold as an inequality ($=$ replaced by $\leq$) for generic weak solutions of Leray-Hopf type. In contrast to the Euler equations and hyperbolic conservation laws, however, there seems to be no mathematical or physical reason to believe that (5.4) should hold with strict inequality.

Let us condense this somewhat loose collection of observations into some general ideas that will form the focus of this survey:

- A great variety of partial differential equations related to continuum dynamics exhibit quantities that can be easily seen, by formal calculation invoking the chain rule of differential calculus, to be conserved in time. For smooth solutions, these calculations are easily justified rigorously.
- These conservation laws can be formulated in a local (e.g., (5.2), (5.3)) or in a global (e.g., (5.1), (5.4)) way.
- On the other hand, less regular solutions may not conserve these quantities. This relates to physically observable effects of anomalous energy dissipation due to turbulence, or to increase of physical entropy due to the Second Law of Thermodynamics.
- It is therefore worth investigating the threshold regularity below which such dissipative effects can occur.

Our discussion here mainly focuses on inviscid models. We begin with the arguably simplest case of linear scalar conservation laws, i.e., linear transport equations, which possess infinitely many conserved quantities. The question of threshold regularity can be interpreted in several ways, depending on whether one wishes to impose regularity conditions on the coefficients alone, or on the coefficients and the solution combined. The first approach leads to the DiPerna-Lions theory of renormalisation, while the second one motivates the commutator estimates of Constantin-E-Titi [12]. This is presented in Sect. 5.2.

The mentioned techniques give bounds from above for the sought threshold regularity (i.e., sufficient conditions for conservation), and can be viewed as restoring the chain rule in regimes of low regularity. On the other side, the construction of examples of non-conservation, and therefore of breakdown of the chain rule, can be much harder. We have seen that classical shocks provide such examples in the context of hyperbolic equations, but in incompressible models, shocks are not available. Instead, convex integration has recently become the method of choice to construct dissipative solutions of the Euler equations. In Sect. 5.3, we outline the method of convex integration in the comparatively simple setting of steady transport, and thus provide counterexamples to renormalisation.

The conservative part of Onsager's Conjecture for the incompressible Euler system forms the topic of Sect. 5.4, including a discussion of recent results concerning bounded domains and the vanishing viscosity limit. The subsequent sections are devoted to various recent extensions of the commutator method to statistical solutions, general conservation laws, and degenerate cases such as the compressible Euler and Navier-Stokes equations with possible vacuum.

5.2 Renormalisation of Transport Equations

The first partial differential equation (PDE) considered in the textbook [25] of L. C. Evans is also the supposedly simplest one: the *transport equation*

$$\partial_t \rho + u \cdot \nabla \rho = 0. \tag{5.5}$$

For further simplicity, let us consider the case of periodic boundary conditions. Then, if $\mathbb{T}^d := \mathbb{R}^d/\mathbb{Z}^d$ is the flat torus, one usually considers the vector field $u : \mathbb{T}^d \times [0, T] \to \mathbb{R}^d$ as given and divergence-free. The scalar field $\rho : \mathbb{T}^d \times [0, T] \to \mathbb{R}$ is the unknown, which could be subject to an initial condition $\rho(\cdot, 0) = \rho^0$.

The transport equation can be interpreted, for instance, as follows: The given field u can be thought of as the known velocity field of an incompressible flow (hence the divergence-free condition), such as water on the surface of the ocean (in which case $d = 2$). The scalar ρ then gives the concentration of, say, a chemical dissolved on the ocean surface, and the chemical is transported by the given flow. As, in this very simple model, the chemical has no effect on the dynamics of the transporting flow, ρ is sometimes called a *passive scalar* or a *passive tracer*.

A remarkable property of the transport equation is that it can be *renormalised*: Suppose an arbitrary C^1 function $\eta : \mathbb{R} \to \mathbb{R}$ is given, then (5.5) can be multiplied by $\eta'(\rho)$ to yield, by means of the chain rule,

$$\partial_t \eta(\rho) + u \cdot \nabla \eta(\rho) = 0, \tag{5.6}$$

so that in fact arbitrary functions of a solution become a solution of the same equation. Moreover, integration in space yields $\frac{d}{dt} \int_{\mathbb{T}^d} \eta(\rho) dx = 0$, and therefore from this renormalisation procedure we obtain infinitely many conserved quantities for (5.5).

However, the chain rule (we used it in the form $\partial_t \eta(\rho) = \eta'(\rho)\partial_t \rho$ and $\nabla \eta(\rho) = \eta'(\rho)\nabla \rho$) is only justified if ρ is C^1, or at least Lipschitz. So the question arises:

Under what conditions on u and/or ρ are solutions of the transport equation renormalised?

First of all, how does the transport even make sense if ρ is not C^1? As usual, one considers a distributional concept of solution, using the fact that $u \cdot \nabla \rho = \operatorname{div}(\rho u)$ by virtue of the divergence-free condition on u. Therefore, a function $\rho \in L^\infty(\mathbb{T}^d \times [0, T])$ is called a *weak solution* of (5.5) with initial data $\rho^0 \in L^\infty(\mathbb{T}^d)$ if, for every $\phi \in C_c^1(\mathbb{T}^d \times [0, T))$, we have

$$\int_0^T \int_{\mathbb{T}^d} \partial_t \phi \rho + \rho \nabla \phi \cdot u dx dt = \int_{\mathbb{T}^d} \phi(x, 0) \rho^0(x) dx.$$

Note this definition makes sense as long as $u \in L^1_{loc}(\mathbb{T}^d \times [0, T))$.

Renormalisation plays an important role not only for the study of (5.5) or of ordinary differential equations, but also for larger systems of PDEs that contain (5.5) or the closely related *continuity equation*

$$\partial_t \rho + \operatorname{div}(\rho u) = 0,$$

where u is no longer assumed divergence-free. Consider two examples: The *isentropic compressible Navier-Stokes system* reads

$$\begin{aligned} \partial_t(\rho u) + \operatorname{div}(\rho u \otimes u) + \nabla p(\rho) &= \operatorname{div} \mathbb{S}(\nabla u), \\ \partial_t \rho + \operatorname{div}(\rho u) &= 0, \end{aligned} \tag{5.7}$$

where the scalar density ρ is now assumed non-negative, p is a given function of density, and $\mathbb{S}$ denotes the Newtonian stress tensor. The velocity u is a vector field $\mathbb{T}^d \times [0, T] \to \mathbb{R}^d$.

Let us ignore the details of the constitutive theory for the moment and imagine $\mathbb{S}$ as the identity, so that the right hand side can be thought of simply as Δu. It is well-known that these equations satisfy a priori bounds given by the energy inequality

$$\begin{aligned} \int_{\mathbb{T}^d} \frac{1}{2} \rho(x, t) |u(x, t)|^2 + P(\rho(x, t)) dx + \int_0^t \int_{\mathbb{T}^d} \mathbb{S}(\nabla u) : \nabla u dx ds \\ \leq \int_{\mathbb{T}^d} \frac{1}{2} \rho^0 |u^0|^2 + P(\rho^0) dx. \end{aligned}$$

Here P is the so-called pressure potential given by

$$P(r) = r\int_1^r \frac{p(s)}{s^2}ds. \tag{5.8}$$

Since $\mathbb{S}(\nabla u) : \nabla u \geq c|\nabla u|^2$, we obtain a bound for u in $L^2(0, T; H^1(\mathbb{T}^d))$, whereas for ρ we obtain only integrability, but no regularity properties. In fact, this is to be expected since (5.7) is parabolic in the momentum equation but only hyperbolic in the mass equation.

The question whether the density can be renormalised plays an important role, e.g., in the Lions-Feireisl theory of weak solutions for (5.7), see [28, 37]. As we just saw, we obtain a priori information on the regularity of the transporting velocity, but not on the transported scalar. In this context, one thus asks: Under what (Sobolev) regularity assumptions on the transporting velocity field u is every bounded weak solution ρ of (5.5) renormalised (i.e. every smooth function of ρ is again a weak solution of (5.5))? This is the subject of the famous DiPerna-Lions theory that we will outline shortly.

As another example, consider an *active scalar equation* of the form

$$\begin{aligned} \partial_t\rho + u\cdot\nabla\rho &= 0,\\ \operatorname{div} u &= 0,\\ u &= T[\rho], \end{aligned}$$

where T is a Fourier multiplier operator of order zero. For instance, if (for $d = 2$) the symbol of T is given by $i\frac{\xi^\perp}{|\xi|}$, we obtain the well-known surface quasi-geostrophic (SQG) equation.

The only thing of interest at the moment, anyway, is the fact that T is bounded from L^p to L^p ($1 < p < \infty$) and therefore, the active scalar ρ and the velocity u have the same (Lebesgue, Besov, Hölder etc.) regularity. In contrast to the Navier-Stokes equations, therefore, we can not (or do not have to) distinguish between the regularities of the scalar and the vector field.

A similar example is given by the *incompressible Euler equations*

$$\begin{aligned} \partial_t u + (u\cdot\nabla)u + \nabla p &= 0\\ \operatorname{div} u &= 0, \end{aligned} \tag{5.9}$$

where in some sense the velocity is transported by itself (this is the effect of *advection*). Again, there is no way to distinguish between the regularities of the transporting and the transported quantity. We will get back to the Euler equations later.

5.2.1 DiPerna-Lions Commutators

Again let us simply consider the transport Eq. (5.5) and assume that ρ, u form a weak solution (where u is still divergence-free and ρ is bounded for simplicity). In cases where some regularity is known for u (loosely speaking, at least one full distributional space derivative), but none for ρ, the theory of DiPerna-Lions is useful.

Let $\chi : \mathbb{T}^d \to \mathbb{R}$ be a standard mollifier, i.e. a smooth non-negative radially symmetric function with compact support in $B_1(0)$ and $\int_{B_1(0)} \chi\, dx = 1$. Set $\chi_\varepsilon(x) = \varepsilon^{-d}\chi\left(\frac{x}{\varepsilon}\right)$, which is supported on $B_\varepsilon(0)$ and still has unit integral. For a function f, We write $f_\varepsilon := f * \chi_\varepsilon$.

Mollifying (5.5), we obtain (ignoring issues of time differentiability)

$$0 = \partial_t \rho_\varepsilon + \operatorname{div}(\rho u)_\varepsilon = \partial_t \rho_\varepsilon + \operatorname{div}(\rho_\varepsilon u) + R_\varepsilon, \tag{5.10}$$

where

$$R_\varepsilon = \operatorname{div}(\rho u)_\varepsilon - \operatorname{div}(\rho_\varepsilon u)$$

is the *commutator*. Let now $\eta \in C^1$ and multiply (5.10) by $\eta'(\rho_\varepsilon)$. Since ρ_ε is smooth, we can apply the chain rule to obtain

$$\partial_t \eta(\rho_\varepsilon) + \operatorname{div}(\eta(\rho_\varepsilon)u) = -\eta'(\rho_\varepsilon)R_\varepsilon$$

(recall that u is divergence-free, so that $\operatorname{div}(\rho_\varepsilon u) = u \cdot \nabla\rho_\varepsilon$). First, it is clear that the terms on the left hand side converge, in the sense of distributions as $\varepsilon \to 0$, to $\partial_t \eta(\rho) + \operatorname{div}(\eta(\rho)u)$. Hence, as $\eta'(\rho_\varepsilon)$ is bounded uniformly in ε, it suffices to show that $R_\varepsilon \to 0$ in $L^1(\mathbb{T}^d \times [0, T])$.

To this end, we compute

$$\begin{aligned}
R_\varepsilon(x,t) &= \operatorname{div}(\rho u) * \chi_\varepsilon(x,t) - \operatorname{div}((\rho * \chi_\varepsilon)u(x,t)) \\
&= -\rho u * \nabla\chi_\varepsilon(x,t) + \rho * \nabla\chi_\varepsilon \cdot u(x,t) \\
&= \int_{B_\varepsilon(x)} \rho(y,t)(u(x,t) - u(y,t)) \cdot \nabla\chi_\varepsilon(x-y)dy \\
&= \varepsilon^{-d-1} \int_{B_\varepsilon(x)} \rho(y,t)(u(x,t) - u(y,t)) \cdot \nabla\chi\left(\frac{x-y}{\varepsilon}\right)dy \\
&= -\int_{B_1(0)} \rho(x+\varepsilon z,t)\frac{u(x+\varepsilon z,t) - u(x,t)}{\varepsilon} \cdot \nabla\chi\,(z)\,dz,
\end{aligned}$$

where in the end we used the transformation $z = \frac{y-x}{\varepsilon}$.

Suppose that $u \in L^1(0,T;W^{1,1}(\mathbb{T}^d))$, then, by standard difference quotient lemmas, $\frac{u(x+\varepsilon z,t)-u(x,t)}{\varepsilon}$ converges, as $\varepsilon \to 0$, to the directional derivative $\nabla u(x)z$ in $L^1(\mathbb{T}^d \times (0,T))$ for fixed z; moreover it is bounded in L^1 uniformly in z and ε. Since $\rho(x+\varepsilon z)$ is in L^∞, uniformly in ε and z, and converges in L^1 to $\rho(x)$ for fixed z, we obtain the strong L^1 convergence

$$\begin{aligned}
R_\varepsilon(x,t) \to -\rho(x,t)\int_{B_1(0)} \nabla u(x,t) : (z \otimes \nabla\chi(z))dz \\
= -\rho(x,t)\partial_j u_i(x,t)\int_{B_1(0)} z_j \partial_i \chi(z)\, dz \\
= \rho(x,t)\delta_{ij}\partial_j u_i(x,t) = \rho(x,t)\operatorname{div} u(x,t) = 0,
\end{aligned}$$

as desired.

As the convergence argument may not be obvious, let us give it in more detail: Write $D_\varepsilon(x,z,t) := \frac{u(x+\varepsilon z,t)-u(x,t)}{\varepsilon}$ and $\rho_\varepsilon(x,z,t) := \rho(x+\varepsilon z,t)$, then

$$\begin{aligned}
&\int_{\mathbb{T}^d\times(0,T)} \left| R_\varepsilon(x,t) + \rho(x,t)\int_{B_1(0)} \nabla u(x,t) : (z\otimes\nabla\chi(z))dz \right| dxdt \\
&\le \|\nabla\chi\|_\infty \left(\int_{\mathbb{T}^d\times(0,T)}\int_{B_1(0)} |\rho_\varepsilon(x,z,t)D_\varepsilon(x,z,t) - \rho_\varepsilon(x,z,t)\nabla u(x,t)z| dzdxdt \right.\\
&\qquad \left. + \int_{\mathbb{T}^d\times(0,T)}\int_{B_1(0)} |\rho_\varepsilon(x,z,t)\nabla u(x,t)z - \rho(x,t)\nabla u(x,t)z| dzdxdt \right) \\
&\le \|\nabla\chi\|_\infty \sup_{\varepsilon,z}\|\rho_\varepsilon\|_\infty \int_{B_1(0)}\int_{\mathbb{T}^d\times(0,T)} |D_\varepsilon - \nabla u z| dxdtdz \\
&\qquad + \|\nabla\chi\|_\infty \sup_{|z|\le 1}\|\nabla u z\|_{L^1} \int_{B_1(0)}\int_{\mathbb{T}^d\times(0,T)} |\rho_\varepsilon - \rho| dxdtdz.
\end{aligned}$$

As mentioned above, for each z, $\int_{\mathbb{T}^d\times(0,T)} |D_\varepsilon - \nabla u z| dxdt$ converges to zero as $\varepsilon \to 0$, and the same is true for $\int_{\mathbb{T}^d\times(0,T)} |\rho_\varepsilon - \rho| dxdt$. Since these expressions are dominated, respectively, by

$$2\sup_{|z|\le 1}\int_{\mathbb{T}^d\times(0,T)} |\nabla u(x,t)z| dxdt, \quad 2\int_{\mathbb{T}^d\times(0,T)} |\rho(x,t)| dxdt,$$

which are constant and hence integrable in z, we conclude by the Dominated Convergence Theorem.

We have thus proved:

Theorem 5.1 (DiPerna-Lions) *If $u \in L^1(0,T;W^{1,1}(\mathbb{T}^d))$ is divergence-free, then every bounded weak solution of* (5.5) *is renormalised in the sense of* (5.6).

Of course this result holds in much greater generality (the assumptions $\rho \in L^\infty$ and $\operatorname{div} u = 0$ can be substantially relaxed). In fact it follows easily from renormalisation that the Cauchy problem for (5.5) admits a unique weak solution.

5.2.2 Constantin-E-Titi Commutators

Let us present a commutator argument that looks very similar to the previous one, but leads to very different conclusions. Assume again u to be divergence-free. Once more we mollify Eq. (5.5) in space,

$$0 = \partial_t \rho_\varepsilon + \operatorname{div}(\rho u)_\varepsilon = \partial_t \rho_\varepsilon + \operatorname{div}(\rho_\varepsilon u_\varepsilon) + S_\varepsilon,$$

with

$$S_\varepsilon = \operatorname{div}(\rho u)_\varepsilon - \operatorname{div}(\rho_\varepsilon u_\varepsilon).$$

Note the only difference compared to (5.10) is that we chose to mollify also u. Multiplying again by $\eta'(\rho_\varepsilon)$, we obtain (noting that u_ε is still divergence-free)

$$\partial_t \eta(\rho_\varepsilon) + \operatorname{div}(\eta(\rho_\varepsilon) u_\varepsilon) = -\eta'(\rho_\varepsilon) S_\varepsilon,$$

so we obtain renormalisation provided we can show the the right hand side converges to zero, in the sense of distributions, as $\varepsilon \to 0$.

To this end, let $\phi \in C_c^1(\mathbb{T}^d \times (0, T))$, so that integration by parts yields

$$\begin{aligned} -\int_0^T \int_{\mathbb{T}^d} \phi \eta'(\rho_\varepsilon) S_\varepsilon dx dt &= \int_0^T \int_{\mathbb{T}^d} \eta'(\rho_\varepsilon) \nabla \phi \cdot ((\rho u)_\varepsilon - \rho_\varepsilon u_\varepsilon) dx dt \\ &\quad + \int_0^T \int_{\mathbb{T}^d} \phi \eta''(\rho_\varepsilon) \nabla \rho_\varepsilon \cdot ((\rho u)_\varepsilon - \rho_\varepsilon u_\varepsilon) dx dt. \end{aligned}$$

We only treat the second integral, as the first one is easier. Assume to this end that η'' is bounded, and consider the pointwise identity

$$(\rho u)_\varepsilon - \rho_\varepsilon u_\varepsilon = -(\rho_\varepsilon - \rho)(u_\varepsilon - u) + \int_{B_\varepsilon(0)} \chi_\varepsilon(y)(\rho(\cdot - y) - \rho)(u(\cdot - y) - u) dy.$$

Thus, one part of the desired estimate is obtained by

$$\int_0^T \int_{\mathbb{T}^d} |\nabla \rho_\varepsilon| |\rho_\varepsilon - \rho| |u_\varepsilon - u| dx dt \le \|\nabla \rho_\varepsilon\|_{L^p} \|\rho - \rho_\varepsilon\|_{L^p} \|u - u_\varepsilon\|_{L^q}$$

for some exponents satisfying $\frac{2}{p}+\frac{1}{q}\leq 1$. The other part can be estimated similarly, so we ignore it.

The question arises under what conditions these norms converge to zero as $\varepsilon\to 0$. To this end, suppose that

$$\lim_{\varepsilon\to 0}\int_0^T\int_{\mathbb{T}^d}\int_{B_\varepsilon(0)}\frac{|\rho(x)-\rho(x-y)|^p}{\varepsilon^{d+\alpha p}}dydxdt=0. \tag{5.11}$$

We aim to show that, under this assumption, $\lim_{\varepsilon\to 0}\varepsilon^{-\alpha}\|\rho-\rho_\varepsilon\|_{L^p}=0$. Indeed, using Jensen's inequality and the definition of χ_ε,

$$\begin{aligned}\varepsilon^{-\alpha p}\|\rho-\rho_\varepsilon\|_{L^p}^p&=\int_0^T\int_{\mathbb{T}^d}\left|\int_{B_\varepsilon(0)}\frac{\rho(x)-\rho(x-y)}{\varepsilon^\alpha}\chi_\varepsilon(y)dy\right|^p dxdt\\&\leq\int_0^T\int_{\mathbb{T}^d}\int_{B_\varepsilon(0)}\frac{|\rho(x)-\rho(x-y)|^p}{\varepsilon^{d+\alpha p}}\chi\left(\frac{y}{\varepsilon}\right)dydxdt,\end{aligned}$$

which converges to zero by virtue of assumption (5.11). Likewise, we have

$$\lim_{\varepsilon\to 0}\varepsilon^{-\beta}\|u-u_\varepsilon\|_{L^q}=0$$

provided

$$\lim_{\varepsilon\to 0}\int_0^T\int_{\mathbb{T}^d}\int_{B_\varepsilon(0)}\frac{|u(x)-u(x-y)|^q}{\varepsilon^{d+\beta q}}dydxdt=0. \tag{5.12}$$

Finally, it is not difficult to get the estimate $\lim_{\varepsilon\to 0}\varepsilon^{1-\alpha}\|\nabla\rho_\varepsilon\|_{L^p}=0$. Assume now that $2\alpha+\beta\geq 1$, then

$$\begin{aligned}&\|\nabla\rho_\varepsilon\|_{L^p}\|\rho-\rho_\varepsilon\|_{L^p}\|u-u_\varepsilon\|_{L^q}\leq\\&\qquad\varepsilon^{1-\alpha}\|\nabla\rho_\varepsilon\|_{L^p}\varepsilon^{-\alpha}\|\rho-\rho_\varepsilon\|_{L^p}\varepsilon^{-\beta}\|u-u_\varepsilon\|_{L^q}\to 0,\end{aligned}$$

and so we arrive at the following result:

Theorem 5.2 *Let* $\frac{2}{p}+\frac{1}{q}\leq 1$ *and* $2\alpha+\beta\geq 1$, *and* $\rho\in L^p(\mathbb{T}^d\times(0,T))$ *and* $u\in L^q(\mathbb{T}^d\times(0,T))$ *be a weak solution of* (5.5) *satisfying* (5.11) *and* (5.12), *where* u *is divergence-free. If* $\eta\in W^{2,\infty}(\mathbb{R})$, *then the renormalised equation holds in the sense of distributions:*

$$\partial_t\eta(\rho)+u\cdot\nabla\eta(\rho)=0.$$

Several remarks are in order. This commutator method relies on the ideas in [12] in the context of Onsager's conjecture (which we will get back to later). In the case of Euler, the transporting field and the transported quantity are identical, which forces $p = q \geq 3$ and $\alpha = \beta \geq \frac{1}{3}$. Mixed regularities for ρ and u were introduced in [29, 36]. Theorem 5.2 was proved in [1], but in a slightly more restrictive functional framework. The paper [1] also relaxes the condition $\eta \in W^{2,\infty}(\mathbb{R})$, and gives an application to active scalar equations. Conditions (5.11) and (5.12) were first formulated in [30] and thus generalised previous works like [10, 12, 24]. They are implied, for instance, by $\rho \in L^p((0,T); C^{\alpha'}(\mathbb{T}^d))$ or $\rho \in L^p((0,T); B^{\alpha'}_{p,\infty}(\mathbb{T}^d))$ for any $\alpha' > \alpha$, where $C^{\alpha'}$ and $B^{\alpha'}_{p,\infty}$ denote the respective Hölder and Besov spaces, and similarly for u.

5.3 Counterexamples to Renormalisation via Convex Integration

It may seem strange that renormalisation could fail for a simple linear equation like (5.5). But in fact, there is a great deal of counterexamples known, starting from the original paper of DiPerna-Lions and extended in various ways in [3, 13, 14, 18, 34, 38–40, 47]. A powerful instrument to produce rough and often pathological solutions to certain PDEs is known as *convex integration*. We present here the main ideas from [14] (full details can be found in that paper). The result discussed here is by no means optimal in terms of regularity of the velocity field, but it does show that renormalisation can fail in basically any conceivable way, and it gives an idea of the general technique of convex integration in a comparatively simple setting.

We consider the stationary problem in 3D, and we aim to show the following:

Theorem 5.3 *Let $f \in \mathcal{D}'(\mathbb{T}^3;\mathbb{R})$ be a distribution such that there exists a bounded continuous solution of the equation $\operatorname{div} w = f$. Then, there exist a bounded vectorfield $u \in L^\infty(\mathbb{T}^3;\mathbb{R}^3)$ and a bounded scalar field $\rho \in L^\infty(\mathbb{T}^3)$ such that*

$$\begin{aligned} \operatorname{div}(\rho u) &= 0 \\ \operatorname{div}(u) &= 0 \\ \operatorname{div}(\rho^2 u) &= f. \end{aligned} \tag{5.13}$$

Note once more that if the chain rule were valid, we would have $\operatorname{div}(\rho^2 u) = 0$. Therefore, f is called the *renormalisation defect*.

We begin by vastly relaxing the problem to the study of triplets (m, u, w) of vectorfields that satisfy

$$\begin{aligned} \operatorname{div}(m) &= 0 \\ \operatorname{div}(u) &= 0 \\ \operatorname{div}(w) &= f. \end{aligned} \tag{5.14}$$

Setting

$$K_C := \left\{ (m, u, w) \in \mathbb{R}^{3\times 3} : \frac{1}{C} \leq |u| \leq C \right.$$
$$\left. \text{and there is } \frac{1}{C} \leq \rho \leq C \text{ such that } m = \rho u,\, w = \rho^2 u \right\}, \tag{5.15}$$

we realise that a solution of (5.14) will be a solution of the original problem (5.13) if $(m, u, w)(x) \in K_C$ for almost every $x \in \mathbb{T}^3$, for some $C > 1$.

Our strategy is now roughly as follows:

1. Start with the "subsolution" $U^0 := (0, 0, w)$, where w has the property $\operatorname{div} w = f$.
2. For almost every $x \in \mathbb{T}^3$, represent $(0, 0, w)$ as a convex combination of the form
$$\sum_{i=1}^{N} \lambda_i (m_i, u_i, w_i)$$
in a way that is compatible with (5.14), and such that the (m_i, u_i, w_i) are in K_C.
3. Find a new triplet of functions (m^1, u^1, w^1) that still satisfies (5.14), and that approximately takes the value (m_i, u_i, w_i) on a volume fraction λ_i near the point x.
4. Iterate the process to obtain a sequence $(m^n, u^n, w^n)_{n\in\mathbf{N}}$ of functions satisfying (5.14), and show that the sequence converges to a solution of (5.13).

Step 1 can be done by assumption. For Step 2, it surely should be clarified what is meant by "compatibility" of the convex combination with constraint (5.14). Clearly, without a differential constraint like (5.14), it is easy to find a continuous map $U^1 := (m^1, u^1, w^1) : \mathbb{T}^3 \to \mathbb{R}^{3\times 3}$ such that, close to any point $x \in \mathbb{T}^3$, U^1 takes values close to $U_i := (m_i, u_i, w_i)$ on a set of volume fraction λ_i; suppose for simplicity that the λ_i and U_i are constant, then such a map could simply be given as (a suitable mollification of) the piecewise constant map

$$U^1(x) = U_n \quad \text{if } x_1 \in \left(\sum_{i=1}^{n-1} \lambda_i, \sum_{i=1}^{n} \lambda_i \right), \qquad n = 1, \ldots, N.$$

Now of course λ_i and U_i do depend on x, but as this is the case continuously, we can cover the domain by small cubes, consider λ_i and U_i constant on each such cube, and glue the various pieces together.

However, we want $U - (0, 0, w)$ to be divergence-free, so that (5.14) remains valid. Suppose, at a given point $x \in \mathbb{T}^3$, $U^0(x)$ can be represented as a convex combination of only two matrices, $U^0(x) = \lambda U_1 + (1 - \lambda)U_2$. Does there exist a divergence-free matrix field U that oscillates between U_1 and U_2?

To this end, we define the *wave cone* for the divergence-free condition as

$$\Lambda := \{\bar{U} \in \mathbb{R}^{3\times 3} : \text{there exists } \xi \neq 0 \text{ such that } \operatorname{div}[h(x\cdot\xi)\bar{U}] \\ = 0 \text{ for any } h : \mathbb{R} \to \mathbb{R}\}. \tag{5.16}$$

Since $\bar{U} \in \Lambda$ is equivalent to $h'(x\cdot\xi)\bar{U}\xi = 0$ for all h and a non-trivial ξ, we get that $\bar{U} \in \Lambda$ if and only if $\det \bar{U} = 0$ or, in other words, the rank of $\bar{U}$ is at most 2. Therefore, there exists an oscillation between U_1 and U_2 if and only if $\operatorname{rank}(U_2 - U_1) \leq 2$.

Inductively, one can then find a condition for divergence-free oscillations between more than two matrices:

Definition 5.1 Suppose $\lambda_i > 0$ for $i = 1, \dots, n$, $\sum_{i=1}^n \lambda_i = 1$, and $U_i \in \mathbb{R}^{3\times 3}$ for $i = 1, \dots, n$. The family of pairs $(\lambda_i, U_i)_{i=1}^n$ satisfies the (inductively defined) **H_n-condition** if

(i) $\operatorname{rank}(U_2 - U_1) \leq 2$ in the case $n = 2$;
(ii) after a relabeling of indices, if necessary, we have $\operatorname{rank}(U_2 - U_1) \leq 2$ and the family $(\tau_i, V_i)_{i=1}^{n-1}$ satisfies the H_{n-1}-condition, where

$$\tau_1 = \lambda_1 + \lambda_2, \quad \tau_i = \lambda_{i+1} \quad \text{for } i = 2, \dots, n-1$$

and

$$V_1 = \frac{\lambda_1}{\tau_1}U_1 + \frac{\lambda_2}{\tau_1}U_2, \quad V_i = U_{i+1} \quad \text{for } i = 2, \dots, n-1$$

in the case $n > 2$.

Moreover we adopt the convention that every pair of the form $(1, U)$ satisfies the H_1-condition.

With this definition, we can thus give a precise meaning to Step (2) above: For almost every $x \in \mathbb{T}^3$, represent $U^0 := (0, 0, w(x))$ as a convex combination of the form $\sum_{i=1}^N \lambda_i(x)U_i(x)$, where $(\lambda_i(x), U_i(x))_{i=1}^N$ satisfy the H_N-condition, and $U_i(x) \in K_C$.

That this can actually be done is the content of the following lemma. Note that, since w is assumed bounded and continuous, we may always assume $|w| \geq 1$, as adding a constant will not affect the property $\operatorname{div} w = f$.

Lemma 5.1 *Let $U = (0, 0, w) \in \mathbb{R}^{3\times 3}$ such that $|w| \geq 1$. Then there exist $(\lambda_i, U_i)_{i=1}^3$ satisfying the H_3-condition, such that*

$$U = \sum_{i=1}^3 \lambda_i U_i,$$

and a number $C > 1$ such that

$$U_i \in K_C \quad \textit{for } i = 1, 2, 3.$$

Proof We split $(0, 0, w)$ into

$$(0, 0, w) = \frac{1}{2}(-w, -w, w) + \frac{1}{2}(w, w, w).$$

If we call the matrices corresponding to the two triplets on the right hand side U_- and U_+, respectively, we first observe that U_- and U_+ are rank-2 connected since $(U_- - U_+)e_3 = 0$. Secondly, $U_+ \in K_C$ for any C such that

$$C \geq |w|.$$

Next, let us further decompose U_-. We make the ansatz

$$(-w, -w, w) = \frac{1}{2}(\rho_1 v_1, v_1, \rho_1^2 v_1) + \frac{1}{2}(\rho_2 v_2, v_2, \rho_2^2 v_2) \tag{5.17}$$

with

$$v_1 = w, \quad v_2 = -3w. \tag{5.18}$$

Then clearly (5.17) is a rank-2 decomposition (in fact even rank-1), and (5.17) and (5.18) result in the conditions

$$\begin{aligned} -\rho_1 + 3\rho_2 &= 2 \\ -\rho_1^2 + 3\rho_2^2 &= -2. \end{aligned} \tag{5.19}$$

A direct calculation shows that these requirements can be satisfied by two numbers $\rho_1, \rho_2 > 1$, and the proof is finished. □

We proceed to Step (3). This consists in finding a continuous divergence-free map $U^1 = (m^1, u^1, w^1) : \mathbb{T}^3 \to \mathbb{R}^{3\times 3}$ that approximately "recovers" the values $U_i(x)$ from the previous step. Given a family $(\lambda_i, U_i)_{i=1}^N$ satisfying the H_N-condition, we call the probability measure

$$\nu := \sum_{i=1}^{N} \lambda_i \delta_{U_i} \in \mathcal{P}(\mathbb{R}^{3\times 3})$$

a *rank-2 laminate*, and if the λ_i and U_i depend on x, then we obtain a probability measure ν_x that depends on $x \in \mathbb{T}^3$. (Such measures are often called *parametrised probability measures* or *Young measures*).

If $K \subset \mathbb{R}^{3\times 3}$ is compact, we denote by $K^{2lc} \supset K$ the *rank-2 lamination convex hull* of K, that is, the set of barycentres of rank-2 laminates supported on K. In other words, if $U \in K^{2lc}$, then U is the convex combination of some matrices in K that satisfy an H_N-condition.

The recovery lemma can then be stated as follows:

Lemma 5.2 *Let $K \subset \mathbb{R}^{3\times 3}$ be compact and $(\nu_x)_{x\in\Omega}$ be a family of probability measures such that*

(a) the measure ν_x is a rank-2 laminate for almost every $x \in \Omega$,
(b) $\operatorname{supp} \nu_x \subset K$ for almost every x.

Assume further that $\psi \in C(\mathbb{R}^{3\times 3}; \mathbb{R})$ is a non-negative function that vanishes on K. Then the barycentre $\bar{\nu}_x = \int_K V d\nu_x(V)$ is well-defined for almost every $x \in \Omega$, and for every $\varepsilon > 0$ there exists a matrix-valued function U such that

(i) $\operatorname{div} U = \operatorname{div} \bar{\nu}$ *in the sense of distributions,*
(ii)

$$\int_\Omega \psi(U(x))dx < \varepsilon,$$

(iii)

$$\|\operatorname{dist}(U(x), K^{2lc})\|_{L^\infty(\Omega)} < \varepsilon,$$

(iv)

$$\int_\Omega |U(x) - \bar{\nu}_x|\, dx < \int_\Omega \int_{\mathbb{R}^{3\times 3}} |V - \bar{\nu}_x|\, d\nu_x(V)dx + \varepsilon. \tag{5.20}$$

Moreover, if $\bar{\nu} \in C(\bar{\Omega})$, then U can be chosen to satisfy $U \in C(\bar{\Omega})$.

The proof of this lemma is largely standard. By translation and localisation arguments, one reduces to the situation where ν is independent of x, and uses nested oscillations along rank-2 lines, as provided by (5.16). Note that all of this requires a localisation technique, i.e. a way to replace a divergence-free matrix field with another one of compact support, such that it is close to the original one on its support. It would be tempting to multiply a given field U by a compactly supported cutoff function $\phi : \mathbb{T}^3 \to \mathbb{R}$, but then ϕU might not be divergence-free; instead, since $\operatorname{div} U = 0$, there exists a potential $\Psi : \mathbb{T}^3 \to \mathbb{R}^{3\times 3}$ such that $\operatorname{curl} \Psi = U$, and one localises considering $\widetilde{U} := \operatorname{curl}(\phi\Psi)$. We omit details.

In this way, we get a new matrix field $U^1 = (m^1, u^1, w^1)$ that is much closer to K_C that the original $U^0 = (0, 0, w)$. However, as Lemma 5.2 still unavoidably contains errors of order ε, we have to repeat the procedure (Step (4)). Since U^1 is no longer of the special form $(0, 0, w)$, we need a more general version of Lemma 5.1:

Lemma 5.3 *Let $\varepsilon > 0$ and $\widetilde{C} > 1$. There exists a strictly increasing continuous function $h : [0, \infty) \to [0, \infty)$, depending only on $\widetilde{C}$, with $h(0) = 0$, and a number $\delta > 0$, depending only on $\widetilde{C}$ and ε, such that for every $1 < C < \widetilde{C} - \varepsilon$ and every $U \in \mathbb{R}^{3\times 3}$ such that $\operatorname{dist}(U, K_C^{2lc}) < \delta$, there exists a rank-2 laminate $\nu = \sum_{i=1}^n \lambda_i \delta_{U_i}$ such that*

$$U = \sum_{i=1}^{n} \lambda_i U_i, \tag{5.21}$$

$$\sum_{i=1}^{n} \lambda_i |U_i - U| \le h\left(\operatorname{dist}(U, K_C)\right), \tag{5.22}$$

and

$$\operatorname{supp} \nu \subset K_{C+\varepsilon}.$$

The proof of this lemma is actually at the core of the whole construction, but we shall not discuss it here since the proof of Lemma 5.1 hopefully already gives a taste for the geometric arguments involved.

To finish up the proof, we need to define a sequence $(U^n)_{n\in\mathbf{N}}$ of divergence-free matrix fields whose average distance to K_C converges to zero, for some $C > 1$.

To this end, let $C_0 > 1$ be as required by Lemma 5.1 applied to $U^0(x)$ for *all* $x \in \bar{\Omega}$ (this is possible since U^0 is bounded). Next, pick a sequence $(C_n)_{n\ge 0}$ that is strictly increasing such that $C_n \nearrow C_0 + 1 =: C$ as $n \to \infty$. We also set $\varepsilon_n := C_{n+1} - C_n$. Then, (ε_n) is a sequence of positive numbers converging to zero.

By Lemma 5.1 there exists for almost every $x \in \mathbb{T}^3$ a rank-2 laminate ν_x^0 of finite order whose expectation is $U^0(x)$ and whose support is contained in K_{C_0}. This completes the definition of U^0 and ν^0.

Suppose now that U^n and ν^n have already been constructed for some $n \ge 0$ in such a way that $\operatorname{supp} \nu^n \subset K_{C_n}$ and (5.14), (5.21), (5.22) are satisfied, that is:

$$\begin{aligned} \operatorname{div}(U^n) &= (0, 0, f)^T, \\ U^n(x) &= \bar{\nu}_x^n, \\ \int_{\mathbb{R}^{3\times 3}} |V - U^n(x)| d\nu_x^n(V) &\le h\left(\operatorname{dist}(U^n, K_{C_{n-1}})\right). \end{aligned} \tag{5.23}$$

The last estimate is claimed only for $n \ge 1$. By Lemma 5.3, where we set $\varepsilon = \varepsilon_{n+1}$ and $\widetilde{C} = C + 1$, there exists $\delta_{n+1} = \delta(\varepsilon_{n+1})$ such that whenever

$$\operatorname{dist}(U, K_{C_n}^{2lc}) < \delta_{n+1},$$

then there exists a rank-2 laminate whose expectation is U and whose support is contained in

$$K_{C_n+\varepsilon_{n+1}} \subset K_{C_{n+1}}. \tag{5.24}$$

Therefore we apply Lemma 5.2 to (ν_x^n) with K_{C_n}, $\varepsilon = \delta_{n+1}$, and

$$\psi = h\left(\operatorname{dist}(\bullet, K_{C_n})\right).$$

This yields a matrix field U_{n+1} satisfying

$$\operatorname{div}(U_{n+1}) = \operatorname{div}\left(\bar{\nu}_x^n\right) = \operatorname{div}(U_n) = (0, 0, f)^T,$$

$$\int_\Omega h\left(\operatorname{dist}(U_{n+1}(x), K_{C_n})\right) dx < \delta_{n+1}, \tag{5.25}$$

and

$$\|\operatorname{dist}(U_{n+1}(x), K_{C_n}^{2lc})\|_{L^\infty(\Omega)} < \delta_{n+1}. \tag{5.26}$$

Therefore, by (5.24), we can indeed find, for every x, a rank-2 laminate ν_x^{n+1} with support in $K_{C_{n+1}}$ satisfying (5.21) and (5.22). This completes the construction of the sequence (U^n).

It is then not hard to show that $U^n \to U$ in $L^1(\mathbb{T}^3; \mathbb{R}^{3\times 3})$, and that consequently $U(x) \in K_C$ for almost every x. Note that, in the construction of our sequence (U^n), the admissible error ε for Lemma 5.2 is chosen as δ_{n+1}, which typically is much smaller than the previous ε. This choice forces $U^{n+1} - U^n$ to oscillate at a much higher frequency that U^n itself. This separation of frequencies, as one might call it, is typical of any convex integration type argument.

5.4 Onsager's Conjecture

5.4.1 Onsager's Conjecture on $\mathbb{T}^3$

Consider again the incompressible Euler equations,

$$\begin{aligned} \partial_t u + (u \cdot \nabla)u + \nabla p &= 0 \\ \operatorname{div} u &= 0. \end{aligned}$$

Here, $u : \mathbb{T}^3 \times [0, T] \to \mathbb{R}^3$ denotes the velocity and $p : \mathbb{T}^3 \times [0, T] \to \mathbb{R}$ the pressure of an ideal (i.e. inviscid) incompressible fluid, like approximately water. Multiplying this system by its velocity and integrating in space gives

$$\frac{d}{dt}\int_{\mathbb{T}^3} |u|^2 dx + \int_{\mathbb{T}^3} (u \cdot \nabla)u \cdot u dx + \int_{\mathbb{T}^3} u \cdot \nabla p dx = 0.$$

The last integral vanishes since u is divergence-free, and for the middle one we compute, by integration by parts,

$$\int_{\mathbb{T}^3}(u\cdot\nabla)u\cdot u\,dx=\int_{\mathbb{T}^3}u_j\partial_j u_i u_i\,dx=-\int_{\mathbb{T}^3}\partial_j u_j u_i u_i\,dx-\int_{\mathbb{T}^3}u_j u_i\partial_j u_i\,dx,$$

and since the first integral on the right hand side again vanishes due to the divergence-free condition, it follows that $\int_{\mathbb{T}^3}(u\cdot\nabla)u\cdot u\,dx=0$, so we are left with the conservation of energy,

$$\frac{d}{dt}\int_{\mathbb{T}^3}|u|^2dx=0. \tag{5.27}$$

In fact, a more careful computation (that omits integration in space) yields the *local energy equality*

$$\partial_t\frac{|u|^2}{2}+\operatorname{div}\left(\left(\frac{|u|^2}{2}+p\right)u\right)=0, \tag{5.28}$$

which is the precise analogue (for $\eta(u)=\frac{1}{2}|u|^2$) of the renormalised Eq. (5.6) in the context of transport equations.

Note once more that the chain rule (or, on a related note, integration by parts) requires u to have at least one full derivative. On the other hand, there are examples known of weak solutions that grossly violate (5.28) and even (5.27), the first one being due to Scheffer [43], whose solutions are no better than L^2_{loc}. The question thus arises whether there exists a threshold regularity that distinguishes dissipative from conservative solutions. In 1949, L. Onsager made the following conjecture [41]:

Conjecture 5.1

(a) If u is a weak solution of the incompressible Euler equations with $u\in C^\alpha$ for an $\alpha>\frac{1}{3}$, then the energy is conserved.
(b) For every $\alpha<\frac{1}{3}$ there exists a weak solution $u\in C^\alpha$ that dissipates energy.

It may seem strange that Onsager, a physicist, would worry about the validity of the chain rule for non-differentiable functions at a time where no counterexamples to the chain rule were known in the first place. In fact, he didn't; rather, his motivation stems from phenomenological turbulence theories (particularly the one of Kolmogorov) that make predictions on the energy spectrum of a fully turbulent fluid, which in turn lead to certain regularity properties of the velocity. This also means that non-conservative solutions of Euler are not always mathematical pathologies, but they are expected in turbulence theory and can be experimentally observed. We refer to [27, 32, 46] for more physically oriented overviews.

Onsager's conjecture has meanwhile largely been proved. Part (b) was completed only recently in [8, 35], based on convex integration techniques whose development started with [15]. Part (a) is more classical—it was initially solved in [12, 26] and then refined in [10, 24, 30].

To formulate Part (a) more precisely, we have the following result:

Theorem 5.4 *Let* $(u, p) \in L^3 \times L^{3/2}(\mathbb{T}^3 \times (0, T))$ *be a weak solution of* (5.9) *such that*

$$\lim_{\varepsilon \to 0} \int_0^T \int_{\mathbb{T}^3} \int_{B_\varepsilon(0)} \frac{|u(x) - u(x-y)|^3}{\varepsilon^4} dydxdt = 0. \tag{5.29}$$

Then, the local energy inequality (5.28) *is satisfied in the sense of distributions.*

To prove this, one proceeds exactly as in the proof of Theorem 5.2, where the velocity now plays the role of both ρ and u, thus forcing $p = q = 3$ and $\alpha = \beta = \frac{1}{3}$.

As before, note that condition (5.29) is implied by $u \in L^3(0, T; C^\alpha(\mathbb{T}^3))$ for some $\alpha > \frac{1}{3}$, or by $u \in L^3(0, T; B^\alpha_{3,\infty}(\mathbb{T}^3))$ for some $\alpha > \frac{1}{3}$ (this is the space considered in [12]), or by $u \in L^3(0, T; B^{1/3}_{3,c_0}(\mathbb{T}^3))$ (this is the space considered in [10]). Another important remark is that the result is actually true in any space dimension (so the exponent $1/3$ has nothing to do with 3D space!). In contrast, Part (b) of Onsager's conjecture is still open in two dimensions.

5.4.2 Bounded Domains

Let now $\Omega \subset \mathbb{R}^3$ be a smooth bounded domain. Recall that the strategy for Theorems 5.2 and 5.4 was to mollify the equation in space and thus to obtain

$$\partial_t u_\varepsilon + \operatorname{div}(u_\varepsilon \otimes u_\varepsilon) + \nabla p_\varepsilon = R_\varepsilon \tag{5.30}$$

for the commutator $R_\varepsilon = \operatorname{div}(u_\varepsilon \otimes u_\varepsilon - (u \otimes u)_\varepsilon)$. It seems like an issue that the mollification of a function cannot be defined in an obvious way on a bounded domain. However, choosing a compactly supported test function $\phi \in C^1_c(\Omega; \mathbb{R})$, u_ε is well-defined on the support of ϕ as long as $\varepsilon < \operatorname{dist}(\operatorname{supp}\phi, \partial\Omega)$. Therefore, testing (5.30) against ϕ, everything is well-defined. Suppose now (5.29) is satisfied on $\operatorname{supp}\phi$ for each $\phi \in C^1_c(\Omega)$; this means that (5.29) holds locally in Ω, but not necessarily uniformly up to the boundary. Then, just as in the proof of Theorem 5.2 we obtain $u_\varepsilon \cdot R_\varepsilon \to 0$ in the sense of distributions, so that the local energy equality holds in the sense of distributions on Ω.

The idea is that the local energy equality (5.28) is stronger than the global one (5.27), since one can deduce the latter from the former simply by integrating in space. On bounded domains, though, one should be a bit more careful.

Given (5.28), integration in space would amount to testing with $\phi \equiv 1$. But we are only allowed to test with $\phi \in C^1_c(\Omega)$. So instead, let $\xi : \mathbb{R}^+_0 \to \mathbb{R}^+_0$ be a smooth cutoff function that is zero near $s = 0$ and one for $s \geq 1$, and let $\xi_\delta(s) := \xi\left(\frac{s}{\delta}\right)$. Set $\phi_\delta(x) := \xi_\delta(\operatorname{dist}(x, \partial\Omega))$, so that $\phi_\delta \in C^1_c(\Omega)$, and test (5.28) against ϕ_δ:

$$\frac{d}{dt}\int_\Omega \phi_\delta \frac{|u|^2}{2} dx = \int_\Omega \nabla\phi_\delta \cdot u \left(\frac{|u|^2}{2} + p\right) dx.$$

The left hand side converges to $\frac{d}{dt}\int_\Omega \frac{|u|^2}{2} dx$ as $\delta \to 0$. If we want (5.27) to hold, therefore, we must show that the right hand side converges to zero.

Since $\partial\Omega$ is smooth, there exists a neighbourhood Γ_0 of $\partial\Omega$ where the orthogonal projection of $x \in \Gamma_0$ to $\partial\Omega$ is uniquely defined. Let's call this projection $\sigma(x) \in \partial\Omega$, and denote by $n(\sigma(x))$ the outer unit normal to $\partial\Omega$ at $\sigma(x)$.

Then, $\nabla\phi_\delta(x)$ is parallel to $n(\sigma(x))$ with absolute value bounded by $\frac{C}{\delta}$, and on the other hand, $\nabla\phi_\delta$ is supported on $\Gamma_\delta := \{\text{dist}(x, \partial\Omega) < \delta\}$, where the volume of Γ_δ is comparable to δ.

Hence, we have

$$\left|\int_\Omega \nabla\phi_\delta \cdot u \left(\frac{|u|^2}{2} + p\right) dx\right| \leq \frac{C}{\delta}\int_{\Gamma_\delta} \left|\frac{|u|^2}{2} + p\right| |u(x) \cdot n(\sigma(x))| dx.$$

From this, it is easy to read off some sufficient conditions for global energy conservation: For instance, one could require u and p to be bounded in some neighbourhood of $\partial\Omega$, $x \mapsto u(x) \cdot n(\sigma(x))$ to be continuous in a neighbourhood of $\partial\Omega$, and $u \cdot n = 0$ on $\partial\Omega$, which is the natural slip boundary condition for Euler anyway.

To summarise:

Theorem 5.5 *Let* (u, p) *be a weak solution of* (5.9) *such that* u *satisfies* (5.29) *locally in* Ω. *Suppose* (u, p) *is bounded in some neighbourhood of* $\partial\Omega$, $x \mapsto u(x) \cdot n(\sigma(x))$ *is continuous in a neighbourhood of* $\partial\Omega$, *and* $u \cdot n = 0$ *on* $\partial\Omega$. *Then, the global energy equality* (5.27) *holds in the sense of distributions.*

This result is taken from [5], where the theorem is stated under much weaker assumptions (in particular on the pressure). Similar results, with some differences on the technical level, were obtained independently in [22]. Earlier results on Onsager's conjecture in domains with boundaries were obtained in [4, 42].

A useful feature of this result is that, near the boundary, it puts conditions only on the *normal* component of the velocity, but not on the tangential one. This gives rise to an application to the viscosity limit.

Recall the Navier-Stokes equations with viscosity $\nu > 0$,

$$\begin{aligned} \partial_t u_\nu + (u_\nu \cdot \nabla) u_\nu + \nabla p_\nu &= \nu \Delta u_\nu \\ \operatorname{div} u_\nu &= 0, \end{aligned} \tag{5.31}$$

which are known to admit (for given initial data in L^2) global weak (so-called Leray-Hopf) solutions in Ω, subject to the no-slip boundary condition $u = 0$ on $\partial\Omega$. It is commonly expected that, for small ν, the Navier-Stokes flow will behave like an Euler flow except on a boundary layer of thickness of order $\sqrt{\nu}$, where it decays steeply to $u = 0$ at $\partial\Omega$. Thus, one expects the normal velocity component to be very small in the boundary layer, but the tangential velocity component to have a gradient of magnitude $\nu^{-1/2}$. Since Theorem 5.5 makes no assumption on the tangential component, the result is consistent with the formation of such a boundary layer. More precisely:

Corollary 5.1 ([5]) *Let* $(u_\nu)_{\nu>0}$ *be a family of Leray-Hopf weak solutions of* (5.31) *with viscosity* ν *and initial data* $u^0 \in L^2(\Omega)$, *and suppose this family satisfies the conditions of Theorem 5.5 uniformly in* ν. *Then, there exists a subsequence* $\nu_k \to 0$ *such that* $u_\nu \to u$ *strongly in* $L^\infty(0, T; L^2(\Omega))$, *where* u *is a weak solution of the Euler equations that conserves energy.*

It is interesting to compare this result to [11, 23], where purely interior assumptions are made on the solution, and anomalous energy dissipation is therefore not excluded.

5.5 Statistical Solutions

As phenomenological theories of turbulence are of a statistical nature, it is natural to place the fundamental PDEs of fluid mechanics in a probabilistic framework. A classical way to do this is given by DiPerna's measure-valued solutions [19], formulated for the incompressible Euler equations by DiPerna and Majda [21]. To illustrate the idea, consider again the incompressible Euler equations (5.9). Usually one wants to find a solution which is a vectorfield $u : \mathbb{T}^3 \times [0, T] \to \mathbb{R}^3$, that is, at each (or at least almost every) point in space and time, one gives the velocity. In contrast, suppose the velocity at a point in space-time is not known exactly, but only as a probability distribution: This can be modelled by a map $\mathbb{T}^3 \times [0, T] \to \mathcal{P}(\mathbb{R}^3)$ from space-time into the set of probability measures on the phase space $\mathbb{R}^3$. Thus, at each point (x, t), there is a probability measure $\mu_{x,t}$ such that the probability that the velocity is in a (measurable) subset $U \subset \mathbb{R}^3$ is given by $\mu_{x,t}(U)$.

Writing $\bar{u}(x, t) := \int_{\mathbb{R}^3} \xi \, d\mu_{x,t}(\xi)$ and $\overline{u \otimes u}(x, t) := \int_{\mathbb{R}^3} \xi \otimes \xi \, d\mu_{x,t}(\xi)$, we say that μ is a measure-valued solution of the Euler equations if

$$\begin{aligned} \partial_t \bar{u} + \overline{u \otimes u} + \nabla p &= 0 \\ \operatorname{div} \bar{u} &= 0 \end{aligned}$$

in the sense of distributions. (Observe that $(u \cdot \nabla)u = \operatorname{div}(u \otimes u)$, which allows to write the equations in divergence form.) We deliberately ignore here the issue of possible concentrations requiring a generalised measure-valued setting, and refer to [21].

Although such solutions are easily shown to exist for any initial datum, and have other useful properties (see e.g. [48]), they are felt by some to contain too little information. Indeed, it is not possible to represent with them two-point correlations, i.e. expressions of the form "the probability that the velocity at point (x, t) is in U_1 and the velocity at (y, t) is in U_2". On a related note, it is not obvious how to make sense of a Besov condition like (5.11) for a measure-valued solution.

To describe the statistics of a flow, Fjordholm et al. [31] propose a new notion of statistical solutions, where solutions to conservation laws are given as *correlation Young measures*. In the context of incompressible Euler, this was studied in [30]. The idea is to describe the two-point statistics of a fluid in terms of a parametrised measure $\mu_{x,y,t} \in \mathcal{P}(\mathbb{R}^4 \times \mathbb{R}^4)$, where $x, y \in \mathbb{T}^3$ are points from the space domain. The measure $\mu_{x,y}(du_1, dp_1, du_2, dp_2)$ can then be interpreted as the joint probability that the velocity and pressure at point x are in $du_1 \times dp_1$ and the velocity and pressure at point y are in $du_2 \times dp_2$.

More precisely, such a correlation measure is a pair (μ^1, μ^2), where $\mu^1_{x,t}$ is a probability measure on $\mathbb{R}^4$ and $\mu^2 : x, t$ is a probability measure on $\mathbb{R}^4 \times \mathbb{R}^4$ for almost every (x, t), such that the following conditions are satisfied:

1. *Symmetry:* If $f \in C_0(\mathbb{R}^3 \times \mathbb{R}^3)$ then
$$\int_{\mathbb{R}^4\times\mathbb{R}^4} f(\xi, \eta) d\mu^2_{x,y,t}(\xi, \eta) = \int_{\mathbb{R}^3\times\mathbb{R}^3} f(\eta, \xi) d\mu^2_{y,x,t}(\eta, \xi)$$
for a.e. $x, y \in \mathbb{R}^3$;
2. *Consistency:* If $f \in C_0(\mathbb{R}^4 \times \mathbb{R}^4)$ is of the form $f(\xi, \eta) = g(\xi)$ for some $g \in C_0(\mathbb{R}^4)$, then
$$\int_{\mathbb{R}^4\times\mathbb{R}^4} f(\xi, \eta) d\mu^2_{x,y,t}(\xi, \eta) = \int_{\mathbb{R}^4} g(\xi) d\mu^1_{x,t}(\xi)$$
for almost every $(x, y) \in \mathbb{R}^3 \times \mathbb{R}^3$.

One can now give a natural definition of μ to be a solution of the Euler equations: Indeed, if (u, p) is a smooth solution and u^i denotes the i-th velocity component, then

$$\partial_t \left(u^i(x)u^j(y)\right) + \sum_k \partial_{x^k} \left(u^i(x)u^k(x)u^j(y)\right) + \sum_k \partial_{y^k} \left(u^i(x)u^k(y)u^j(y)\right)$$
$$+ \partial_{x^i} \left(p(x)u^j(y)\right) + \partial_{y^j} \left(u^i(x)p(y)\right) = 0,$$

as can be seen by adding the equation evaluated at x and multiplied by $u^j(y)$ to the equation evaluated at y and multiplied by $u^i(x)$. Thus, replacing the occurrences of the velocity and pressure by the correlation measure, one is led to the equation

$$\partial_t \langle \mu^2_{x,y} u^i_1 u^j_2 \rangle + \sum_k \partial_{x^k} \langle \mu^2_{x,y} u^i_1 u^k_1 u^j_2 \rangle + \sum_k \partial_{y^k} \langle \mu^2_{x,y} u^i_1 u^k_2 u^j_2 \rangle$$
$$+ \partial_{x^i} \langle \mu^2_{x,y} p_1 u^j_2 \rangle + \partial_{y^j} \langle \mu^2_{x,y} u^i_1 p_2 \rangle = 0,$$

where we wrote $\langle \mu^2_{x,y} u_1^i u_2^j \rangle = \int_{\mathbb{R}^4 \times \mathbb{R}^4} u_1^i u_2^j d\mu_{x,y}(u_1, p_1, u_2, p_2)$ etc. (hence the dummy variables u_1, u_2, p_1, p_2 stand for $u(x), u(y), p(x), p(y)$, respectively).

The natural extension of the Besov-type assumption (5.29) is then

$$\lim_{\varepsilon \to 0} \int_0^T \int_{\mathbb{T}^3} \int_{B_\varepsilon(0)} \frac{\langle \mu^2_{x,x-y}; |u_1 - u_2|^3 \rangle}{\varepsilon^4} dydxdt = 0,$$

which is the measure-valued analogue of (5.29).

In [30] we show that the energy is (locally) conserved under this assumption.

5.6 General Conservation Laws

A crucial observation made in [29] is that the commutator estimates remain valid even when the nonlinearities involved are not quadratic. For instance, for the isentropic compressible Euler system

$$\begin{aligned} \partial_t(\rho u) + \operatorname{div}(\rho u \otimes u) + \nabla p(\rho) = 0, \\ \partial_t \rho + \operatorname{div}(\rho u) = 0, \end{aligned} \tag{5.32}$$

one formally has the local conservation of energy:

$$\partial_t \left(\frac{\rho |u|^2}{2} + P(\rho) \right) + \operatorname{div} \left[\left(\frac{\rho |u|^2}{2} + P(\rho) + p(\rho) \right) u \right] = 0,$$

where P is the pressure potential as defined in (5.8). To prove this rigorously for weak solutions with a commutator argument, one needs to estimate, e.g.,

$$p(\rho)_\varepsilon - p(\rho_\varepsilon).$$

If $\rho \mapsto p(\rho)$ is twice differentiable in the closure of the range of ρ, Taylor expansion yields

$$p(\rho_\varepsilon) \sim p(\rho) + p'(\rho)(\rho_\varepsilon - \rho) + \frac{1}{2} p''(\rho)(\rho_\varepsilon - \rho)^2$$

as well as

$$p(\rho(y)) \sim p(\rho(x)) + p'(\rho(x))(\rho(y) - \rho(x)) + \frac{1}{2} p''(\rho(x))(\rho(y) - \rho(x))^2.$$

Multiplication of the latter by $\chi_\varepsilon(x - y)$, integration w.r.t. y, and subtraction of both equations yields

$$|p(\rho_\varepsilon) - p(\rho)_\varepsilon| \lesssim \|p\|_{C^2}(\rho_\varepsilon - \rho)^2,$$

so we have reduced the problem again to a quadratic nonlinearity. In this way, one arrives at the following result:

Theorem 5.6 *Let ρ, u be a solution of* (5.32) *in the sense of distributions. Assume ρ and ρu both satisfy* (5.11) *and v satisfies* (5.12)*, where also a time shift is considered*[1]*. Suppose further*

$$0 \leq \underline{\rho} \leq \rho \leq \overline{\rho} \text{ a.a. in}(0, T) \times \mathbb{T}^d,$$

for some constants $\underline{\varrho}$, $\overline{\varrho}$, and

$$2\alpha + \beta > 1, \quad \alpha + 2\beta > 1, \quad p = q = 3.$$

Assume moreover that $p \in C^2[\underline{\varrho}, \overline{\varrho}]$, and, in addition

$$p'(0) = 0 \text{ as soon as } \underline{\varrho} = 0.$$

Then the energy is locally conserved, i.e.

$$\partial_t \left(\frac{1}{2}\rho|u|^2 + P(\rho)\right) + \operatorname{div}\left[\left(\frac{1}{2}\rho|u|^2 + p(\rho) + P(\rho)\right) u\right] = 0$$

in the sense of distributions on $(0, T) \times \mathbb{T}^d$.

Some remarks are in order. The result bears resemblance to Theorem 5.2, the most important difference arguably being the symmetry between the regularity and integrability indices α, β, p, q; this comes from the fact that, unlike for the transport equation, there are now several commutators to control, in some of which the terms $\rho, \rho u$ appear twice and u appears once, and vice versa. Secondly, we need to control the Besov-type regularity now also in time, since there is the nonlinear term ρu under the time derivative. (A possible way to avoid this is to write (5.32) in conservative variables, thus replacing ρu by m, and to obtain the energy equality upon testing with $\frac{m_\varepsilon}{\rho_\varepsilon}$ rather than u_ε. This idea, carried out in [36] for the inhomogeneous incompressible Euler equations, succeeds in avoiding any assumption on time regularity, but leads to trouble with vacuum states.)

[1] This means that (5.11) turns into

$$\lim_{\varepsilon \to 0} \int_0^T \int_{\mathbb{T}^d} \int_{B_\varepsilon(0)} \frac{|\rho(x,t) - \rho(x-y, t-\tau)|^p}{\varepsilon^{d+1+\alpha p}} dy d\tau dx dt = 0,$$

where the ball with radius ε is considered in space-time. Analogously, one needs to alter (5.12).

The assumption $p \in C^2$ in the closure of the range of ρ will typically require *absence of vacuum*: A common choice for the pressure is the polytropic law $p(\rho) = \rho^\gamma$, $\gamma > 1$, and often $\gamma \le 5/3$ (the exponent $5/3$ corresponds to a monoatomic gas). As such a function p is twice differentiable only away from $\rho = 0$, the assumption to justify the Taylor expansion boils down to absence of vacuum states. Note that this property (namely, $\rho \ge c > 0$) is not necessarily propagated in time: Suppose the initial density satisfies $\rho_0 \ge c > 0$, then the maximum principle for transport equations [20] implies that ρ will remain bounded away from zero if the divergence of the velocity is bounded. This, however, is not necessarily satisfied for weak solutions of (5.32) (not even for the compressible Navier-Stokes system). We will get back to the vacuum problem in the next section.

More generally, the Taylor expansion strategy applies to essentially any system of conservation laws that possesses an entropy [6, 7, 33]. To illustrate the point, let

$$\partial_t u + \partial_x f(u) = 0$$

be a scalar conservation law in one dimension, so that $u : \mathbb{R} \times [0, T] \to \mathbb{R}$ is the unknown and f is a given smooth flux function. Let $\eta : \mathbb{R} \to \mathbb{R}$ be any convex function (an *entropy*) and $q : \mathbb{R} \to \mathbb{R}$ a corresponding *entropy flux*, meaning that $q' = f'\eta'$. Then it is easy to see that, again by the chain rule, the *entropy equality*

$$\partial_t \eta(u) + \partial_x q(u) = 0$$

holds, at least if u is smooth. (The entropy equality is really the same thing as the renormalised Eq. (5.6) for transport equations, or the local energy equality (5.28) for the Euler equations.)

Using the arguments outlined above, one can then show that this is the case if (5.29) holds. Thus, the exponent $1/3$ appears universally, simply because the leading order in the Taylor expansion of a smooth function which does not commute with a mollification is the second order. Remarkably, shocks provide an easy example that the 1/3-condition (5.29) is optimal.

5.7 Degenerate Cases

The reduction of an arbitrary nonlinearity to a quadratic one relies crucially on the boundedness of second derivatives for the nonlinearities and entropies involved. There are at least two interesting cases when this condition fails: First, when a transport equation is to be renormalised with $\eta(\rho) = |\rho|^p$ with $p < 2$, and secondly when one considers the compressible Euler system with possible vacuum, for pressure laws $p(\rho) = \rho^\gamma$ with $1 < \gamma < 2$ (these are the physically interesting ones).

These problems have been studied in [1] and [2], respectively. We give here a brief outline of the latter. Recall the discussion of the compressible Euler system in the previous section. One idea is to approximate p locally uniformly by a function p^δ with bounded second derivatives (up to $\rho = 0$ of course). The main error term introduced by this additional layer of approximation takes the form

$$\int_{\mathbb{T}^3} u_\varepsilon \cdot \nabla[p^\delta(\rho)_\varepsilon - p(\rho)_\varepsilon]dx = -\int_{\mathbb{T}^3} \operatorname{div} u_\varepsilon[p^\delta(\rho)_\varepsilon - p(\rho)_\varepsilon]dx,$$

and thus converges to zero, uniformly in ε, provided $\operatorname{div} u$ is a bounded measure. While for the Euler equations this condition (which is quite popular in the theory of hyperbolic conservation laws, see [9]) cannot be guaranteed a priori, for the compressible Navier-Stokes equations (5.7) it follows directly from the energy estimate. We therefore obtain that the compressible Navier-Stokes equations conserve energy, even with possible vacuum, under the assumptions of Theorem 5.6 except the C^2 condition on the pressure law. This yields a nice complement to the results in [49].

If one is not prepared to make such an assumption on $\operatorname{div} u$, one needs to consider two commutator terms involving the pressure, which are therefore sensitive to the condition that p'' be bounded. These two commutators, which appear in the course of the computation, are

$$R^1_\varepsilon := \int_{\mathbb{T}^3} \operatorname{div} u_\varepsilon(p(\rho_\varepsilon) - p(\rho)_\varepsilon)dx,$$

and

$$R^2_\varepsilon := \int_{\mathbb{T}^3} \operatorname{div}(\rho_\varepsilon u_\varepsilon - (\rho v)_\varepsilon)P'(\rho_\varepsilon)dx.$$

It is not difficult to show $R^1_\varepsilon \to 0$ as long as ρ, u satisfy (5.11), (5.12) (in the space-time sense of Footnote 5.6) with $p = q = 3$ and

$$\gamma\alpha + \beta > 1,$$

which is a stronger version of the previous condition $2\alpha + \beta > 1$. (To show convergence of R^1_ε under this condition, one uses "Taylor expansion to order γ", see [2, Lemma 4.2].)

The other commutator R^2_ε is more delicate. We compute

$$\begin{aligned} R^2_\varepsilon &= \int_{\mathbb{T}^3} \operatorname{div}(\rho_\varepsilon u_\varepsilon - (\rho u)_\varepsilon)P'(\rho_\varepsilon)dx \\ &= -\int_{\{\rho_\varepsilon>0\}} (\rho_\varepsilon u_\varepsilon - (\rho u)_\varepsilon) \cdot P''(\rho_\varepsilon)\nabla\rho_\varepsilon dx \\ &\sim \int_{\{\rho_\varepsilon>0\}} (\rho_\varepsilon - \rho)(u_\varepsilon - u) \cdot \rho_\varepsilon^{\gamma-2}\nabla\rho_\varepsilon dx, \end{aligned}$$

then split the domain of integration into $B^\varepsilon := \{0 < \rho_\varepsilon < \varepsilon^\alpha\}$ and $C^\varepsilon := \{\rho_\varepsilon \geq \varepsilon^\alpha\}$, and only consider integration over B^ε here (the C^ε part is easier):

$$\begin{aligned}
&\left|\int_{B^\varepsilon} (\rho_\varepsilon - \rho)(u_\varepsilon - u) \cdot \rho_\varepsilon^{\gamma-2} \nabla \rho_\varepsilon dx\right| \\
&\leq \int_{B^\varepsilon} \left|\frac{\rho_\varepsilon - \rho}{\rho_\varepsilon}\right| |u_\varepsilon - u| \left|\rho_\varepsilon^{\gamma-1}\right| |\nabla \rho_\varepsilon| dx \\
&\leq \varepsilon^{\alpha(\gamma-1)} \int_{\mathbb{T}^3} \left|\frac{\rho_\varepsilon - \rho}{\rho_\varepsilon}\right| |u_\varepsilon - u| |\nabla \rho_\varepsilon| dx \qquad (5.33) \\
&\leq \varepsilon^{\alpha(\gamma-1)} \|u_\varepsilon - u\|_{L^3} \|\nabla \rho_\varepsilon\|_{L^3} \left\|\frac{\rho_\varepsilon - \rho}{\rho_\varepsilon}\right\|_{L^3} \\
&\lesssim \varepsilon^{\gamma\alpha+\beta-1} \left\|\frac{\rho_\varepsilon - \rho}{\rho_\varepsilon}\right\|_{L^3}.
\end{aligned}$$

Now all is well as long as $\left\|\frac{\rho_\varepsilon - \rho}{\rho_\varepsilon}\right\|_{L^3}$ is bounded uniformly in ε; however, in general, this may be false: If $p > 1$, then there are smooth non-negative functions ρ such that $\left\|\frac{\rho_\varepsilon - \rho}{\rho_\varepsilon}\right\|_{L^p} \to \infty$ as $\varepsilon \searrow 0$ [2, Subsection 4A]. Only for $p = 1$ are we able to control this term [2, Lemma 4.3].

This suggests we should use a Hölder $\infty - \infty - 1$ estimate instead of a $3 - 3 - 3$ estimate in (5.33); for this, in turn, we need to require ρ and v to be Hölder continuous. We thus arrive at the following result:

Theorem 5.7 *Replace $2\alpha + \beta$ with $\gamma\alpha + \beta$ as well as* (5.11) *and* (5.12) *with $\rho \in C^\alpha(\mathbb{T}^3 \times (0, T))$ and $u \in C^\beta(\mathbb{T}^3 \times (0, T))$ in the assumptions of Theorem 5.6, and allow for $1 < \gamma < 2$, $\rho \geq 0$. Then the energy is conserved.*

References

1. I. Akramov, E. Wiedemann, Renormalization of active scalar equations. Nonlinear Anal. **179**, 254–269 (2019)
2. I. Akramov, T. Dębiec, J.W.D. Skipper, E. Wiedemann, Energy conservation for the compressible Euler and Navier-Stokes equations with vacuum. Anal. PDE **13**, 789–811 (2020)
3. G. Alberti, S. Bianchini, G. Crippa, Structure of level sets and Sard-type properties of Lipschitz maps: results and counterexamples. Ann. Sc. Norm. Super. Pisa Cl. Sci. **5**, 863–902 (2013)
4. C. Bardos, E. Titi, Onsager's conjecture for the incompressible Euler equations in bounded domains. Arch. Ration. Mech. Anal. **228**, 197–207 (2018)
5. C. Bardos, E. Titi, E. Wiedemann, Onsager's conjecture with physical boundaries and an application to the vanishing viscosity limit. Commun. Math. Phys. **370**, 291–310 (2019)
6. C. Bardos, P. Gwiazda, A. Świerczewska-Gwiazda, E.S. Titi, E. Wiedemann, On the extension of Onsager's conjecture for general conservation laws. J. Nonlinear Sci. **29**, 501–510 (2019)

7. C. Bardos, P. Gwiazda, A. Świerczewska-Gwiazda, E.S. Titi, E. Wiedemann, Onsager's conjecture in bounded domains for the conservation of entropy and other companion laws. Proc. A **475**, 20190289, 18pp. (2019)
8. T. Buckmaster, C. De Lellis, L. Székelyhidi, Jr., V. Vicol, Onsager's Conjecture for admissible weak solutions. Commun. Pure Appl. Math. **72**, 229–274 (2019)
9. G.-Q. Chen, H. Frid, Divergence-measure fields and hyperbolic conservation laws. Arch. Ration. Mech. Anal **147**, 89–118 (1999)
10. A. Cheskidov, P. Constantin, S. Friedlander, R. Shvydkoy, Energy conservation and Onsager's conjecture for the Euler equations. Nonlinearity **21**, 1233–1252 (2008)
11. P. Constantin, V. Vicol, Remarks on high Reynolds numbers hydrodynamics and the inviscid limit. J. Nonlinear Sci. **28**, 711–724 (2018)
12. P. Constantin, E. Weinan, E.S. Titi, Onsager's conjecture on the energy conservation for solutions of Euler's equation. Commun. Math. Phys. **165**, 207–209 (1994)
13. G. Crippa, N. Gusev, S. Spirito, E. Wiedemann, Non-uniqueness and prescribed energy for the continuity equation. Commun. Math. Sci. **13**, 1937–1947 (2015)
14. G. Crippa, N. Gusev, S. Spirito, E. Wiedemann, Failure of the chain rule for the divergence of bounded vector fields. Ann. Sc. Norm. Super. Pisa Cl. Sci. **17**, 1–18 (2017)
15. C. De Lellis, L. Székelyhidi Jr., The Euler equations as a differential inclusion. Ann. Math. **170**, 1417–1436 (2009)
16. C. De Lellis, L. Székelyhidi Jr., On admissibility criteria for weak solutions of the Euler equations. Arch. Ration. Mech. Anal **195**, 225–260 (2010)
17. C. De Lellis, L. Székelyhidi Jr., Dissipative continuous Euler flows. Invent. Math. **193**, 377–407 (2013)
18. N. Depauw, Non unicité des solutions bornées pour un champ de vecteurs BV en dehors d'un hyperplan. C. R. Math. Acad. Sci. Paris **337**, 249–252 (2003)
19. R.J. DiPerna, Measure-valued solutions to conservation laws. Arch. Ration. Mech. Anal **88**, 227–270 (1985)
20. R.J. DiPerna, P.-L. Lions, Ordinary differential equations, transport theory and Sobolev spaces. Invent. Math. **98**, 511–547 (1989)
21. R.J. DiPerna, A.J. Majda, Oscillations and concentrations in weak solutions of the incompressible fluid equations. Commun. Math. Phys **108**, 667–689 (1987)
22. T.D. Drivas, H.Q. Nguyen, Onsager's conjecture and anomalous dissipation on domains with boundary. SIAM J. Math. Anal. **50**, 4785–4811 (2018)
23. T.D. Drivas, H.Q. Nguyen, Remarks on the emergence of weak Euler solutions in the vanishing viscosity limit. J. Nonlinear Sci. **29**, 709–721 (2019)
24. J. Duchon, R. Robert, Inertial energy dissipation for weak solutions of incompressible Euler and Navier-Stokes equations. Nonlinearity **13**, 249–255 (2000)
25. L.C. Evans, *Partial Differential Equations*. Graduate Studies in Mathematics, vol. 19, 2nd edn. (American Mathematical Society, Providence, 2010)
26. G.L. Eyink, Energy dissipation without viscosity in ideal hydrodynamics, I. Fourier analysis and local energy transfer. Phys. D **78**, 222–240 (1994)
27. G.L. Eyink, K.R. Sreenivasan, Onsager and the theory of hydrodynamic turbulence. Rev. Modern Phys. **78**, 87–135 (2006)
28. E. Feireisl, On compactness of solutions to the compressible isentropic Navier-Stokes equations when the density is not square integrable. Comment. Math. Univ. Carolin. **42**, 83–98 (2001)
29. E. Feireisl, P. Gwiazda, A. Świerczewska-Gwiazda, E. Wiedemann, Regularity and energy conservation for the compressible Euler equations. Arch. Ration. Mech. Anal. **223**, 1375–1395 (2017)
30. U.S. Fjordholm, E. Wiedemann, Statistical solutions and Onsager's conjecture. Phys. D **376/377**, 259–265 (2018)
31. U.S. Fjordholm, S. Lanthaler, S. Mishra, Statistical solutions of hyperbolic conservation laws: foundations. Arch. Ration. Mech. Anal. **226**, 809–849 (2017)

32. U. Frisch, *Turbulence. The legacy of A. N. Kolmogorov* (Cambridge University Press, Cambridge, 1995)
33. P. Gwiazda, M. Michálek, A. Świerczewska-Gwiazda, A note on weak solutions of conservation laws and energy/entropy conservation. Arch. Ration. Mech. Anal. **229**, 1223–1238 (2018)
34. P. Isett, V. Vicol, Hölder continuous solutions of active scalar equations. Ann. PDE **1**, Art. 2 (2015)
35. P. Isett, A proof of Onsager's conjecture. Ann. Math. **188**, 871–963 (2018)
36. T.M. Leslie, R. Shvydkoy, The energy balance relation for weak solutions of the density-dependent Navier-Stokes equations. J. Differ. Equ. **261**, 3719–3733 (2016)
37. P.-L. Lions, *Mathematical Topics in Fluid Mechanics, vol. 2. Compressible Models*. Oxford Lecture Series in Mathematics and its Applications, vol. 10 (Oxford Science Publications/The Clarendon Press/Oxford University Press, New York, 1998)
38. S. Modena, G. Sattig, Convex integration for the transport equation with full dimensional concentration. Ann. Inst. H. Poincaré Anal. Non Linéaire **37**, 1075–1108 (2020)
39. S. Modena, L. Székelyhidi Jr., Non-uniqueness for the transport equation with Sobolev vector fields. Ann. PDE **4**, Art. 18 (2018)
40. S. Modena, L. Székelyhidi Jr., Non-renormalized solutions to the continuity equation. Calc. Var. Partial Differ. Equ. **58**, 208, 30 pp. (2019)
41. L. Onsager. Statistical hydrodynamics. Nuovo Cimento **6**, 279–287 (1949)
42. J. Robinson, J.L. Rodrigo, J.W.D. Skipper, Energy conservation in the 3D Euler equations on $\mathbb{T}^2 \times \mathbb{R}_+$ for weak solutions defined without reference to the pressure. Asymptot. Anal. **110**, 185–202 (2018)
43. V. Scheffer, An inviscid flow with compact support in space-time. J. Geom. Anal. **3**, 343–401 (1993)
44. A. Shnirelman, On the nonuniqueness of weak solution of the Euler equation. Commun. Pure Appl. Math. **50**, 1261–1286 (1997)
45. A. Shnirelman. Weak solutions with decreasing energy of incompressible Euler equations. Commun. Math. Phys. **210**, 541–603 (2000)
46. R. Shvydkoy, Lectures on the Onsager conjecture. Discrete Contin. Dyn. Syst. Ser. S **3**, 473–496 (2010)
47. R. Shvydkoy, Convex integration for a class of active scalar equations. J. Am. Math. Soc. **24**, 1159–1174 (2011)
48. E. Wiedemann, Weak-strong uniqueness in fluid dynamics, in *Partial Differential Equations in Fluid Mechanics*. London Mathematical Society Lecture Note Series, vol. 452 (Cambridge University Press, Cambridge, 2018), pp. 289–326
49. C. Yu, Energy conservation for the weak solutions of the compressible Navier-Stokes equations. Arch. Ration. Mech. Anal. **225**, 1073–1087 (2017)

GPSR Compliance

The European Union's (EU) General Product Safety Regulation (GPSR) is a set of rules that requires consumer products to be safe and our obligations to ensure this.

If you have any concerns about our products, you can contact us on ProductSafety@springernature.com

In case Publisher is established outside the EU, the EU authorized representative is:

Springer Nature Customer Service Center GmbH
Europaplatz 3
69115 Heidelberg, Germany

Batch number: 10755332

Printed by Printforce, the Netherlands